Handbook

of

Simplified Solid-State

Circuit Design

REVISED AND ENLARGED

③ 7-23-1992

(전력 수요 사상 최고)

(2000 만 KW) 한국전체?

(Corona P/Library 에서)

(3D - 기피현상)

Dangerous
Dirty
Difficult

개인 능력 극대화 시킬 수 있다.

Second Edition

Handbook

of

Simplified Solid-State

Circuit Design

REVISED AND ENLARGED

JOHN D. LENK

Consulting Technical Writer

PRENTICE-HALL, INC., *Englewood Cliffs, N.J.   07632*

Library of Congress Cataloging in Publication Data

LENK, JOHN D
    Handbook of simplified solid state circuit design.

    Includes index.
    1. Transistor circuits. 2. Electronic circuit
design. I. Title.
TK7871.9.L364 1978   621.3815 '3 '0422   77-23555
ISBN   0-13-381707-5

This Reward paperback edition
published 1979 by Prentice-Hall, Inc.

Printed in the United States of America

10 9 8 7 6 5 4

PRENTICE-HALL INTERNATIONAL, INC., *London*
PRENTICE-HALL OF AUSTRALIA PTY. LIMITED, *Sydney*
PRENTICE-HALL OF CANADA, LTD., *Toronto*
PRENTICE-HALL OF INDIA PRIVATE LIMITED, *New Delhi*
PRENTICE-HALL OF JAPAN, INC., *Tokyo*
PRENTICE-HALL OF SOUTHEAST ASIA PTE. LTD., *Singapore*
WHITEHALL BOOKS LIMITED, *Wellington, New Zealand*

To my wife Irene, daughter Karen,
Mr. Lamb, Ianna, Mortimer,
and the Magic Time

CONTENTS

chapter three

OPERATIONAL AMPLIFIERS
WITH INTEGRATED CIRCUITS   137

chapter four

RADIO-FREQUENCY CIRCUITS   227

chapter five

WAVEFORMING AND WAVESHAPING CIRCUITS   323

chapter six

POWER-SUPPLY CIRCUITS   367

PREFACE

This second edition of the HANDBOOK OF SIMPLIFIED SOLID-STATE CIRCUIT DESIGN carries through all of the features that made the first edition so successful. That is, the second edition provides a simplified, step-by-step approach to solid-state circuit design. All the chapters in the second edition have been expanded or enlarged to include new material. Existing data has been up-dated to reflect current design trends. Also, much of the existing material from the first edition has been revised for clarification and/or simplification. *No previous design experience is required* to use the design data and techniques described in this second edition.

As in the original, the basic approach of the second edition is to start all design problems with approximations or rules-of-thumb for the selection of components on a trial value basis, assuming a specific design goal and a given set of conditions. Then, using these approximate values in experimental test circuits, the desired results (gain, frequency response, impedance match, etc.) are produced by varying the test component values.

The second edition concentrates on simple, practical approaches to circuit design, not on circuit analysis or model circuits. Theory is kept to a minimum, and appears only where required to understand the design steps. Thus, the reader need not memorize elaborate theories or understand abstract mathematics to use the design data. With any solid-state circuit, it is possible to apply certain guidelines for the selection of component values. These rules can be stated in basic equations, requiring only simple arithmetic for their solution.

The component values will depend upon transistor characteristics, available power sources, the desired performance (voltage amplification, stability, etc.) and external circuit conditions (input/output impedance match, input signal amplitude, etc.). The transistor characteristics are to be found in the manufacturer's data, or from actual test. The circuit characteristics can then be determined, based on reasonable expectation of the transistor characteristics. Often, the final circuit is a result of many tradeoffs between desired performance and available characteristics. The second edition discusses the problem of tradeoffs from a simplified, practical standpoint.

It is assumed that the reader is familiar with solid-state basics at a level found in the author's HANDBOOK FOR TRANSISTORS (Prentice-Hall, Inc., Englewood Cliffs, N.J. 1976). It is especially important the reader be able to interpret solid-state datasheets. However, *no direct reference to any of the author's previous books is required to understand and use this second edition.*

The first chapter of this book is devoted to basic design rules, especially those related to transistor-bias techniques. This is essential since proper bias must be used in every transistor circuit. The second edition provides an expanded Chapter 1 that includes practical details of mounting techniques for metal-packaged power semiconductors.

Chapter 2 is devoted entirely to audio circuit design. The second edition provides an expanded section on transformerless audio amplifiers, as well as a revised section on transformer-coupling in audio circuits.

Chapter 3 covers operational amplifiers. The second edition assumes that today's designer will use commercial IC op-amps as the basic element for all op-amp design applications. With this in mind, the chapter concentrates on circuits external to the IC package that modify the op-amp to produce a given design function or characteristic (modification of frequency response, gain, etc.) and to interpretation of IC op-amp datasheets and/or test results.

Chapter 4 describes design of RF circuits, particularly RF power amplifiers which must be designed "from scratch". (Generally, RF circuits are not available in IC or package form.) The second edition is expanded to cover y-parameters (how y-parameters fit into simplified design), and includes examples of using y-parameters to design both VHF and UHF power amplifiers. An introduction to the Smith Chart is also included.

Chapter 5 covers waveforming and waveshaping circuits. The section on RF oscillators has been revised extensively to reflect current design trends.

Chapter 6 describes solid-state power supply circuits, including converters and regulators. A new section on switching regulators has been provided in the second edition.

Many professionals have contributed their talent and knowledge to the revision and enlargement of the new edition. The author gratefully acknowledges that the tremendous effort to make this second edition such a comprehensive work is impossible for one person, and he wishes to thank all who have contributed directly and indirectly. The author wishes to give special thanks to the following: the Semiconductor Products Division of Motorola Inc., the Semiconductor Products Department of General Electric, the Components Group of Texas Instruments, and the Solid State Division of Radio Corporation of America. The author also wishes to thank Mr. Joseph A. Labok of Los Angeles Valley College.

JOHN D. LENK

Handbook

of

Simplified Solid-State

Circuit Design

REVISED AND ENLARGED

voltage. However, *any electronic power source* is subject to some voltage variation. Therefore:

Always allow for some variation in source voltage when an electronic power supply is used.

Another factor that affects maximum voltage is temperature. Note that in Fig. 1-1 the maximum voltage is specified at 25°C. Usually, breakdown will occur at a lower voltage when temperature is increased. The topic of how operating temperatures affect transistor design is discussed further in Sec. 1-4.

In Fig. 1-1, the maximum base–emitter voltage is listed as V_{EBO} of 1 V. Again, this is a test voltage rather than an operating design voltage. Usually, the base–emitter junction has some current flowing at all times. The voltage drop across the junction is about 0.2 to 0.4 V for germanium and 0.5 to 0.7 V for silicon transistors. The lower voltage drops (0.2 V or 0.5 V) will produce some current flow, while the higher drops (0.4 V or 0.7 V) will produce heavy current flow. In Sec. 1-6, bias circuits designed to produce the desired drop are discussed. Either the higher or lower drops can be used, depending upon the desired results.

In general, the lower drops will produce less current drain and lower no-signal power dissipation. The higher drops may result in operation on a more linear portion of the transistor characteristics. For the purposes of standardization, the lower voltage drops will be used throughout this book. However, if desired, the higher drops can be used as an alternative by slight changes of the bias circuit values given in the design examples.

In practical design, it is often necessary to select a bias (base–emitter voltage) on the basis of *input signal* rather than on some arbitrary point of the transistor's characteristic curve. Keep in mind that the input signal to a transistor can come from an external source or a previous stage, or in the case of an oscillator or operational amplifier, can be *feedback*. Therefore:

Always consider any input signal that may be applied to the base–emitter junction, in addition to the normal operating bias.

1-2.2. Collector Current

In Fig. 1-1, the collector current is listed as I_C of 25 mA, at 25°C. As is discussed later, collector current will increase with temperature (and temperature increases as current increases). Therefore:

Do not operate any transistor at or near its maximum current rating.

Of course, if you could be absolutely certain that the transistor would dissipate any temperature increases (a practical impossibility), the circuit could be designed to operate near the maximum current.

In practical design applications, it is the *power* dissipated in the collector circuit (rather than a given current) that is of major concern. For example,

TYPICAL TRANSISTOR SPECIFICATION 2N332

ABSOLUTE MAXIMUM RATINGS (25°C.)

Voltages:

Collector to base (emitter open)	V_{CBO}	**45 volts**
Emitter to base (collector open)	V_{EBO}	**1 volt**

Collector current I_C **25 ma**

Power*

Collector dissipation (25°C.)	P_C	**150 mw**
Collector dissipation (125°C.)	P_C	**50 mw**

Temperature range:

Storage	T_{STG}	**−65°C. to 200°C.**
Operating	T_A	**−55°C. to 175°C.**

ELECTRICAL CHARACTERISTICS (25°C.)
(Unless otherwise specified, $V_{CB} = 5$ v; $I_E = -1$ ma; $f = 1$ kc)

		min.	nom.	max.	
Small signal characteristics:					
Current transfer ratio	h_{fe}	9	15	20	
Input impedance	h_{ib}	30	53	80	ohms
Reverse voltage transfer ratio	h_{rb}	.25	1.0	5.0	$\times 10^{-4}$
Output admittance	h_{ob}	0.0	.25	1.2	μmhos
Power gain					
($V_{CE} = 20$ v; $I_E = -2$ ma; $f = 1$ kc;					
$R_G = 1$ K ohms; $R_L = 20$ K ohms)	G_e		35		db
Noise figure	NF		28		db
High frequency characteristics:					
Frequency cutoff					
($V_{CB} = 5$ v; $I_E = -1$ ma)	f_{ab}		15		mc
Collector to base capacity					
($V_{CB} = 5$ v; $I_E = -1$ ma; $f = 1$ mc)	C_{ob}		7		μμf
Power gain (common emitter)					
($V_{CB} = 20$ v; $I_E = -2$ ma; $f = 5$ mc)	G_e		17		db
D-c characteristics:					
Collector breakdown voltage					
($I_{CBO} = 50$ μa; $I_E = 0$; $T_A = 25$°C.)	BV_{CBO}	45			volts
Collector cutoff current					
($V_{CB} = 30$ v; $I_E = 0$; $T_A = 25$°C.)	I_{CBO}		.02	2	μa
($V_{CB} = 5$ v; $I_E = 0$; $T_A = 150$°C.)	I_{CBO}			50	μa
Collector saturation resistance					
($I_B = 1$ ma; $I_C = 5$ ma)	R_{SC}		80	200	ohms
Switching characteristics:					
($I_{B_1} = 0.4$ ma; $I_{B_2} = -0.4$ ma;					
$I_C = 2.8$ ma)					
Delay time	t_d		.75		μsec
Rise time	t_r		.5		μsec
Storage time	t_s		.05		μsec
Fall time	t_f		.15		μsec

*Derate 1mw/°C increase in ambient temperature.

Fig. 1-1. Typical transistor data sheet (2N332).

transistor selection, operating point, etc.) is covered. This is followed by reference to the equations (on the working schematic) and procedures for determining component values that will produce the desired results.

Finally, a design example is given. A specific design problem is stated. The value of each circuit component is calculated, in step-by-step procedures, using the rules of thumb established in the design considerations.

Where applicable, procedures for testing the completed circuit are given in detail (generally at the end of each chapter).

1-2. Interpreting Data Sheets

Most of the basic design information for a particular transistor can be obtained from the data sheet. There are some exceptions to this rule. For extreme-high-frequency work, and in digital work where switching characteristics are of particular importance, it may be necessary to test a transistor under simulated operating conditions.

In any event, it is always necessary to interpret data sheets. Each manufacturer has its own system of data sheets. It would be impractical to discuss all data-sheet formats here. Instead, we shall discuss the typical information found on data sheets, and see how this information affects simplified design.

Figure 1-1 is the data sheet for a 2N332 transistor. This transistor is an industrial type, suitable for low-power audio or RF (up to about 10 MHz), as well as switching.

1-2.1. Maximum Voltage

The first specifications listed are those of *maximum voltage*. In Fig. 1-1, the maximum collector voltage is listed as V_{CBO} of 45 V. Actually, this is a test voltage rather than an operating design voltage. (V_{CBO} usually indicates collector–base breakdown voltage, with the emitter circuit open. Transistors do not operate in this way in circuits.) However, for design purposes, the 45-V figure can be considered as the *absolute maximum* collector voltage. Keep the following in mind on maximum collector voltage: Except for RF circuits used in transmitters, most transistors will be operated with their collector at some voltage value (V_C) less than the source voltage (V_{CC}). For example, in a typical class A circuit, the collector voltage will be half the source voltage, at the normal operating point. However, the collector voltage will rise to or near the source voltage when the transistor is at or near cutoff. Therefore:

Never design any circuit in which the collector is connected to a source higher than the maximum voltage rating, even through a resistance.

The next maximum voltage design problem to be considered is the type of source voltage. A battery power source will not deliver more than its rated

1

BASIC DESIGN RULES

In this chapter, we shall establish a set of basic design rules. Generally, these rules will apply to all circuits discussed in remaining chapters. You need not commit these rules to memory. They will be referred to frequently in the remaining chapters. However, it is strongly recommended that you read this chapter in its entirety before attempting to design any solid-state circuit, even the simplest diode detector.

1-1. How To Use This Book

Once you have read this chapter, you may go directly to the section that describes the design procedures for a particular circuit. Use the Contents or Index to locate the first page of the section. All circuits of a given type (oscillators, low-frequency amplifiers, RF amplifiers, etc.) have been grouped into separate chapters. In turn, a separate section has been assigned to each circuit within the chapter. The sections are either complete within themselves or make reference to another specific section (by section number).

The same format or pattern is used in each section (where practical). First, a working schematic is presented for the circuit, together with a brief description of the operational theory. Where practical, the working schematic also includes the operational characteristics of the circuit (in equation form), as well as the rule-of-thumb relationship of circuit values (also in equation form).

Next, design considerations such as desired performance, use with external circuits, and available (or required) power sources are discussed. Each major design factor (supply voltage, amplification, operating frequency, assume that the collector operates at 45 V and 25 mA. This results in a power dissipation of over 1 W, far above the 150 mW specified for 25°C.

1-2.3. Power and Temperature Range

The power-dissipation capabilities of a transistor in any circuit are closely associated with the temperature range. As shown in Fig. 1-1, power dissipation is 150 mW at 25°C, 50 mW at 125°C, and must be *derated* 1 mW for each degree (°C) increase in ambient temperature. Because of the importance of temperature to power dissipation, the subject is discussed fully in Sec. 1-4.

1-2.4. Small-Signal Characteristics

Small-signal characteristics can be defined as those where the ac signal is small compared to the dc bias. For example, h_{fe} or *forward current transfer ratio* (also known as *ac beta* or *dynamic beta*), is properly measured by noting the change in collector alternating current for a given change in base alternating current, without regard to static base and collector currents.

Small-signal characteristics do not provide a truly sound basis for practical design. As discussed in related chapters, the performance of a transistor in a working circuit can be controlled by the circuit component values (within obvious limits, of course). There are two basic reasons for this approach.

First, not all manufacturers list the same small-signal characteristics on their data sheets. To further complicate matters, manufacturers call the same characteristic by different names (or even use the same name to identify different characteristics).

Second, the small-signal characteristics listed in data sheets are based on a set of *fixed operating conditions*. If the conditions change (as they must in any practical circuit), the characteristics will change. For example, beta changes drastically with temperature, frequency, and *operating point*. Therefore:

Use small-signal characteristics as a starting point for simplified design, not as hard-and-fast design rules.

1-2.5. High-Frequency Characteristics

High-frequency characteristics are especially important in the design of RF networks. As is discussed in later chapters, networks (such as RF stages in a transmitter) provide the dual function of frequency selection (tank circuit) and impedance matching between the transistor and a load. Unfortunately, the high-frequency information provided in many data sheets (such as Fig. 1-1) is not adequate for simplified design. To properly match impedances, both the resistive (real part) and reactive (imaginary part) components must be considered. The reactive component (either inductive or capacitive) changes with frequency. Therefore, it is necessary to know the reactance values over a wide range of frequencies, not at some specific test frequency (unless you happen to be designing for that test frequency only). The best way to show how resistance and reactance vary in relation to frequency for a particular transistor is by means of curves. Fortunately, manufacturers who are trying to sell their transistors for high-frequency use generally provide a set of curves showing the characteristics over the anticipated frequency range.

1-2.6. Direct-Current Characteristics

Direct-current characteristics, while important in the design of basic bias circuits, do not have too critical an effect on the operation of the final operating circuit. The dc characteristics shown in Fig. 1-1 are primarily test values rather than design parameters. The important point to remember regarding such dc characteristics is that they serve as a starting point for bias design, and that they will change with temperature.

1-2.7. Switching Characteristics

Switching characteristics are important in the design of all pulse circuity (multivibrators, gates, etc.). The switching times shown in Fig. 1-1 are defined in Fig. 1-2. These time factors (delay, storage, rise, fall) determine the operating limits for switching circuits. For example, if a transistor gate with a 20-ns rise time is used to pass a 15-ns pulse, the pulse will be hopelessly distorted. Likewise, if there is a 1.5-μs delay added to a 1-μs pulse an absolute minimum of 2.5-μs is required before the next pulse can occur. This means a maximum pulse-repetition rate (PRR) of 400 kHz ($1/2.5^{-6} = 400,000$ Hz). Actually, the prf is lower since there is some "off" time between pulses.

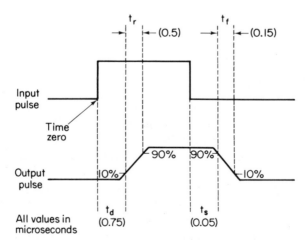

Fig. 1-2. Definition of switching-time characteristics.

1-3. Determining Parameters at Different Frequencies

Data sheets specify most parameters at some given frequency. Often, it is desired to know the parameter at another frequency. The following paragraphs describe methods for converting from one parameter to another at different frequencies.

1-3.1. Basic Parameter Relationships

One or both of two major parameters, h_{fb} (alpha) and h_{fe} (beta), are found on most data sheets. (Table 1-1 is a glossary of terms used in this section.)

Table 1-1. GLOSSARY OF TRANSISTOR PARAMETER SYMBOLS

Symbol	*Definition*
h_{fb}	Common-base ac forward current gain (alpha).
h_{fbo}	Value of h_{fb} at 1 kHz.
h_{fe}	Common-emitter ac forward current gain (beta).
h_{feo}	Value of h_{fe} at 1 kHz.
f_{ab}	Common-base current-gain cutoff frequency. Frequency at which h_{fb} has decreased to a value 3 dB below h_{fbo} (where $h_{fb} = 0.707h_{fbo}$).
f_{ae}	Common-emitter current-gain cutoff frequency. Frequency at which h_{fe} has decreased to a value 3 dB below h_{feo} (where $h_{fe} = 0.707h_{feo}$).
f_T	Gain-bandwidth product. Frequency at which $h_{fe} = 1$ (0 dB).
G_{pe}	Common-emitter power gain.
f_{max}	Maximum frequency of oscillation. Frequency at which $G_{pe} = 1$ (0 dB).
K	Phase-shift factor. (Phase shift of current in transistor base.)

1-3.1.1. Common-Base Parameters

The quantity h_{fbo} (the value of h_{fb} at 1 kHz) will remain constant as frequency is increased until a top limit is reached. After the top limit, h_{fb} begins to drop rapidly. The frequency at which a significant drop in h_{fb} occurs provides a basis for comparison of the expected frequency performance of different transistors. This frequency is known as f_{ab} and is defined as that frequency at which h_{fb} is 3 dB below h_{fbo}.

A curve of h_{fb} versus frequency for a transistor with an f_{ab} of 1 MHz is shown in Fig. 1-3. This curve has the following significant characteristics:

1. At frequencies below f_{ab}, h_{fb} is nearly constant and approximately equal to h_{fbo}.
2. h_{ab} begins to decrease significantly in the region of f_{ab}.

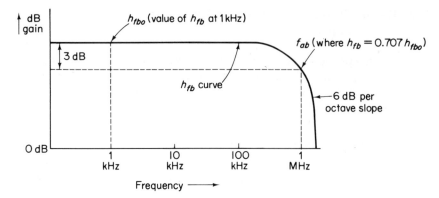

Fig. 1-3. Typical curve of common-base current gain versus frequency. Courtesy Motorola.

3. Above f_{ab}, the rate of decrease of h_{fb} (with increasing frequency) approaches 6 dB/octave.

The curve of common-base current gain versus frequency for any transistor has the same characteristics, and the same general appearance, as the curve of Fig. 1-3.

1-3.1.2. Common-Emitter Parameters

The common-emitter parameter which corresponds to f_{ab} is f_{ae}, the common-emitter current-gain cutoff frequency. This f_{ae} is the frequency at which h_{fe} (beta) has decreased 3 dB below h_{feo}. A typical curve of h_{fe} versus frequency for a transistor with an f_{ae} of 100 kHz is shown in Fig. 1-4. The curve of Fig. 1-4 has the same significant characteristics as those described for Fig. 1-3. That is, h_{fe} is considered to be decreasing at a rate of 6 dB/octave, at f_{ae}.

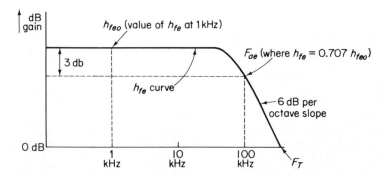

Fig. 1-4. Typical curve of common-emitter current gain versus frequency. Courtesy Motorola.

These characteristics allow such a curve to be constructed for a particular transistor by knowing only h_{feo} and f_{ae}. With the curve constructed, h_{fe} at any frequency could be determined. Furthermore, if f_{ae} is not known, a curve could also be constructed if h_{feo} and h_{fe} at *any frequency above f_{ae}* were known. Thus, to find h_{fe} at any frequency, it is necessary to know only h_{feo} (which is available on most data sheets), and either f_{ae} or h_{fe}, at some frequency greater than f_{ae}.

1-3.1.3. Gain-Bandwidth Product

Gain-bandwidth product (f_T) is sometimes specified on data sheets instead of f_{ae} (or instead of h_{fe} at some frequency greater than f_{ae}). This f_T is the frequency at which gain drops to unity (0 dB). f_T can be approximated when f_{ae} is multiplied by h_{feo}. ($f_T = f_{ae} \times h_{feo}$.)

On those data sheets where h_{fe} is specified at some frequency greater than f_{ae}, f_T can be approximated when the specified frequency is multiplied by the specified h_{fe}.

It should be noted that f_T is a *common-emitter parameter* and should not be used with common-base calculations. It should also be noted that common-emitter f_T is approximately equal to the common-base parameter of f_{ab}. Usually, f_T is slightly less than f_{ab}.

1-3.1.4. Maximum Operating Frequency

Although common-emitter current gain is equal to 1 at f_T, there may still be considerable power gain at f_T due to different input and output impedance levels. Thus, f_T is not necessarily the highest useful frequency of operation for a transistor in the common-emitter mode. An additional parameter, the maximum frequency of oscillation ($f_{\max}$), is found.

The term $f_{\max}$ is the frequency at which common-emitter *power gain* is equal to 1. A plot of common-emitter power gain versus frequency has the same characteristics as the voltage-gain plot. The curve will appear similar to that of Fig. 1-4.

The maximum frequency of oscillation, $f_{\max}$, may be found by measuring power gain at some frequency on the 6 dB/octave (slope) portion of the power gain versus frequency curve and multiplying the square root of the power gain (in magnitude) by the frequency of measurement.

The problem here is that data sheets do not always specify if the power-gain figure is on the slope of the power gain versus frequency curve. However, there are some clues which can be used to estimate the location of the power-gain figure on the curve.

If two power-gain figures are given, the high-frequency figure can be considered to be on the slope, and can be used to find $f_{\max}$. For example, the data sheet of Fig. 1-1 shows a power gain of 35 dB (magnitude 3000) at 1 kHz, and 17 dB (magnitude 50) at 5 MHz. The 17-dB (magnitude 50) at 5 MHz figure can be used to find f_{max} by

$$f_{max} = \text{frequency of measurement } \sqrt{\text{power gain in magnitude}}$$
$$= 5\sqrt{50}$$
$$= 35 \text{ MHz (approximately)}$$

If a power-gain figure is given for a *frequency higher than f_{ae}*, it can be considered as being on the slope and can be used to find f_{max}.

1-3.2. Conversion between Parameters

One of the problems with data sheets is the mixing of parameters. For example, the data sheet for a 2N332 shows a nominal common-emitter (beta) current transfer ratio h_{fe} of 15, and a nominal common-base (alpha) cutoff frequency f_{ab} of 15 MHz. Likewise, the data sheet of a 2N337 shows a "typical" current transfer ratio h_{fe} of 55 (magnitude) and a "typical" alpha cutoff frequency of 30 MHz. (Actually, the data sheet lists alpha cutoff frequency as f_{hfb}.) Many other data sheets, and transistor specification books, have similar forms of parameter mixing. Therefore, it is necessary to convert between parameters for design purposes. Usually, it is necessary to convert from common base to common emitter, but the reverse can also be true. The following rules summarize the conversion process:

To find beta when alpha is given:

$$\text{beta} = \frac{\text{alpha}}{1 - \text{alpha}}$$

To find alpha when beta is given:

$$\text{alpha} = \frac{\text{beta}}{1 + \text{beta}}$$

To find h_{feo} when h_{fbo} is given:

$$h_{feo} = \frac{h_{fbo}}{1 - h_{fbo}}$$

To find h_{fbo} when h_{feo} is given:

$$h_{fbo} = \frac{h_{feo}}{1 + h_{feo}}$$

To find f_{ae} when f_{ab} is given:

$$f_{ae} = K(1 - h_{fbo})f_{ab}$$

To find f_{ab} when f_{ae} is given:

$$f_{ab} = \frac{f_{ae}}{K(1 - h_{fbo})}$$

Note that the constant K refers to a phase-shift factor. This constant ranges between 0.8 and 0.9 for all current transistors, except MADT types, which have a phase-shift factor of 0.6.

1-3.3. Determining Common-Emitter Parameters at Different Frequencies

The common-emitter configuration is used in 90 per cent of present-day transistor circuits. There are three frequency-related common-emitter parameters of particular importance to design. These are: h_{fe} at some frequency other than specified on the data sheet, f_T and f_{max}.

The following rules summarize the procedures for finding these three parameters, when they are not listed directly on the data sheet.

To find h_{fe} at a particular frequency:

h_{fe} is approximately equal to h_{feo} when the frequency of interest is less than f_{ae}.

h_{fe} is approximately equal to $0.7h_{feo}$ when the frequency of interest is near f_{ae}.

h_{fe} decreases at a rate of 6 dB/octave and is approximately equal to f_T/frequency when the frequency of interest is above f_{ae}.

h_{fe} is equal to 1 (unity gain) at f_T.

To find f_T:

When h_{feo} and f_{ae} are given, $f_T = h_{feo} \times f_{ae}$.

When h_{fbo} and f_{ab} are given, $f_T = h_{fbo} \times f_{ab} \times K$.

To find f_{max}: As discussed in Sec. 1-3.1.4, f_{max} is a common-emitter parameter and requires that the common-emitter power gain be established at some frequency on the sloping portion of the gain versus frequency curve. With such a gain established:

$$f_{max} = \text{frequency of measurement} \sqrt{\text{power gain in magnitude}}$$

1-3.4. Examples of Relating Data-Sheet Frequency Parameters to Design Problems

Assume that a 2N337 transistor (Sec. 1-3.2) is to be used in RF transmitter circuits. One circuit must provide a voltage gain of at least 10 (magnitude) at 2.5 MHz. Another circuit must operate at 25 MHz with some gain. Both circuits must operate in the common-emitter configuration. Assume that the K factor (phase shift) is 0.9. Each problem requires that f_T be found. There are two approaches to finding f_T.

First approach: Find f_T using f_{ae} and h_{feo}. Since the data sheet gives a common-emitter h_{feo} of 55 (magnitude) but a common-base cutoff of 30 MHz, it is necessary to find f_{ae}.

$$f_{ae} = K(1 - h_{fbo})f_{ab} \quad \text{or} \quad 0.9 \times 0.018 \times 30 = 0.49 \text{ MHz}$$

With f_{ae} found, then,

$$f_T = h_{feo} \times f_{ae} \quad \text{or} \quad 55 \times 0.49 = 26.9 \text{ MHz}$$

Second approach: Find f_T using f_{ab} and h_{fbo}. Since the data sheet gives a common-emitter h_{feo} of 55 (magnitude) and a common-base cutoff of 30 MHz, it is necessary to find h_{fbo}.

$$h_{fbo} = \frac{h_{feo}}{1 + h_{feo}} \quad \text{or} \quad \frac{55}{56} = 0.982$$

With h_{fbo} found, then,

$$f_T = K \times h_{feo} \times f_{ab}$$
$$= 0.9 \times 0.982 \times 30$$
$$= 26.5 \text{ MHz}$$

With either approach: $h_{fe} = f_T/$frequency of interest, or 26.9/2.5; h_{fe} is greater than 10 at 2.5 MHz, and the 2N337 could be used in the first circuit.

Also, the f_T is greater than 25 MHz; the 2N337 would provide some gain at 25 MHz and could be used for the second circuit.

Note: No matter what system is used to find transistor parameters at different frequencies, remember that the parameters are voltage- and current-dependent, and the operating point must be considered in all cases. For example, the high-frequency h_{fe} measured at one collector voltage and current must not be used to calculate f_T directly at another voltage or current, without considering the possible effects of the different operating point.

1-4. Temperature-Related Design Problems

There are two basic temperature-related problems in design work. First, data sheets specify transistor parameters at a given temperature. Many of these parameters will change with temperature. Since transistors rarely operate at the exact temperature shown on the data sheet, it is important to know the parameters at the actual operating temperature. Although practically all parameters change with temperature, the three most critical parameters (from a design standpoint) are: current gain, collector leakage, and power dissipation. To compound this problem, changes in parameters can affect transistor temperature. For example, an increase in current gain or power dissipation will result in a temperature increase.

Second, in addition to knowing the effects of temperature on parameters, it is important to know how *heat sinks* or transistor mounting can be used to offset the effects of temperature. For example, if a transistor is used with a heat sink, or is mounted on a metal chassis that acts as a heat sink, an increase in temperature (from any cause) can be dissipated into the surrounding air.

The following paragraphs describe methods for approximating important transistor parameters at temperatures other than those shown on data sheets. Methods for determining the proper dissipation characteristics of heat sinks are also discussed.

1-4.1. Effects of Temperature on Collector Leakage

Collector leakage, I_{cbo}, increases with temperature. As a rule of thumb:
I_{cbo} *doubles with every 10°C increase in temperature for germanium transistors.*

I_{cbo} *doubles with every 15°C increase in temperature for silicon transistors.*

Collector leakage also increases with voltage applied to the collector. For example, the 2N332 (Fig. 1-1) data sheet lists an I_{cbo} of 2 μA at 25°C and of 50 μA at 150°C. However, the 25°C figure is with a voltage of 30 V (V_{cb}), while the 150°C figure is with 5 V. If the temperature was raised from 25°C to 150°C with 30 V at the collector, I_{cbo} would be approximately 500 μA.

Therefore, always consider the possible effects of a different collector voltage when approximating I_{cbo} at temperatures other than those on the data sheet.

1-4.2. Effects of Temperature on Current Gain

Current gain, h_{fe}, increases with temperature. As a rule of thumb:
Current gain doubles when the temperature is raised from 25°C to 100°C, for germanium transistors.

Current gain doubles when the temperature is raised from 25°C to 175°C, for silicon transistors.

It is obvious that silicon transistors are less temperature-sensitive than germanium transistors. Data sheets usually specify a maximum operating temperature, or ambient temperature. If this temperature is not given, or is unknown, the rule of thumb is:
Do not exceed 100°C for germanium transistors or 200°C for silicon transistors.

1-4.3. Effects of Temperature on Power Dissipation

The power-dissipation capabilities of a transistor must be carefully considered when designing any circuit. Of course, in small-signal circuits, the power dissipation is usually less than 1 W, and heat sinks are not needed. In such circuits, the only concern is that the rated power dissipation (as shown on the data sheet) not be exceeded. As a rule of thumb:
Do not exceed 90 per cent of the maximum power dissipation, as shown on the data sheet, for small-signal circuits.

As with other characteristics, transistor manufacturers specify maximum power dissipation in a variety of ways on their data sheets. Some manufacturers provide *safe-operating-area curves* for temperature and/or power dissipation of transistors. Other manufacturers specify *maximum power dissipation*, in relation to a given ambient temperature or a given case temperature. Still others specify a *maximum junction temperature* or a *maximum case temperature.*

1-4.3.1. Thermal Resistance

Transistors designed for power applications usually have some form of *thermal resistance* specified to indicate the power-dissipation capability. Thermal resistance can be defined as *the increase in temperature of the transistor junction (with respect to some reference) divided by the power dissipated,* or °C/W.

Power-transistor data sheets usually specify thermal resistance at a given temperature. For each increase in temperature from this specified value, there will be a change in the temperature-dependent characteristics of the transistor. Since there will be a change in temperature with changes in power dissipation of the transistor, the junction-to-ambient temperature will also change, resulting in a characteristic change. Therefore, the transistor characteristics can change with ambient temperature changes and with changes produced by variation in power dissipation.

In power transistors, thermal resistance is normally measured from the transistor pellet (or junction) to the case. This results in the term θ_{JC}. On those transistors where the case is bolted directly to the mounting surface with an integral threaded bolt or stud, the term θ_{MB} (thermal resistance to mounting base) or θ_{MF} (thermal resistance to mounting flange) is used. These terms take into consideration only the thermal paths from junction to case (or mount). For power transistors in which the pellet is mounted directly on a header or pedestal, the total internal thermal resistance from junction to case (or mount) varies from 50°C/W to less than 1°C/W.

1-4.3.2. Thermal Runaway

The main problem in operating a transistor at or near its maximum power dissipation limits is a condition known as *thermal runaway.* When current passes through a transistor junction, heat will be generated. If not all of this heat is dissipated by the case (an impossibility), the junction temperature will increase. This, in turn, will cause more current to flow through the junction, even though the voltage, circuit values, etc., remain the same. In turn, this will cause the junction temperature to increase even further, with a corresponding increase in current flow. If the heat is not dissipated by some means, the transistor will burn out and be destroyed.

1-4.3.3. Operating Transistors without Heat Sinks

If a transistor is not mounted on a heat sink, the thermal resistance from case to ambient air, θ_{CA}, is so large in relation to that from junction to case (or mount) that the total thermal resistance from junction to ambient air, θ_{JA}, is primarily the result of the θ_{CA} term.

Table 1-2 lists values of case-to-ambient air thermal resistances for a number of popular transistor cases. It will be seen that the large, heavy-duty cases such as TO–3 will have a small temperature increase (for a given wattage) in comparison to such cases as the TO–5. That is, the heavy-duty cases will dissipate the heat into the ambient air.

Table 1-2. CASE-TO-AMBIENT THERMAL RESISTANCE
FOR TYPICAL TRANSISTOR CASES

Case	$\theta_{CA}\,(^{\circ}C/W)$
TO–3	30
TO–5	150
TO–8	75
TO–18	300
TO–36	25
TO–39	150
TO–46	300
TO–60	70
TO–66	60

The information in Table 1-2 can be used to approximate the maximum power dissipation of transistors (without heat sinks) when such information is not shown on the data sheet. Assume that a germanium transistor with a TO–5 case is to be used and it is desired to find the absolute maximum power dissipation (without a heat sink).

The θ_{CA} factor for a TO–5 case is 150 (Table 1-2). As discussed in Sec. 1-4.2, germanium transistors should not be operated above 100°C. Assuming a 25°C ambient temperature, the transistor temperature must not be allowed to increase more than 75°C maximum:

$$\text{maximum power dissipation} = \frac{\text{maximum allowable temperature increase}}{\theta_{CA}}$$

$$= \frac{75}{150} = 0.5 \text{ W}$$

Therefore, the TO–5 case could dissipate 0.5 W and permit only a 75°C increase in temperature. However, as discussed in Sec. 1-4.2, current gain will double when the temperature is raised from 25°C to 100°C. Assuming that the voltage remains constant, the maximum dissipation allowable would be 0.25 W. This would be an *absolute maximum* figure, assuming a germanium transistor, TO–5 case, and an ambient temperature of 25°C. For practical design purposes, the 0.25-W figure would be safe if the case were mounted on a metal chassis, which would act as a heat sink.

1-4.3.4. Operating Transistors with Heat Sinks

After about 1 W (or less) it becomes impractical to increase the size of the case to make the case-to-ambient thermal resistance term comparable to the junction-to-case term. For this reason, most power transistors are designed for use with an external heat sink. Sometimes, the chassis or mounting area serves as the heat sink. In other cases, a heat sink is attached to the case. Either way, the primary purpose of the heat sink is to increase the effective heat-dissipation area of the case and to provide a low-heat-resistance path from case to ambient.

To properly design (or select) a heat sink for a given application, the thermal resistance of both the transistor and heat sink must be known. For this reason, power-transistor data sheets will specify the θ_{JA} that must be combined with the heat-sink thermal resistance to find the total power-dissipation capability. (Note that some power-transistor data sheets specify a *maximum case temperature* rather than θ_{JA}. As discussed in Sec. 1-4.3.5, maximum case temperature can be combined with heat-sink thermal resistance to find maximum power dissipation.)

Heat-sink ratings. Commercial fin-type heat sinks are available for various transistor case sizes and shapes. (Refer to Sec. 1-8.) Such heat sinks are especially useful when the transistors are mounted in Teflon sockets, which provide no thermal conduction to the chassis or printed circuit board. Commercial heat sinks are rated by the manufacturer in terms of thermal resistance, usually in terms of °C/W. When heat sinks involve the use of washers, the °C/W factor usually includes the thermal resistance between the transistor case and sink, θ_{CA}. Without a washer, only the sink-to-ambient, θ_{SA}, thermal resistance factor is given. Either way, the thermal resistance factor represents temperature increase (in °C) divided by wattage dissipated.

For example, if the heat-sink temperature rises from 25°C to 100°C (a 75°C increase) when 25 W is dissipated, the thermal resistance is $75 \div 25 = 3$. This could be listed on the data sheet as θ_{SA}, or simply as 3°C/W.

All other factors being equal, the heat sink with the lowest thermal resistance (°C/W) is best; that is, a heat sink with 1°C/W is better than a 3°C/W heat sink. Of course, the heat sink must fit the transistor case and the space around the transistor. Except for these factors, selection of a suitable heat sink should be no particular problem.

Calculating heat-sink capabilities. The thermal resistance of a heat sink can be calculated if the following factors are known: material, mounting provisions, exact dimensions, shape, thickness, surface finish, and color. Even if all these factors are known, the thermal resistance calculations are approximate.

As a *very approximate* rule of thumb:

$$\text{heat-sink thermal resistance (in } °C/W) = \sqrt{\frac{1500}{\text{area}}}$$

where the area (total area exposed to the air) is in square inches, material is $\frac{1}{8}$-in-thick aluminum, and the shape is a flat disc.

From a practical design standpoint, it is better to accept the manufacturer's specifications for a heat sink. The heat-sink thermal resistance actually consists of two series elements: the thermal resistance from the case to the heat sink that results from conduction (θ_{CS}), and the thermal resistance from the heat sink to the ambient air caused by convection and radiation (θ_{SA}).

Practical heat-sink considerations. To operate a transistor at its full power capabilities, there should be no temperature difference between the case and ambient air. This will occur only when the thermal resistance of the heat sink is zero, and the only thermal resistance is that between the junction and case. It is not practical to manufacture a heat sink with zero resistance. However, the greater the ratio θ_{JC}/θ_{CA}, the nearer the maximum power limit (set by θ_{JC}) can be approached.

When transistors are to be mounted on heat sinks, some form of electrical insulation must be provided between the case and the heat sink (unless a grounded collector circuit is used). Because good electrical insulators usually are also good thermal insulators, it is difficult to provide electrical insulation without introducing some thermal resistance between case and heat sink. The best materials for this application are mica, beryllium oxide (Beryllia), and anodized aluminum. The properties of these three materials for case-to-heat sink insulation of a TO–3 case are compared in Table 1-3.

Table 1-3. COMPARISON OF INSULATING WASHERS FOR HEAT SINKS

Material	Thickness (ins)	θ_{CS} (°C/W)	Capacitance (pF)
Beryllia	0.063	0.25	15
Anodized aluminum	0.016	0.35	110
Mica	0.002	0.4	90

For small, general-purpose transistors with a TO–5 case, a beryllium-oxide washer can be used to provide insulation between the case and a metal chassis or printed circuit board. The use of a zinc-oxide-filled silicon compound (such as Dow Corning 340 or Wakefield 1201) between the washer and chassis, together with a moderate amount of pressure from the top of the transistor, helps to decrease thermal resistance. If the transistor is mounted within a heat sink, a beryllium cup should also be used between the transistor and heat sink. Figure 1-5 shows both types of mounting. Section 1-8 describes some mounting techniques for power transistors.

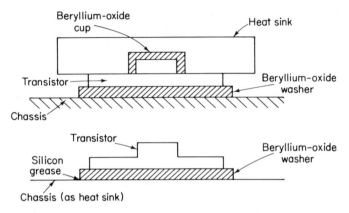

Fig. 1-5. Typical mounting arrangements for transistor heat sinks.

Note that a "Capacitance" column appears in Table 1-3. Any isolation between collector and the chassis (as is produced by the washer between the case and heat sink) will also result in capacitance between the two metals. This capacitance can be a problem and must be considered in RF circuit design. However, the problem of collector-to-chassis capacitance is usually noticed only at frequencies above 100 MHz.

1-4.3.5. Calculating Power Dissipation under Steady-State Operating Conditions

For design purposes, the no-signal dc collector voltage and current can be used to calculate power dissipation, when a transistor is operated under steady-state conditions. Actually, there are other currents that result in power dissipation (collector–base leakage, current emitter–base current). However, these can be ignored, and the power dissipation (in watts) can be considered as the dc collector voltage times the collector current.

Once the power dissipation has been calculated, the maximum *power dissipation capability* must be found. Under steady-state conditions, the maximum dissipation capability is dependent upon three factors: the sum of the series thermal resistances from the transistor junction to ambient air, the maximum junction temperature, and the ambient temperature.

Following are some examples of how power dissipation can be calculated. Assume that it is desired to find the *maximum power dissipation* of a transistor (in watts). The following conditions are specified: a maximum junction temperature of 200°C (typical for a silicon power transistor), a junction-to-case thermal resistance (θ_{JC}) of 2°C/W, a heat sink with a thermal resistance (θ_{SA}) of 3°C/W, and an ambient temperature of 25°C.

Note that the thermal resistance of the heat sink includes any thermal resistance produced by the washer between the transistor case and heat sink. (If this factor were not known, it would be a good rule of thumb to add 0.5°C/W thermal resistance for any washer between case and heat sink. As shown in Table 1-3, this thermal resistance is high and allows a safe tolerance.)

First, find the total junction-to-ambient thermal resistance θ_{JA}:

$$\theta_{JA} = \theta_{JC} + \theta_{SA} \quad \text{or} \quad 5 = 2 + 3$$

Next, find the maximum permitted power dissipation:

$$\text{maximum power} = \frac{\text{maximum junction temperature} - \text{ambient temperature}}{\theta_{JA}}$$

$$= \frac{200 - 25}{5} = 35 \text{ W (maximum)}$$

If the same transistor is used without a heat sink, but under the same conditions and with a TO–3 case, the maximum power can be calculated as follows. First, find the total junction-to-ambient thermal resistance θ_{JA}:

$$\theta_{JA} = \theta_{JC} + \theta_{CA} \quad \text{or} \quad 32 = 2 + 30$$

The value of 30 for a TO–3 case is obtained from Table 1-2. Next, find the maximum permitted power dissipation using:

$$\frac{\text{maximum junction temperature} - \text{ambient temperature}}{\theta_{JA}}$$

$$= \frac{200 - 25}{32} = 5 \text{ W (approximately)}$$

Some power-transistor data sheets specify a *maximum case temperature* rather than a maximum junction temperature. Assume that a maximum case temperature of 130°C is specified instead of a maximum junction temperature of 200°C. In that event, subtract the ambient temperature from the maximum permitted case temperature:

$$130° - 25° = 105°C$$

Then divide the case temperature by the heat-sink thermal resistance:

$$\frac{105°C}{3°C} = 35 \text{ W maximum power}$$

Note that the case temperature of any transistor can be calculated by:

maximum case temperature

$$= \text{maximum power} \times (\theta_{SA}) + \text{ambient temperature}$$

1-4.3.6. Calculating Power Dissipation under Pulse Operating Conditions

When transistors are operated by pulses, the maximum permitted power dissipation is much greater than with steady-state operation. The following are some examples of power dissipation calculations for both nonrepetitive and repetitive pulses.

For a *single, nonrepetitive pulse*, the *transient* thermal resistance must be calculated. Usually, the transient thermal resistance is shown on the safe-operating-area curves in the form of a *power multiplier*. The power multiplier is given for a specific case temperature and given pulse width.

For example, a 2N3055 transistor has power multipliers of 2.1 for 100-ms pulses, 3.0 for 1-ms pulses, and on up to 7.7 for 30-μs pulses. The steady-state (or direct current) maximum power is multiplied by these factors to find the maximum permitted single pulse power. Assuming that the maximum permitted steady-state power is 100 W for a given set of conditions, the single pulse maximum power would be 300 W if a 1-ms pulse were used.

Usually, the data sheets will specify the power mutipliers for a given case temperature. If the case temperature is increased, the factor must be derated. The temperature derating factor is found by:

$$\text{derating factor} = \frac{\text{case temperature} - \text{ambient temperature}}{\text{maximum junction temperature} - \text{ambient temperature}}$$

For example, assume that a transistor has a maximum junction temperature of 200°C, that the case temperature under these conditions is 130°C, that the ambient temperature is 25°C, and that the maximum steady-state power is 35 W. Further assume that the transistor data sheet specifies a power multiplier of 3.0 for 1-ms pulses, with a case temperature of 25°C.

The derating factor is

$$\frac{130°C - 25°C}{200°C - 25°C} = \frac{3}{5}$$

To find the maximum single pulse power:

$$\text{maximum power} = \text{multiplier} \times (1 - \text{derating factor})$$
$$\times \text{steady-state power}$$
$$= 3 \times (1 - \tfrac{3}{5}) \times 35 = 42 \text{ W}$$

These calculations for single-pulse operation are based on the assumption that the heat-sink capacity is large enough to prevent the heat-sink temperature from rising between pulses.

For *repetitive pulses*, both the case and heat-sink temperatures will rise. This temperature increase must be taken into account when determining the maximum power dissipation.

The maximum permitted power dissipation for a transistor operated with repetitive pulses is calculated by:

maximum permitted power

$$= \frac{\text{power multiplier} \times \text{maximum junction} - \text{ambient temperature}}{\theta_{JC} + \text{power multiplier} \times \text{duty cycle percentage} \times \theta_{JA}}$$

Assume that it is desired to find the maximum permitted power of the same transistor described for steady-state operation, but now operated by 1-ms pulses repeated at 100-Hz intervals. The conditions are: power multiplier 3.0, θ_{JC} 2°C/W, θ_{JA}-5°C/W (including heat sink), maximum junction temperature of 200°C, ambient temperature of 25°C, and a duty cycle of 10 per cent (1 ms on and 9 ms off):

$$\text{maximum permitted power} = \frac{3(200 - 25)}{2 + 3(0.1)5} = 150 \text{ W}$$

The case temperature of a transistor operated by repetitive pulses is:

case temperature = peak pulse power × duty cycle percentage

$$\times \; \theta_{JA} \text{ ambient temperature}$$

Peak pulse power is obtained by multiplying the collector voltage by the collector current (assuming that the transistor is operated in a grounded-emitter or grounded-base configuration, and that the transistor is switched full-on and full-off by the pulses).

For example, assume an ambient temperature of 25°C, a duty cycle of 10 per cent, a total θ_{JA} of 5°C/W, and peak pulses of 120 W (say, a collector voltage of 60, and a collector current of 2 A). The case temperature is:

$$\text{case temperature} = 120 \times 0.1 \times 5 + 25 = 85°C$$

1-5. Using Load Lines in Simplified Design

Load lines are useful design tools in interpreting data-sheet curves. Many data sheets show transistor characteristics by means of curves that are reproductions of displays obtained with an oscilloscope-type curve tracer. The collector current–voltage curves shown in Fig. 1-6 are typical. Such curves are obtained by applying a series of stepped base currents, then sweeping the collector voltage over a given range. Several curves are made in rapid succession at different base currents. Gain can be found by noting the difference in collector current for a given change in base current, while maintaining a fixed collector voltage.

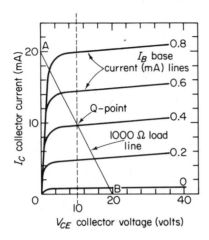

Fig. 1-6. Typical load line drawn on characteristic curves of common-emitter transistor.

It is also possible to examine the combined effect of base current and a collector load resistor of given value by superimposing a load line on the curves. For example, assume that a collector load of 1000 Ω is used with the transistor displayed in Fig. 1-6. When the collector current reaches 20 mA (base current approximately 0.8 mA), the collector voltage drops to zero. Likewise, when the collector current drops to zero, collector voltage rises to 20 V (base current zero). If a load line is connected between these two extreme points (marked A and B on Fig. 1-6), the instantaneous collector voltage and collector current can be obtained for any base current along the line. This permits an operating point to be selected by graphic means. For example, assume that it is desired to operate the collector at 10 V with no signal applied. Draw a line up from the 10-V point and note that it intercepts the load line at about 0.4-mA base current (and about 10-mA collector current). The transistor circuit bias resistors are then chosen to provide a 0.4-mA base current.

There are some schools of design that use the load-line method to select final component values. While this method is convenient and does provide some fairly accurate approximations, there is one drawback. The data-sheet curves are "typical" for a transistor of a given type and represent an average gain (or, in some cases, the minimum gain). Also, as previously discussed, transistor gain is temperature- and frequency-dependent. Therefore, the selection of load values, bias values, and operating point on the basis of static gain curves is subject to error. If the transistor's beta shifts, the operating point must shift, requiring different bias and load values.

As is discussed in later chapters, the problem of variable gain can be overcome by means of *feedback to stabilize the gain.* The feedback method of design makes the circuit characteristics dependent upon the *relationship of circuit values* rather than on transistor gain characteristics.

1-6. Basic Bias Schemes

All transistors (and other solid-state devices) require a form of bias. In a typical transistor the collector–base circuit must be reverse-biased. That is, current should not flow between collector and base. Any collector–base current that does flow is a result of leakage or breakdown. Breakdown must be avoided by proper design. Leakage is an undesirable (but almost always present) condition that must be reckoned with in practical design.

The emitter–base circuit of a typical transistor must be forward-biased. That is, current should flow between base and emitter during normal operation. Some current flows all the time in certain circuits (class A amplifiers and most oscillators). In other circuits (class C amplifiers, switches, etc.) current flows only in the presence of an operating signal or trigger. Either way, the emitter–base circuit must be biased so that current can flow.

The desired bias is accomplished by applying voltages to the corresponding transistor elements, usually through resistances. The following sections describe several basic bias circuits. These circuits (or variations of them) represent most of the bias methods used in transistor circuits.

Keep one point in mind when studying the following bias schemes: The purpose of the basic bias circuit is to establish collector–base–emitter voltage and current relationships *at the operating point of the circuit.* (The operating point is also known as the quiescent point, Q-point, no-signal point, idle point, or static point.) Since transistors rarely operate at this static point, the basic bias circuits are to be used as a reference or starting point for design. The actual circuit configuration, and (especially) the bias circuit values, must be selected on the basis of *dynamic circuit conditions* (desired output voltage swing, expected input signal level, etc.).

1-6.1. Basic Bias Design Considerations

The first step in bias design is to determine the characteristics of both the circuit and the transistor to be used. For example, will the circuit be used as an amplifier, oscillator, or switch? What class of operation (A, AB, B, or C) will be required? How much gain (if any) will be required? What power supply voltages will be available? What transistor is to be used? Must input and or output impedances be set at some arbitrary value?

In all the following basic bias circuit examples, it is assumed that some gain will be required (such as with an amplifier or oscillator circuit), that a 2N332 *NPN* transistor is to be used, that the power-supply voltage is 20 V, and that collector current must flow at all times (class A operation). Note that the same basic bias circuits can be used (but with different values) to produce other classes of operation.

Once these preliminary factors have been decided, the design data should be recorded. This will require reference to the transistor data sheet (Fig. 1-1), as well as some arbitrary decisions.

Available (or desired) power-supply voltage. An arbitrary 20 V has been selected. The voltage must be below the collector–emitter breakdown voltage (BV_{CEO}). The rated collector–emitter breakdown voltage for a 2N332 is 45 V, so the arbitrary 20 V is well within tolerance.

Maximum collector-current rating. The maximum rated collector current for a 2N332 is 25 mA (from the data sheet).

Maximum power dissipation. The maximum rated power for a 2N332 is 150 mW at 25°C and 50 mW at 125°C. *When the exact operating temperature is not known, use from one-half to one-fourth of the 25°C rating.* An arbitrary 50 mW is chosen for all of the following bias-circuit examples.

Leakage current (I_{CBO}). The maximum rated leakage current for a 2N332 is 2 μA at 25°C and 50 μA at 150°C. *When the exact operating temperature is not known, use 10 times the 25°C rating.* An arbitrary 20 μA is chosen for the examples.

1-6.2. Operating Load Current and Resistance

Once the basic design characteristics have been tabulated, the next step is to calculate the operating load current and resistance. In some circuits, an arbitrary load resistance must be used (to achieve a given circuit output impedance, for example). In other circuits, an arbitrary load current is required (to obtain a given power output). In the following examples, it is assumed that both load current and load resistance will be chosen solely to obtain a given operating point (no-signal collector voltage). Usually, the no-signal collector voltage is one-half the supply voltage. In the examples, this is 10 V (20 ÷ 2 = 10).

Two major factors determine operating load current for a transistor: *leakage current* and *maximum rated current.* Obviously, the load current cannot exceed the maximum rated current for the transistor (25 mA). Likewise, the load current must not be less than the leakage current (20 μA), or no current will flow.

If the circuit is to be operated from a battery, an operating load current near the low end should be selected to minimize battery drain. However, *the load current should be no less than 10 times the leakage current.* Since the leakage current is 20 μA, the *minimum load current* should be 0.2 mA.

At the high end, the maximum load current should not exceed the *maximum power-dissipation voltage.* Using the arbitrary 50 mW, divided by 10 V ($I = P \div E$), the maximum load current is 5 mA.

The operating load current should be midway between these two extremes: $5 - 0.2 = 4.8$, divided by 2 or 2.4 mA. The rule for selecting an operating load current midway between the two limits should be followed in all design work, unless there is specific need for a given collector voltage with a given load resistance.

Once the operating load current has been established, the next step is to calculate the load resistance value. When operating at frequencies up to about 100 kHz, the load resistance value can be calculated on a dc basis. At higher frequencies, it may be necessary to select a load resistance on the basis of impedance. This is discussed in later chapters.

When the operating load current is flowing at the selected no-signal point, the collector load resistance should drop the collector voltage to one-half the supply voltage. This is typical for all class A circuits. In the example, the source of 20 V is dropped to 10 V, through the load resistance. The desired 10 V divided by 2.4 mA ($R = E \div I$) produces a load resistance of 4166 Ω. The nearest standard value is 4300 Ω.

1-6.3. Bias with Emitter Feedback

Once the desired (or arbitrarily selected) load current and load resistance have been determined, the final step in the basic circuit design is to select the resistances for the emitter and base that will produce the correct operating point. Although there are many bias networks, each with its own advantages and disadvantages, one major factor must be considered for any network. *The basic network must maintain the desired base current in the presence of temperature (and frequency, in some cases) changes.* This is often referred to as *bias stability.*

There are several methods to provide temperature and frequency compensation for bias circuits. One of the most effective is with *emitter feedback.* Note that all the following bias circuits use some form of emitter resistance to produce feedback.

The use of an emitter-feedback resistance in any bias circuit can be summed up as follows: Base current (and, consequently, collector current) depends upon the differential in voltage between base and emitter. If the differential voltage is lowered, less base current (and, consequently, less collector current) will flow. The opposite is true when the differential is increased. All current flowing through the collector (ignoring collector–base leakage) also flows through the emitter resistor. The voltage drop across the emitter resistor is therefore dependent (in part) on the collector current.

Should the collector current increase for any reason, then both the emitter current and the voltage drop across the emitter resistor will also increase. This *negative feedback* tends to decrease the differential between base and emitter, thus lowering the base current. In turn, the lower base current tends to decrease the collector current and to offset the initial collector-current increase.

1-6.4. *Determining Base Current*

If the exact base current required to produce a given collector current is known, bias-circuit design is a simple matter. Unfortunately, the *exact* relationship between base and collector currents (or gain) is never known. Nor can data sheets be trusted for exact gain information. As has been discussed, gain is temperature- and frequency-dependent, as well as dependent upon circuit values.

There are two basic ways to find the *approximate base current* that will produce a given operating point. The first method involves a *load line* drawn on static collector characteristic curves, as shown in Fig. 1-7. As indicated, a base current of about 0.2 mA produces a collector current of 2.4 mA. This indicates a beta of about 12. Note that the load line is drawn between the source voltage (20 V) and the maximum permitted current (5 mA).

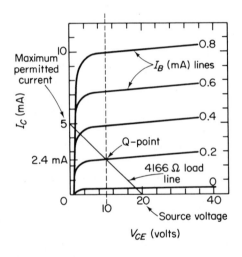

Fig. 1-7. Load line drawn on static characteristic curves to find approximate base current and Q-point.

The second method is to divide the desired collector current by the nominal beta as shown on the data sheet (Fig. 1-1). In this case, the nominal beta is 15. Using the 2.4-mA collector current, the base current is 2.4 ÷ 15, or 0.16 mA.

No matter what method is used, keep in mind that the values are approximate. Therefore, the selected bias resistance values are just as approximate. In addition, all resistors have some tolerance for their values (usually 5 or 10 per cent).

In practice, the calculated bias resistance values can be tried with a breadboard circuit. Then the resistance values are trimmed to produce the desired results. In the case of a bias circuit, "desired results" are a given collector voltage, current, and load. When the basic bias circuit is used in another circuit, "desired results" can be overall circuit functions (such as

gain for an amplifier, output voltage for an oscillator, etc.). This aspect of
bias adjustment is discussed in later chapters.

In all of the following examples, it is assumed that the base current is
0.2 mA.

1-6.5. Bias Circuit A

Figure 1-8 is the diagram for bias circuit A. The basic characteristics for
this circuit are noted below.

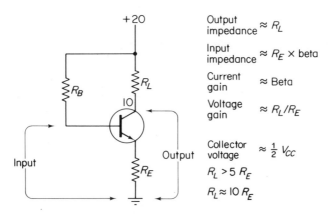

Fig. 1-8. Bias circuit A.

The output impedance is approximately equal to R_L (at frequencies up to
about 100 kHz).

The input impedance is approximately equal to beta times R_E (at frequencies up to about 100 kHz). Since input impedance is dependent upon beta,
the input impedance is subject to wide variation.

The circuit current gain is approximately equal to the ac beta of the
transistor.

The voltage gain is approximately equal to the ratio R_L/R_E.

Bias circuit A offers the greatest possible voltage gain, but the least
stability, of all the basic bias circuits described here.

The value of R_L is determined by the desired collector voltage and current,
or by an arbitrary need for a given output impedance, as discussed in Sec.
1-6.2.

The value of R_E is dependent upon a tradeoff between stability and gain.
An increase in the value of R_E, in relation to R_L, increases stability and
decreases gain. The value of R_E should be between 100 and 1000 Ω and
should not be greater than one-fifth of R_L. An R_E that is one-tenth of R_L can
be considered as typical. Assuming that R_L is 4300 Ω (Sec. 1-6.3), R_E should
be 430 Ω.

The value of R_B is selected to provide a given base current at the Q-point (no-signal point). For an *NPN* silicon transistor, it can be assumed that the base voltage will be 0.5 V more positive than the emitter. (The base of a *PNP* silicon transistor will be 0.5 V more negative than the emitter.) The emitter voltage is found by noting the drop across R_E ($E = IR$). Both the base current and collector current flow through R_E (although the base current can usually be ignored). Therefore, the current through R_E is $2.4 + 0.2$ or 2.6 mA. The drop across R_E is 2.6×430 or 1.2 V (approximately). With R_E at $+1.2$ V, the base voltage should be 1.7 V ($1.2 + 0.5$).

Since R_B is connected to the source of 20 V, the drop across R_B must be $20 - 1.7$, or 18.3 V. With a base current of 0.2 mA through R_B, a resistance of 91,500 Ω is required to produce a drop of 18.3 V ($R = E \div I$). The nearest standard resistor is 91,000 Ω. This is close enough for a trial value of R_B.

Some transistor data sheets show the collector current for a given base current, or given base–emitter voltage. When such information is available on data sheets, it is shown in the form of curves such as in Fig. 1-9. Note

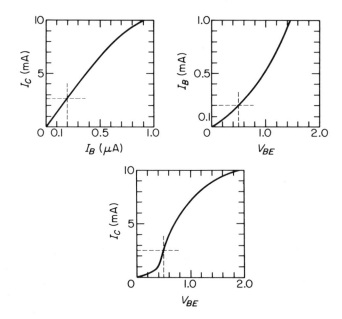

Fig. 1-9. Typical transfer characteristic curves.

that these curves show a collector current of 2.4 mA with a base current of 0.2 mA and with a base–emitter differential of 0.5 V. These curves, usually called *transfer characteristic curves*, make it possible to calculate a more accurate value of R_B. However, in practice, the calculated value of R_B is only a trial value. In the absence of transfer characteristic curves, the value of 0.5 V for the base–emitter differential is a good rule of thumb for all silicon transistors.

1-6.6. Bias Circuit B

Figure 1-10 is the diagram for bias circuit B. The basic characteristics for bias circuit B are the same as for circuit A, except that stability is increased. The increase in stability is brought about by connecting base resistance R_B to the collector rather than to the source. If collector current increases for any reason, the drop across R_L will increase, lowering the voltage at the collector. This will lower the base voltage and current, thus reducing the collector current. The feedback effect is combined with that produced by the emitter resistor R_E (Sec. 1-6.3) to offset any variations in collector current. However, gain for bias circuit B is only slightly less than for circuit A.

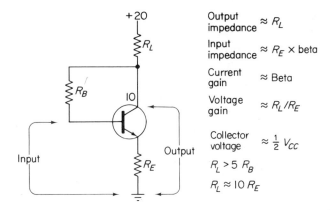

Fig. 1-10. Bias circuit B.

The values for R_L and R_E in circuit B are the same as for circuit A. The value of R_B is found in the same way but will be a different value. Since R_B is connected to the collector (at a theoretical 10 V), the drop across R_B must be 10 − 1.7, or 8.3 V. With a base current of 0.2 mA through R_B, a resistance of 41,500 Ω is required to produce a drop of 8.3 V. The nearest standard resistor is 43 kΩ.

1-6.7. Bias Circuit C

Figure 1-11 is the diagram for bias circuit C. The basic characteristics are noted below.

The output impedance is approximately equal to R_L (at frequencies up to about 100 kHz).

The input impedance is approximately equal to R_B (at frequencies up to about 100 kHz). Actually, the input impedance is equal to R_B in parallel

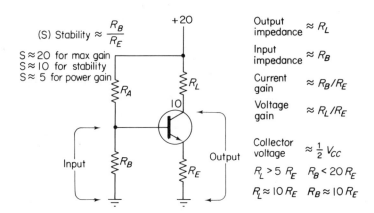

Fig. 1-11. Bias circuit C.

with $R_E \times$ (beta $+ 1$). However, unless the beta is very low, the $R_E \times$ (beta $+ 1$) factor will be much greater than R_B. Therefore, the value of R_B (or slightly less) can be considered as the stage or circuit input impedance and will be so considered when bias circuit C or any of its variations appear in examples throughout this book.

The current gain is approximately equal to the ratio R_B / R_E.

The voltage gain is approximately equal to the ratio R_L / R_E.

Bias circuit C offers more stability than either bias circuit A or B, but with a tradeoff of lower gain and lower input impedance.

The value of R_L is determined by the desired collector voltage and current, or by an arbitrary need for a given output impedance, as discussed in Sec. 1-6.2.

The value of R_E is dependent upon a tradeoff between stability and gain. An increase in the value of R_E, in relation to R_L, increases stability and decreases voltage gain. An increase in the value of R_E, in relation to R_B, increases stability and decreases current gain. The value of R_E should be between 100 and 1000 Ω, and should not be greater than one-fifth of R_L. An R_E that is one-tenth of R_L can be considered as typical. Assuming that R_L is 4300 Ω (Sec. 1-6.3), R_E should be 430 Ω.

The value of R_B is dependent upon tradeoffs between the value of R_E, current gain, stability, and the desired input impedance. As a basic rule, R_B should be approximately 10 times R_E. A higher value of R_B will increase current gain and decrease stability. Input impedance of the circuit will be approximately equal to R_B (actually slightly less). Therefore, if the input impedance is of special importance in the circuit, the value of R_B must be selected on that basis. This may require a different value of R_E to maintain stability and current gain relationships. Of course, any change in R_E will result in a change of the voltage gain (assuming that R_L remains the same).

The value of R_A is selected to provide a given base voltage at the Q-point. Assume the same base–emitter relationships as in the previous example (base at 1.7 V, emitter at 1.2 V, 0.5 V differential). Also assume that R_B is 10 times R_E (430 × 10 = 4300 Ω).

Since R_B is connected to the base, the drop across R_B is 1.7 V. Therefore, the current through R_B is 1.7 ÷ 4300, or 0.4 mA (approximately). This current combines with the base current (0.2 mA) to produce a total current of 0.6 mA through R_A.

Since R_A is connected to the source of 20 V, the drop across R_A must be 20 − 1.7, or 18.3 V. With a combined current of 0.6 mA through R_A, a resistance of 30,500 Ω is required to produce a drop of 18.3 V. The nearest standard resistor is 30,000 Ω.

In addition to stability, the major advantage of bias circuit C is that the input and output impedances, as well as voltage and current gain, are not dependent upon transistor beta. Instead, *circuit characteristics are dependent upon circuit values.*

1-6.8. Bias Circuit D

Figure 1-12 is the diagram of bias circuit D. The basic characteristics for bias circuit D are the same as for circuit C, except that temperature stability is increased. The increase in temperature stability is brought about by connecting diode D between the base and R_B. Diode D (forward-biased) is of the same material (silicon) as the base–emitter junction and is maintained at the same temperature. Thus, the voltage drops across diode D and the base–emitter junction are the same and remain the same with changes in temperature.

The values for R_L, R_E, and R_B in circuit D are the same as for circuit C. The value of R_A will be slightly different. Since the drop across diode D is

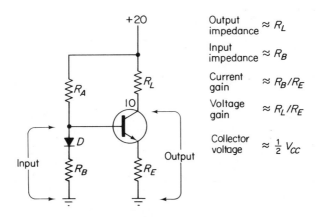

Fig. 1-12. Bias circuit D.

the same as across the base–emitter junction (0.5 V), the drop across R_B is the same as across R_E (1.2 V). Therefore, the current through R_B is 1.2 ÷ 4300, or 0.28 mA. This current combines with the base current (0.2 mA) to produce a total current of 0.48 mA through R_A.

Since R_A is connected to the source of 20 V, the drop across R_A must be 20 − 1.7, or 18.3 V. With a combined current of 0.48 mA through R_A, a resistance of 38,125 Ω is required to produce a drop of 18.3 V. The nearest standard resistor is 39,000 Ω.

As in the case of bias circuit C, bias circuit D's characteristics are not dependent upon transistor beta. In practice, diode D is mounted near the transistor so that the base–emitter junction and diode D are at the same temperature.

1-6.9. Bias Circuit E

Figure 1-13 is the diagram of bias circuit E. The basic characteristics for bias circuit E are the same as for circuit C. However, bias circuit E is used in those special applications that require a negative and positive voltage, each with respect to ground, to control base current.

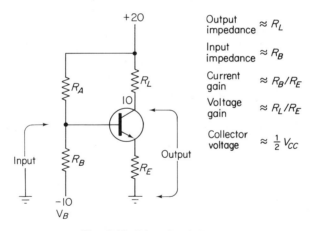

Fig. 1-13. Bias circuit E.

The values for R_L, R_E, and R_B in circuit E are the same as for circuit C. The value of R_A will be different due to the large amount of current through R_B. Since R_B is connected to a negative source V_B, the drop across R_B is V_B + the drop across R_E (1.2 V) + the base–emitter drop (0.5 V). Assume that V_B is −10 V. Then the drop across R_B is 10 + 1.2 + 0.5, or 11.7 V. Therefore, the current through R_B is 11.7 ÷ 4300, or 2.7 mA. This current combines with the base current (0.2 mA) to produce a total current of 2.9 mA through R_A.

Since R_A is connected to the source of 20 V, the drop across R_A must be 20 − 1.7, or 18.3 V. With a combined current of 2.9 mA through R_A, a resistance of 6310 Ω is required to produce a drop of 18.3 V. The nearest standard resistor is 6200 Ω.

1.6.10. Bias Circuit F

Figure 1-14 is the diagram of bias circuit F. This circuit is used in those special applications where it is necessary to supply collector–emitter current from both a positive and negative source. Since the transistor is *NPN*, the collector is connected to the positive source through R_L, while the emitter is connected to the negative source through R_E. If both sources are approximately equal, it is difficult to design a circuit that will produce any voltage gain (unless emitter-bypassing techniques are used as described in Chapter 2). The collector and emitter currents are approximately equal (ignoring base current). Therefore, if R_L drops the positive source to half (say from 20 to 10 V), then R_E would have to drop the entire negative source (from 20 to 0 V), and R_E would be approximately twice the resistance of R_L. This would produce a voltage loss, all other factors being equal.

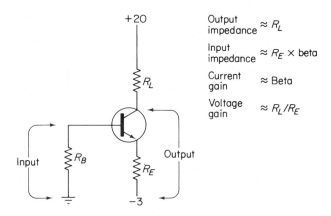

Output impedance	$\approx R_L$
Input impedance	$\approx R_E \times$ beta
Current gain	$\approx$ Beta
Voltage gain	$\approx R_L/R_E$

Fig. 1-14. Bias circuit F.

The basic characteristics for bias circuit F are essentially the same as for bias circuit A (Sec. 1-6.5), except that voltage gain will be low due to the lower R_L/R_E ratio.

In the following example, it is assumed that the conditions are the same as for previous bias circuit examples, except that R_E is returned to a −3-V source, rather than ground. That is, base current is 0.2 mA, collector–emitter current is 2.4 mA, positive source is 20 V, R_L drops the positive 20 V to approximately 10 V, and R_L is 4300 Ω, and the base and emitter are to be maintained at 1.7 and 1.2 V, respectively, at the Q-point.

The voltage drop across R_E must equal the desired $+1.2$ V, plus the -3 V from the negative source, or 4.2 V total. Both collector and base current flow through R_E, resulting in a total current of 2.6 mA, as in previous examples. A resistance of 1615 Ω is required for R_E to drop 4.2 V with 2.6 mA flowing.

The voltage drop across R_B must equal the desired $+1.7$ V. With a base current of 0.2 mA, a resistance of 8500 Ω is required for R_B.

1-7. Practical Design Techniques

This book assumes that the reader is familiar with the use of test equipment, hand tools, and basic shop practice. For those readers with little practical design experience, and as a refresher for the experienced technician, the following notes summarize basic shop and laboratory practice, as related to *general* solid-state design. Specialized solid-state design techniques such as test procedures for a particular type of circuit (amplifier, multivibrator, etc.) are discussed in the related chapter.

1-7.1. Assembling Experimental Circuits

Ideally, experimental circuits should be soldered together and arranged in the same approximate physical position as they will occupy in the final circuit. This makes for realistic test procedures to check design. Soldering experimental circuits also creates some problems. During the design process, it is inconvenient to solder and unsolder the same parts over and over again. In the case of some parts (transistors and diodes in particular), the continued application of the heat involved in soldering can result in permanent damage.

For this reason, it is recommended that only the final design circuit be soldered for testing. All experimental circuits should be assembled on screw-type terminal strips, as shown in Fig. 1-15. (Do not cut the component leads.) The terminal strips can be screwed or bolted to a flat wooden board (breadboard). Avoid the use of a metal work surface to prevent accidental shorts. Use barrier-type terminal strips with two terminals, one on each side. It is possible to secure three or four component leads to each terminal. Thus, six to eight leads can be connected to a given point. This is far more than usually required for any experimental circuit. Make certain to tighten the terminal screws down on the component leads to ensure good contact.

There are many commercial breadboard schemes available, such as the Vector "plugboards" or "Vectorboards," with push-in terminals. These commercial methods are especially helpful where several stages must be assembled in breadboard form.

Any parts to be used frequently should be provided with alligator clips. This will eliminate the need for soldering.

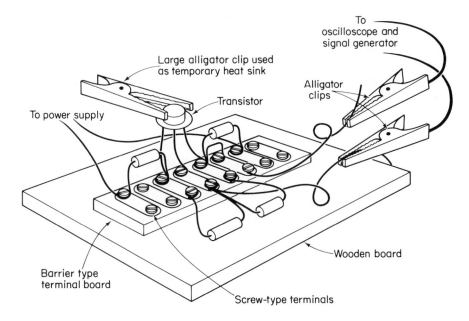

Fig. 1-15. Practical method for assembling experimental circuits.

Ideally, experimental circuits should be assembled in a room free from stray 60-Hz signals (such as a screen room). When this is not practical, the underside of the workbench can be covered with screen material (grounded). Try to avoid running long test leads. Where RF interference is a particular problem, the Vector "High-Frequency" breadboard kit may prove helpful.

1-7.2. Avoiding Thermal Problems

Overheating of transistors (from any cause) must be avoided. Ideally, the temperature of a transistor in an experimental circuit should be checked frequently with a thermometer. (Needless to say, this is rarely done.) Unless your fingers are very sensitive, they should be able to withstand about 50°C continuously. Any increase beyond that point, say, up to about 60°C, will be too painful to touch for more than an instant. Since 50°C is a good working temperature for germanium transistors, and a conservative temperature for silicon, any transistor that is not too hot to hold is safe.

The ideal ambient temperature for assembling experimental circuits is about 25°C (since this is the "normal" rating shown on most transistor data sheets). Of course, if a circuit is intended to operate at extreme temperatures in final form, it should be tested at those extreme temperatures after the final design circuit is complete.

As a quick test for thermal stability, check the experimental circuit characteristics (for bias, gain, etc.) at the ambient temperature. Then hold a soldering tool or "heat gun" near the transistor case for a short time to raise case temperature, while continuing to check the same characteristics. *Avoid prolonged contact between soldering tool and case.*

If the experimental circuit involves a power transistor that must be biased and tested with high currents, mount the transistor on a heat sink *before* applying any *power.*

A large alligator clip clamped onto the case of a small transistor makes a good *temporary* heat sink. Use more than one clip if necessary. If the alligator clip becomes very hot to the touch, disconnect the circuit *immediately.*

When soldering the final test circuit, avoid holding the hot tool on transistor and diode leads any longer than required.

1-7.3. Test Equipment for Design

The test equipment required for design, as well as the test equipment capabilities (frequency range, impedance, etc.), should be the same as for the comprehensive testing of the final circuit. For example, if the final circuit is an amplifier that must have a flat response from 30 Hz to 100 kHz, the design circuit should be tested with a signal generator having a flat output over that range.

The following equipment and parts should be available for design of any circuit.

At least two volt-ohmmeters (*VOM*). The VOM is the most useful test instrument for transistor circuit design. All the transistor bias circuits described thus far can be analyzed with an electronic voltmeter or a very sensitive VOM. A number of manufacturers produce VOMs specifically designed for transistor circuit analysis and design. (The Simpson Model *250* is a typical example.) These VOMs have very low voltage scales (0 to 0.05 V, 0 to 0.250 V) to measure the small voltage differences that often exist between elements of a transistor (especially between emitter and base). Such VOMs also have a very low voltage drop (50 mV) on the current ranges.

Regulated direct-current power source. Ideally, the voltage should be adjustable from 0 to 50 V, at least from 0 to 30 V. If possible, the *current limit* should also be adjustable. With an adjustable current supply, the maximum current can be set at a safe level for the transistor in the experimental circuit. This will prevent damage to the transistor due to an overload. Some power supplies also have a short-circuit protection feature (current drops to zero in case of a short across the supply output terminals). A power supply with both positive and negative outputs (from the same reference, or two separate sources) is most convenient.

Four resistance substitution boxes. Most bias circuits can be assembled with four resistors. These resistance values can be simulated by decade resistance boxes, one for each required resistor. If decade boxes are not available, potentiometers or other variable resistances can be used. Either way, the resistances are set to the calculated values, the experimental circuit is checked, and the resistances are readjusted to get the desired circuit operation (bias relationship, gain, feedback, etc.). When the circuit performs as desired, the final resistance values can be read off the decade box dials. If potentiometers or variable resistances are used instead of decade boxes, it will be necessary to disconnect the resistance value from the circuit, and measure the value with an ohmmeter.

Two capacitor substitution boxes. Two capacitor boxes allow for at least one input and one output capacitor. In addition, there should be several standard-value capacitors available, preferably with alligator clips. Emitter-bypass capacitors usually require large values (up to 1000 μF). Such capacitance values are available in low-voltage electrolytics (5 to 50 V).

Miscellaneous parts. The need for several silicon diodes, as well as Zener diodes of different voltage ratings, is certain to arise in most design work. Likewise, a battery or group of batteries with 1.5-V taps can be used most effectively to check bias circuits. For example, an external battery can provide a good fixed voltage source to check an experimental circuit's operation.

Signal generator. As discussed, the characteristics of a signal generator in design work must duplicate those for test of the final circuit. As a rule of thumb, the signal generator should have a low output impedance (500 Ω or less) since most transistor circuits have low input impedances. The generator output voltage should be variable from 0 to 10 V. Ideally, the output should be available in precision steps (such as 1 mV, 10 mV, 100 mV, and 1 V). If not, a voltage divider can be made up as shown in Fig. 1-16. In use, the generator is adjusted for 10 V peak to peak and each of the outputs checked with a precision voltmeter. Accuracy of the step voltage divider will depend upon accuracy of the resistance values. Precision resistors with 1 per cent accuracy should be sufficient.

Oscilloscope. The requirements for an oscilloscope used in design work are about the same as for any other shop or laboratory applications. It is helpful (but not absolutely essential) that the oscilloscope have a *differential input*. This makes it possible to measure small voltage drops across transistor elements. More important, a differential input minimizes pickup of stray noise (such as 60-Hz hum). Noise pickup is a particular problem in any experimental circuit. The circuit leads must be exposed to RF, hum, etc. A differential input responds only to a difference in voltage between two points. Noise, hum, RF, etc., appear on all leads at once, producing no differential voltage. (For this reason, noise and hum are referred to as *common-mode signals*. The ability of a differential input to reject such signals

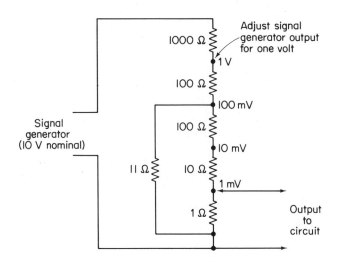

Fig. 1-16. Voltage divider with low output impedance.

is known as *common-mode rejection* and is usually expressed in decibels.) If an oscilloscope is being considered for design application, a high common-mode rejection (CMR) ratio is best.

A dual-trace oscilloscope can also be helpful in experimental design. The dual-trace feature makes it possible to observe both the input and output of an experimental circuit simultaneously. This is useful when adjusting bias values for a given performance.

1-7.4. Troubleshooting Design Circuits

All the circuits described here should work well, if the design rules are followed. However, it may be necessary to troubleshoot an experimental circuit. The first step in troubleshooting any transistor circuit is to check the basic bias circuit. If bias relationships are correct, it should be possible to isolate other circuit problems (either in design or because of defective components).

The remainder of this section is devoted to methods for analyzing faults in basic bias circuits, and for the check of components. Troubleshooting circuits is discussed in the test sections of related chapters.

1-7.4.1. Normal Bias Relationships

Figure 1-17 shows the basic bias connections for both *PNP* and *NPN* transistor circuits. As previously discussed, the basic bias circuit could include a diode for temperature compensation, or could omit one of the resistance values. However, the relationships remain the same. The purpose of Fig. 1-17

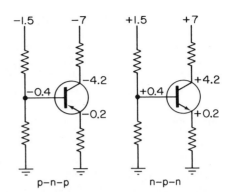

Fig. 1-17. Basic connections for *PNP* and *NPN* transistors with normal bias relationships.

is to establish "normal" voltage relationships. With a "normal" pattern established, it is relatively simple to find an "abnormal" condition.

In practically all transistor circuits, the emitter–base junction must be forward-biased to get electron flow through a transistor. In a *PNP* transistor, this means that the base must be made more negative (or less positive) than the emitter. Under these conditions, the emitter–base junction will draw current and cause a heavy current to flow from collector to emitter. In an *NPN* transistor, the base must be more positive (or less negative) than the emitter in order for current to flow from emitter to collector. Note that throughout this book, the term "current flow" refers to *electron flow* and is from negative to positive.

The following general rules can be helpful in practical analysis of transistor voltages in experimental circuits.

1. The middle letter in *PNP* or *NPN* always applies to the base.
2. The first two letters in *PNP* or *NPN* refer to the *relative* bias polarities of the *emitter* with respect to either the base or collector. For example, the letters *PN* (in *PNP*) indicate that the emitter is positive with respect to both the base and collector. The letters *NP* (*NPN*) indicate that the emitter is negative with respect to both the base and collector.
3. The collector–base junction is always reverse-biased.
4. The emitter–base junction is always forward-biased, or at a voltage level that can be forward-biased by a trigger signal.
5. A *base input voltage* that opposes or decreases the forward bias also decreases the emitter and collector currents.
6. A base input voltage that aids or increases the forward bias also increases the emitter and collector currents.
7. The dc *electron flow* is always against the direction of the arrow on the emitter.
8. If electron flow is into the emitter, electron flow will be out from the collector.
9. If electron flow is out from the emitter, electron flow will be into the collector.

Using these rules, "normal" transistor voltage can be summed up this way:

1. For a *PNP* transistor, the base is negative, the emitter is not quite as negative, and the collector is far more negative.
2. For an *NPN* transistor, the base is positive, the emitter is not quite as positive, and the collector is far more positive.

1-7.4.2. Measuring Transistor Voltage

There are two schools of thought on how to measure transistor voltages. Some prefer to measure transistor voltages from element to element and note the *difference in voltage*. For example, in the circuits of Fig. 1-17, an 0.2-V differential would exist between emitter and base. (This is typical for a germanium transistor.) Likewise, a 3.8-V differential would exist between base and collector. The element-to-element method of measuring transistor voltages will quickly establish the existence of forward and reverse bias.

The other method of measuring transistor voltages is to measure from the element to a *common* or *ground*. For example, all the voltages for the *PNP* transistor of Fig. 1-17 are negative with respect to ground. (The positive test lead of the meter must be connected to ground, and the negative test lead is connected to each element, in turn.)

This method of labeling transistor voltage is sometimes confusing to those not familiar with transistors, since it appears to break the rules. (In a *PNP* transistor, both the emitter and collector should be positive, yet all the elements are negative.) However, the rules still apply.

In the case of the *PNP* transistor of Fig. 1-17, the emitter is at −0.2 V, whereas the base is at −0.4 V. The base is *more negative* than the emitter. Therefore, the emitter is *positive with respect to the base*, and the base–emitter junction is forward-biased (normal).

On the other hand, the base is at −0.4 V, whereas the collector is at −4.2 V. The base is *less negative* than the collector. Therefore, the base is *positive with respect to the collector*, and the base–collector junction is reverse-biased (normal).

1-7.4.3. Troubleshooting with Transistor Voltages

The first step in troubleshooting any transistor circuit is to make a *visual check* of all connections. (This cures more problems in experimental circuits than the reader may realize.) Also, the circuit should be checked by someone other than the person who made the connections, if possible. It is very easy to overlook one's own mistakes. After the connections have been checked, measure the circuit voltages.

Assume that an experimental *PNP* circuit was assembled, and the voltages were found to be similar to those in Fig. 1-18. Except in one case, these voltages would indicate a defect. It is obvious that the transistor is not

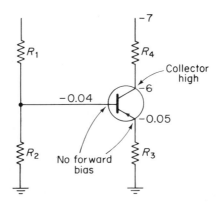

Abnormal voltages: collector high, emitter low, emitter-base not forward biased

Fig. 1-18. *PNP* transistor circuit with abnormal voltages.

forward-biased because the emitter is more negative than the base (reverse bias for a *PNP* transistor). The only circuit where this might be normal is one that requires large "trigger" signal to turn it on (such as a class C amplifier).

The first clue in Fig. 1-18 that something is wrong is that the collector voltage is almost as large as the collector source (at R_4). This means that very little current is flowing through R_4 in the collector–emitter circuit. The transistor could be defective. The trouble is more likely caused by a problem in bias. The emitter voltage depends mostly on the current through R_3, so unless the value of R_3 is completely out of limits (this would be unusual), the problem is one of incorrect bias on the base.

The next step in this case is to measure the bias source voltage at R_1. (Note that R_1 could be connected to the same source as R_4, or to the collector, or to another source that has a common reference to the source at R_4. In this example, it is assumed that R_1 is connected to a 1.5-V source, as shown in the "normal" circuit of Fig. 1-17.)

If the source is as shown in Fig. 1-19 (as 0.5 V), the problem is obvious: an incorrect external bias voltage. This will probably show up as a defect in the power supply, and will appear as an incorrect voltage in other circuits.

If the source voltage is correct, as shown in Fig. 1-20 (1.5 V), the cause of the trouble will probably be an incorrect selection of values for R_1 or R_2, or a defect in the transistor.

The next step is to remove all voltages from the experimental circuit and measure the resistance of R_1 and R_2. If either value is incorrect, the corresponding resistor must be replaced. If both values are correct, the transistor must be tested and/or substituted.

Do not attempt to measure resistance values in transistor circuits (experimental or otherwise) with the resistors still connected. This practice may

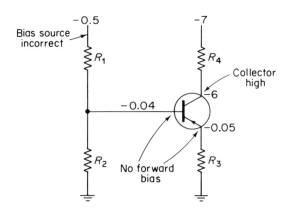

Fig. 1-19. *PNP* transistor circuit with incorrect bias source.

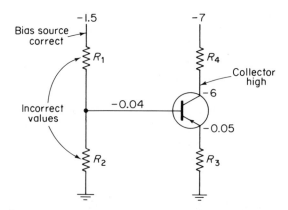

Fig. 1-20. *PNP* transistor with incorrect bias resistance values.

be correct for vacuum-tube circuits but not with transistor circuits. One reason for this is that the voltage produced by the ohmmeter battery could damage some transistors. Even if the voltages are not dangerous, the chance for an error is greater with a transistor circuit because the transistor junctions will pass current in one direction. This can complete a circuit through other resistors and produce a series or parallel combination, thus producing a false indication.

For example, assume that an ohmmeter is connected across R_2 (in Figs. 1-17 through 1-20) with the ohmmeter battery's positive terminal connected to ground. Because R_3 is also connected to ground, the positive terminal is connected to the ground end of R_3. Since the battery's negative terminal is connected to the transistor base, the emitter–base junction is forward-biased and current flows. In effect, R_3 is now in parallel with R_2, and the ohmmeter reading is incorrect.

1-7.5. Transistor and Diode Tests with a Meter

It is possible that the failure of an experimental circuit is due to the failure of the transistor or diodes. Substitution is a good method for checking components. If a new component works, it removes all doubt. There are four basic tests required for transistors (when they are checked as individual components): gain, leakage, breakdown, and switching time. All these tests are best made with commercial transistor testers and oscilloscopes. However, it is possible to test a transistor with an ohmmeter. These simple tests will determine whether the transistor is leaking and shows some gain. Usually, a transistor will operate in a circuit if the transistor shows some gain and is not showing any excessive leakage. Likewise, an ohmmeter will provide a quick check of a diode's ability to pass current in one direction only.

1-7.5.1. Diode Continuity Test

A simple resistance measurement, or continuity check, can be used to test a diode's ability to pass current in one direction only. An ohmmeter can be used to measure the forward and reverse resistance of a diode, as shown in Fig. 1-21.

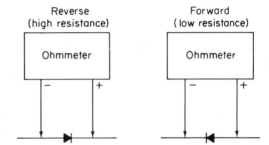

Fig. 1-21. Basic ohmmeter continuity test of diodes for front-to-back ratio.

A good diode will show high resistance in the reverse direction and low resistance in the forward direction. If the resistance is high in both directions, the diode is probably open. A low resistance in both directions usually indicates a shorted diode.

It is possible for a defective diode to show a difference in forward and reverse resistance. The important factor in making a diode resistance test is the *ratio* of forward-to-reverse resistance (often known as the *back-to-front ratio*). The actual ratio will depend upon the type of diode. However, as a rule of thumb, a small-signal diode has a ratio of several hundred to one, and a power rectifier can operate satisfactorily with a ratio of 10:1.

Diodes used in power circuits are usually not required to operate at high frequencies. Such diodes may be tested effectively with direct current or low-frequency alternating current. Diodes used in other circuits, even in audio equipment, must be capable of operation at higher frequencies and should be so tested.

1-7.5.2. Transistor Leakage Test

For the purpose of this test with an ohmmeter, a transistor can be considered as two diodes connected back to back. Therefore, each diode should show low forward resistance and high reverse resistance. These resistances can be measured with an ohmmeter as shown in Fig. 1-22. The same ohmmeter range should be used for each pair of measurements (base to emitter, base to collector, collector to emitter). On low-power transistors, there may be a few ohms indicated from collector to emitter. Avoid using the R × 1 range or an ohmmeter with a high internal-battery voltage. Either of these conditions can damage a low-power transistor.

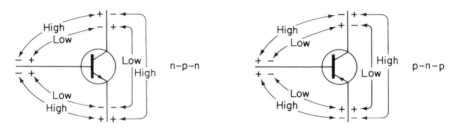

Fig. 1-22. Transistor leakage tests with an ohmmeter.

If both forward and reverse readings are very high, the transistor is open. Likewise, if any of the readings show a short or very low resistance, the transistor is shorted. Also, if the forward and reverse readings are the same (or nearly equal), the transistor is defective.

A typical forward resistance is 300 to 700 Ω. Typical reverse resistances are 10 to 60 kΩ. Actual resistance values will depend upon ohmmeter range and battery voltage. Therefore, the ratio of forward to reverse resistance is the best indicator. Almost any transistor will show a ratio of 30:1. Many transistors show ratios of 100:1 or greater.

1-7.5.3. Transistor Gain Test

Normally, there will be little or no current flow between emitter and collector until the base–emitter junction is forward-biased. This fact can be used to provide a basic gain test of a transistor using an ohmmeter. The test circuit is shown in Fig. 1-23. In this test, the R × 1 range should be used. Any internal-battery voltage can be used, provided it does not exceed the maximum collector–emitter breakdown voltage.

In position A of switch S_1, there is no voltage applied to the base, and the base–emitter junction is not forward-biased. Therefore, the ohmmeter should read a high resistance. When switch S_1 is set to B, the base–emitter circuit is forward-biased (by the voltage across R_1 and R_2), and current flows in the emitter–collector circuit. This is indicated by a lower resistance reading on

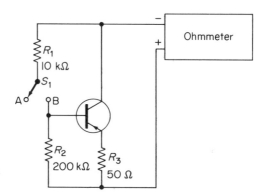

Fig. 1-23. Simple transistor gain test using an ohmmeter.

the ohmmeter. A 10:1 resistance ratio is typical for an audio-frequency transistor.

1-8. Practical Transistor Mounting Techniques

As discussed in Sec. 1-4, proper mounting procedures must be followed if the interface thermal resistance between the transistor package (or case) and heat sink is to be minimized. Proper mounting will provide cooler and more reliable operation of transistors. This section describes mounting techniques for metal-packaged power transistors (and other semiconductors). Included in this section are discussions concerning preparation of the mounting surface, use of thermal compounds, and fastening techniques. Typical interface thermal resistance is given for a number of packages.

1-8.1. Basic Thermal Resistance Concepts

The basic thermal resistance model used in this section is shown in Fig. 1-24. The equivalent electrical circuit may be analyzed using the following equation:

$$T_J = P_D(R_{\theta JC} + R_{\theta CS} + R_{\theta SA}) + T_A$$

where T_J = junction temperature
P_D = power dissipation
$R_{\theta JC}$ = semiconductor thermal resistance (junction to case)
$R_{\theta CS}$ = interface thermal resistance (case to heat sink)
$R_{\theta SA}$ = heat-sink thermal resistance (heat sink to ambient)
T_A = ambient temperature

As shown in Fig. 1-24, and discussed in Sec. 1-4, the thermal resistance junction to ambient is *the sum of the individual components.* Each component must be minimized if the lowest junction temperature is to result. The value of interface thermal resistance $R_{\theta CS}$ is affected by the mounting procedure.

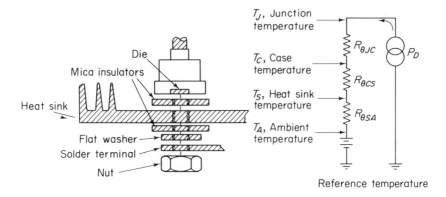

Fig. 1-24. Basic thermal-resistance model showing thermal-to-electrical analogy for a semiconductor. Courtesy Motorola.

Figure 1-25 shows some typical values of $R_{\theta CS}$ for various mounting situations. The following sections discuss practical methods to get lowest junction temperature under all mounting conditions.

1-8.2. Preparation of Mounting Surface

In general, the heat sink should have a flatness and finish comparable to that of the transistor. In lower-power applications, the heat-sink surface is satisfactory if it *appears* flat against a straight edge and is free of deep scratches. In high-power applications, a more detailed examination of the surface is required.

Surface flatness is determined by comparing the variance in height (Δh) of the test specimen to that of a reference standard, as shown in Fig. 1-26. Flatness is normally specified as a fraction of the *total indicator reading* (TIR). The mounting-surface flatness (Δh/TIR) is satisfactory if the deviation (Δh) is less than 0.001 TIR.

Surface finish is the average of the deviations both above and below the mean value of surface height. A finish in the range of 30 to 60 μin is satisfactory. A finer finish is costly and does not significantly lower contact resistance.

Most commercially available cast or extruded heat sinks require spot-facing. In general, milled or machined surfaces are satisfactory if prepared with tools in good working condition.

It is also necessary that the surface be free from all foreign material, film, and oxide (freshly bared aluminum forms an oxide layer in a few seconds). Unless used immediately after machining, it is good practice to polish the mounting area with No. 000 steel wool, followed by an acetone or alcohol rinse. Thermal grease should be applied immediately thereafter.

| | Package type and data | | | $(R_{\theta CS})$ | Interface thermal resistance (°C/W) | | | | | | |
| Motorola Case # | JEDEC Outline # | Description | Recomended hole and drill size | Torque in-lb | Metal-to-metal | | With insulator | | | | Motorola kit # |
					Dry	Lubed	Dry	Lubed	Type		
56	DO-4	10-32 Stud 7/16" Hex	0.188, #12	15	0.41	0.22	1.24	1.06	3 mil Mica		MH745
58	DO-5	1/4-28 Stud 11/16" Hex	0.25, #1	30	0.38	0.20	0.89	0.70	5 mil Mica		MH746
43	DO-21	Pressfit, 1/2"	See fig 1-29	†	0.15	0.1	—	—	—		
1,3,4	TO-3	Diamond	0.14, #28	†	0.2 / — / —	0.1 / — / —	1.45 / 0.8 / 0.4	0.8 / 0.4 / 0.35	3 mil teflon / 2 mil Mica / Annodized Aluminum		MK10 / MK15 / MK20
80	TO-66	Diamond	0.14, #28	†	—	0.5					
179		1 1/4" Square assembly	0.14, #28*	†	0.24	0.1					
219	TO-83	1/2"-20 Stud	0.5, 0.5	130	—	0.1					
246	TO-94	1-1/16" Hex									

* Can be tapped for 10-24 machine screw.
† Limited by mounting screw capability.

Fig. 1-25. Typical values for interface thermal-resistance and other package data. Courtesy Motorola.

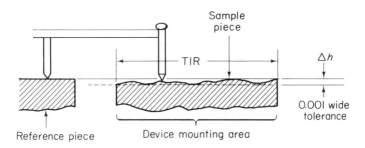

Fig. 1-26. Surface flatness of device mounting area. Courtesy Motorola.

Many aluminum heat sinks are *black-anodized* for appearance, durability, performance, and economy; however, anodizing is an electrical and thermal *insulator* which offers resistance to heat flow. Thus, anodizing should be removed from the mounting area.

Another aluminum finish is *irridite*, or chromate acid dip, which offers low resistance because of its thin surface. But, for best performance, the irridite finish must be cleaned of oils and films that collect in the manufacture and storage of the sinks.

Some heat sinks are *painted* after manufacture. Paint of any kind has a high thermal resistance (compared to metal). For that reason it is essential that paint be removed from the heat-sink surface where the transistor is attached.

1-8.3. Thermal Compounds

To improve contact between transistor and heat sink, thermal joint compounds (silicon greases) are used to fill air voids between the mating surfaces. A typical compound has a resistivity of about 60°C-in/W, compared to about 1200°C-in/W for air. Thus, the thermal resistance of voids, scratches, and imperfections filled with a joint compound will be about one-twentieth of the original value.

Joint compounds are a formulation of fine zinc particles in a silicon oil which maintains a greaselike consistency with time and temperature. There are two commonly used methods for applying the compounds. With one technique, the compound is applied in a very thin layer with a spatula or lintless brush, wiping lightly to remove excess material. The other technique involves applying a small amount of compound to the center of the contact area, then using the mounting and rotating pressure to spread the compound. Any excess compound is then removed after the mounting is completed.

The excess compound is wiped away using a cloth moistened with acetone or alcohol.

Recommended joint compounds are

1. Astrodyne, Conductive Compound 829.
2. Dow Corning, Silicon Heat Sink Compound 340.
3. Emerson & Cuming, Inc., Eccotherm, TC-4.
4. General Electric, Insulgrease.
5. George Risk Industries, Thermal Transfer Compound XL500.
6. IERC, Thermate.
7. Thermalloy, Thermacote.
8. Wakefield, Thermal Compound Type 1201.

1-8.4. Fastening Techniques

Each of the various types of transistor packages in use requires different fastening techniques. Mounting details for *stud, flat-base, press-fit,* and *disc-type* transistors are shown in Figs. 1-27, 1-28, 1-29, and 1-30, respectively. The following notes supplement these illustrations.

With any of the mounting schemes, the screw threads should be free of grease to prevent inconsistent torque readings when tightening nuts. Maximum allowable torque should always be used to reduce $R_{\theta CS}$. However, care

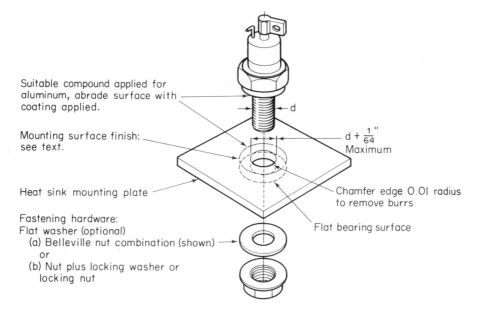

Fig. 1-27. Mounting details for stud-mounted semiconductors. Courtesy Motorola.

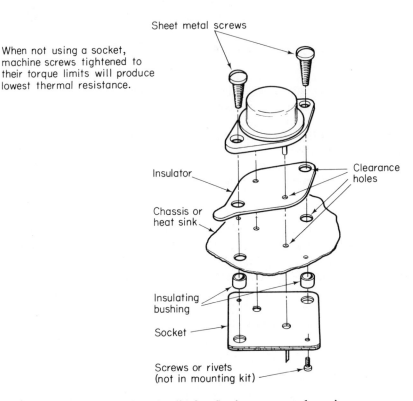

When not using a socket,
machine screws tightened to
their torque limits will produce
lowest thermal resistance.

Fig. 1-28. Mounting details for flat-base-mounted semiconductors (TO-66 shown). Courtesy Motorola.

must be exercised to not exceed the torque rating of parts. Excessive torque
applied to disc- or stud-mounted parts could cause damage to the semiconductor die. To prevent galvanic action from occurring when devices are
used on aluminum heat sinks in a corrosive atmosphere, many devices are
nickel- or gold-plated. Consequently, precautions must be taken not to mar
the finish.

With press-fit devices (Fig. 1-29), the hole edge must be chamfered as
shown to prevent shearing off the knurled edge of the device during press-in.
The pressing force should be applied evenly on the shoulder ring to avoid
tilting or canting of the device case in the hold during the pressing operation.
Also, joint compound should be used to ease the device into the hole. Typically, the pressing force will vary from 250 to 1000 lb, depending upon the
heat-sink material. Recommended hardnesses are: copper, less than 50 on
the Rockwell F scale; aluminum, less than 65 on the Brinell scale. A heat
sink as thin as $\frac{1}{8}$ in. may be used, but the $R_{\theta CS}$ will increase in proportion

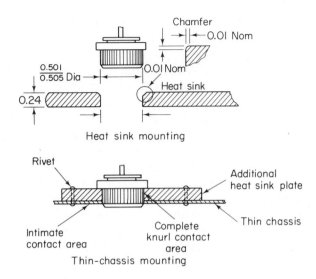

Heat sink mounting

Thin-chassis mounting

Fig. 1-29. Mounting details for press-fit semiconductors. Courtesy Motorola.

to the reduction in contact area. A thin chassis requires the addition of a back-up plate.

With the disc-type mounting (Fig. 1-30), a self-leveling type of mounting clamp is recommended to assure that the contacts are parallel and that there

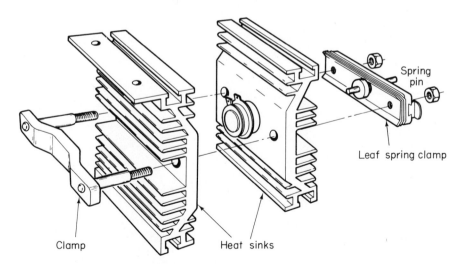

Fig. 1-30. Mounting details for disc-type semiconductors. Courtesy Motorola.

is an even distribution of pressure on each contact area. A swivel-type clamp or a narrow leaf spring in contact with the heat sink is usually acceptable.

The clamping force should be applied smoothly, evenly, and perpendicular to the disc-type package to prevent deformation of the device, or the sink mounting surfaces, during installation. The spring used should provide a mounting force within the range recommended by the transistor manufacturer. Typical clamping forces for disc-type transistors are 800 to 2000 lb.

Installation of a disc-type device between two heat sinks should be done in a manner to permit one heat sink to move with respect to the other. Such movement will avoid stresses being devloped due to thermal expansion, which could damage the transistor. Also, when two or more transistors are to be operated electrically in parallel, one of the heat sinks can be common to both (or all) transistors. Individual heat sinks must be provided against the other mounting surfaces of the transistors so that the mounting force applied in each case will be independently adjustable.

2

AUDIO AMPLIFIERS

2-1. Basic Audio Amplifier

Figure 2-1 is the working schematic of a basic, single-stage, audio amplifier. Note that the basic audio circuit is similar to bias circuit C (Sec. 1-6.7, Fig. 1-11), except that input and output coupling capacitors C_1 and C_2 are added. These capacitors prevent direct-current flow to and from external circuits. Note that a bypass capacitor C_3 is shown connected across the emitter resistor R_E. Capacitor C_3 is required only under certain conditions, as discussed in Sec. 2-2.

Input to the amplifier is applied between base and ground, across R_B. Output is taken across the collector and ground. The input signal adds to, or subtracts from, the bias voltage across R_B. Variations in bias voltage cause corresponding variations in base current, collector current, and the drop across collector resistor R_L. Therefore, the collector voltage (or circuit output) follows the input signal waveform, except that the output is inverted in phase. (If the input swings positive, the output swings negative, and vice versa.)

Variations in collector current also cause variations in emitter current. This results in a change of voltage drop across the emitter resistor R_E, and a change in the base–emitter bias relationship. As previously discussed in Chapter 1, the change in bias that results from the voltage drop across R_E tends to cancel the initial bias change caused by the input signal, and serves as a form of negative feedback to increase stability (and limit gain). This form of emitter feedback (current feedback) is known as *stage feedback*, or *local*

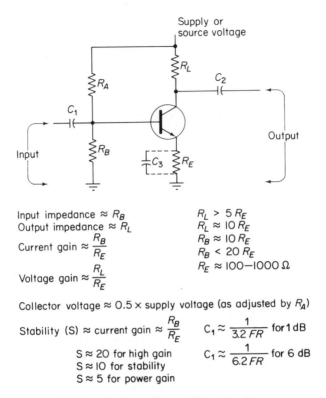

Fig. 2-1. Basic audio-amplifier circuit.

feedback, since only one stage is involved. As discussed in later chapters and sections, *overall feedback*, or *loop feedback*, is sometimes used where several stages are involved.

The outstanding characteristic of the circuit in Fig. 2-1 is that circuit characteristics (gain, stability, impedance) are determined (primarily) by circuit values rather than transistor characteristics (beta).

2-1.1. Design Considerations

The circuit is shown with an *NPN* transistor. Reverse the power supply polarity if a *PNP* transistor is used.

If a maximum source voltage is specified in the design problem, the maximum peak-to-peak output voltage is set. For class A operation, the collector is operated at approximately one-half the source voltage. This permits the maximum positive and negative swing of output voltage. The peak-to-peak output voltage cannot exceed the source voltage. Generally, the absolute maximum peak-to-peak output can be between 90 and 95 per cent of the source. For example, if the source is 20 V, the collector will operate at 10 V

(Q-point), and swing from about 1 V to 19 V. However, there is less distortion if the output is one-half to one-third of the source.

If a source voltage is not specified, two major factors should determine the value: the maximum collector voltage rating of the transistor, and the desired output voltage (or the desired collector voltage at the operating point). The maximum collector voltage rating must not be exceeded. As discussed in Chapter 1, choose a source voltage that does not exceed 90 per cent of the maximum rating. This allows a 10 per cent safety factor. Any desired output voltage (or collector Q-point voltage) can be selected, within these limits.

If the circuit is to be battery-operated, choose a source voltage that is a multiple of 1.5 V.

If a peak-to-peak output voltage is specified, add 10 per cent (to the peak-to-peak value) to find the absolute minimum source voltage.

If a collector Q-point voltage is specified, double the collector Q-point voltage.

For minimum distortion, use a source that is 2 to 3 times the desired output voltage.

If the input and/or output impedances are specified, the values of R_B and R_L are set, as shown in Fig. 2-1. However, there are certain limitations for R_B and R_L imposed by the tradeoff between gain and stability. For example, for a stage current gain of 10 and nominal stability, R_B should be 10 times R_E. R_B should never be greater than 20 times R_E (for maximum current gain and minimum stability), nor should R_B be less than 5 times R_E (for minimum current gain and maximum stability). Since R_E is between 100 and 1000 Ω in a typical class A audio circuit, R_B (and the input impedance) should be between 500 (100 × 5) and 20,000 Ω (1000 × 20).

The value of R_L should also be chosen on the basis of a tradeoff between stability and gain, with R_L never exceeding 20 times R_E, and never being less than 5 times R_E. Since R_E is between 100 and 1000 Ω, R_L (and the output impedance) should also be between 500 and 20,000 Ω.

If the input and/or output impedances are not specified, try to match the impedances of the previous stage and following stage, where practical. This will provide maximum power transfer.

The values of coupling capacitors C_1 and C_2 are dependent upon the *low-frequency limit* at which the amplifier is to operate, and on the resistances with which the capacitors operate. As frequency increases, capacitive reactance decreases and the coupling capacitors become (in effect) a short to the signal. Therefore, the high-frequency limit need not be considered in audio circuits. Capacitor C_1 forms a high-pass *RC* filter with R_B. Capacitor C_2 forms another high-pass filter with the input resistance of the following stage (or the load). This condition is shown in Fig. 2-2. The input voltage is applied across the capacitor and resistor in series. The output voltage is taken across the resis-

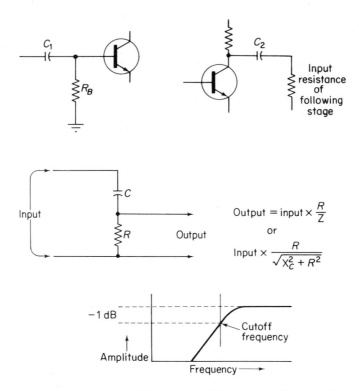

Fig. 2-2. Formation of high-pass RC filter by coupling capacitors and related resistances.

tance. The relation of input voltage to output voltage is:

$$\text{output voltage} = \text{input voltage} \times R \div Z$$

where R is the dc resistance value and Z is the impedance obtained by the vector combination of series capacitive reactance and dc resistance.

When the reactance drops to approximately one-half of the resistance, the output will drop to approximately 90 per cent of the input (or approximately a 1-dB loss). Using the 1-dB loss as the low-frequency cutoff point, the value of C_1 or C_2 can be found by

$$\text{capacitance} = \frac{1}{3.2FR}$$

where capacitance is in farads, F is the low-frequency limit in hertz, and R is resistance in ohms.

The *minimum ac beta of the transistor* should be higher than the desired gain, even though gain for the circuit is set by circuit values. Since the circuit of Fig. 2-1 is designed for maximum gains of $20:1$, any transistor with a minimum beta of 20 should be satisfactory.

2-1.2. Design Example

Assume that the circuit of Fig. 2-1 is to be used as a single-stage voltage amplifier. The desired output is 3 V (peak-to-peak) with a 2000-Ω impedance. The input impedance is to be 1000 Ω. The input signal is 0.3 V (peak to peak). This requires a voltage gain of 10. The low-frequency limit is 30 Hz, with a high-frequency limit of 100 kHz. Minimum distortion is desired. (The circuit should not be overdriven.) The source voltage and transistor type are not specified, but the circuit will be battery-operated.

Supply voltage and operating point. The 3-V output could be obtained with a 4.5-V battery. However, for minimum distortion, the supply should be 2 or 3 times the desired output, or between 6 and 9 V. A 9-V battery will provide the maximum insurance against distortion. Therefore, the collector Q-point voltage should be 4.5 V (9 V $\div$ 2 = 4.5 V).

Load resistance and collector current. The value of R_L should provide the output impedance of 2000 Ω. With a 4.5-V drop across R_L, the collector current is 4.5 $\div$ 2000, or 2.25 mA.

Emitter resistance, current, and voltage. To provide a voltage gain of 10, the value of R_E should be one-tenth of R_L or 2000 $\div$ 10 = 200 Ω. [If tests prove gain slightly below 10, try reducing the value of R_E to the next lower standard value (180 Ω).] The current through R_E will be the collector current of 2.25 mA, plus the base current. Assuming a stage current gain of 10, the base current is 2.25 $\div$ 10, or 0.225 mA. The combined currents through R_E are 2.25 + 0.225, or 2.475 mA. This will produce a drop of 0.495 V across R_E. For practical design, this could be rounded off to 0.5 V.

Input resistance and current. The value of R_B should provide the input impedance of 1000 Ω. The value of R_B should also be at least 5 times R_E, which makes the 1000 $\div$ 200-Ω relationship correct. This relationship provides maximum circuit stability. The base voltage will be 0.5 V higher than the emitter voltage, or 0.5 + 0.5 = 1.0 V. With a 1-V drop across R_B, the current through R_B is 1 $\div$ 1000 = 1 mA.

Base resistance and current. The value of R_A should be sufficient to drop the 9-V source to 8 V, so that the base will be 1 V above ground. The current through R_A will be the current through R_B of 1 mA, plus the base current of 0.225 mA, or 1.0 + 0.225 = 1.225 mA. The resistance required to produce an 8-V drop with 1.225 mA is 8 $\div$ 1.225, or 6530 Ω. Use a 6500-Ω standard resistance as the trial value (or use a decade-box setting of the precise 6530 Ω).

Coupling capacitors. The value of C_1 forms a high-pass filter with R_B. The high limit of 100 kHz can be ignored. The low-frequency limit of 30 Hz requires a capacitance value of 1 $\div$ (3.2 $\times$ 30 $\times$ 1000), or 10 μF (approximately). This provides an approximate 1-dB drop at the low-frequency limit of 30 Hz. If a greater drop can be tolerated, the capacitance value of C_1 can be lowered. The value of C_2 is found in the same manner, except the resistance value must be the load resistance (2000 Ω). Therefore, C_2 requires a value of $1 \div (3.2 \times 30 \times 2000)$, or 5.2 μF (6 μF for practical design). The voltage values of C_1 and C_2 should be 1.5 times the maximum voltage involved, or $9 \times 1.5 = 13.5$ V (15 V for practical design).

Transistor selection. Some circuits must be designed around a given transistor. In such cases, the source voltage, collector current, power dissipation, etc., must be adjusted accordingly. In this example, a 2N337 transistor is available, and can be used, provided that it meets the circuit requirements.

The following is a comparison of 2N337 characteristics and circuit requirements. (Note that any other transistor can be used if it meets the same requirements.)

2N337 characteristics	Circuit requirements
collector voltage (max) 45 V	9-V source
collector current (max) 20 mA	2.25 mA nominal (at saturation, collector current would be 4.5 mA, $9 \div 2000 = 4.5$)
power dissipation (max) 125 mW	approximately 10 mW (4.5 V $\times$ 2.25 mA)
(The 2N337 must be derated 1 mW/°C above 25°C ambient temperature. See Sec. 1-4.)	
ac beta (as h_{fe}) (min) 19 (type) 55	10, *minimum gain at low frequency*
beta (as h_{fe} at 100 kHz) 25	10, minimum gain at 100 kHz

(Section 1-3.4 shows that the h_{fe} at 2.5 MHz is greater than 10. Therefore, the gain will be greater than 10 at 100 kHz.)

2-2. Basic Audio Amplifier with Emitter Bypass

Figure 2-1 shows (in phantom) a bypass capacitor C_3 across emitter resistor R_E. This arrangement permits R_E to be removed from the circuit as far as the signal is concerned, but leaves R_E in the circuit (in regard to direct current). With R_E removed from the signal path, the voltage gain is approximately $R_L \div$ dynamic resistance of the transistor, and the current gain is approximately ac beta of the transistor. Thus, the use of an emitter-bypass capacitor permits the highly temperature-stable dc circuit to remain intact while providing a high signal gain.

2-2.1. Design Considerations

An emitter-bypass capacitor creates some problems. Transistor input impedance changes with frequency, and from transistor to transistor, as does beta. Therefore, current and voltage gains can only be approximated. When the emitter resistance is bypassed, the circuit input impedance is approximately beta times transistor input impedance. Therefore, circuit input impedance is even more subject to variation, and is unpredictable. The author does not recommend a bypassed emitter, except in those rare cases where high voltage gain must be obtained from a single stage.

The value of C_3 can be found by:

$$\text{capacitance} = \frac{1}{6.2FR}$$

where capacitance is in farads, F is low-frequency limit in hertz, and R is input impedance of the transistor in ohms.

2-2.2. Design Example

Assume that C_3 is to be used as an emitter bypass for the circuit described in the previous design example (Sec. 2-1.2) to increase voltage gain. All the circuit values remain the same, as does the low-frequency limit of 30 Hz. Assume that the transistor (2N337) dynamic input resistance is 50 Ω (obtained from the data sheet). This would provide a voltage gain of approximately 2000 ÷ 50, or 40 V. The desired 3-V output could then be obtained with 0.075-V input rather than the 0.3-V input required for the previous example.

The low-frequency limit of 30 Hz requires a C_3 capacitance value of $1 ÷ (6.2 \times 30 \times 50)$, or 107 μF (110 μF for practical design). This value will provide a reactance across R_E that is less than the transistor input impedance and will effectively short the emitter (signal path) to ground. The voltage value of C_3 should be 1.5 times the maximum voltage involved, or $1.0 \times 1.5 = 1.5$ V (3 V for practical design).

2-3. Basic Audio Amplifier with Partially Bypassed Emitter

Figure 2-3 is the working schematic of a basic, single-stage, audio amplifier with a partially bypassed emitter resistor. This design is a compromise between the basic design without bypass (Sec. 2-1) and the fully bypassed emitter (Sec. 2-2). The dc characteristics of both the unbypassed and partially bypassed circuits are essentially the same. All circuit values (except C_3 and R_C) can be calculated in the same way for both circuits. As shown in Fig. 2-3, the voltage and current gains for a partially bypassed amplifier are greater than an unbypassed circuit but less than for the fully bypassed circuit.

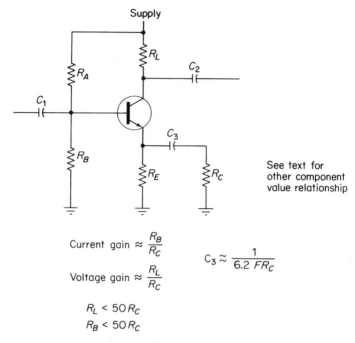

$$\text{Current gain} \approx \frac{R_B}{R_C}$$

$$C_3 \approx \frac{1}{6.2 \ FR_C}$$

$$\text{Voltage gain} \approx \frac{R_L}{R_C}$$

$$R_L < 50 \, R_C$$
$$R_B < 50 \, R_C$$

Fig. 2-3. Basic audio amplifier with partially bypassed emitter.

2-3.1. Design Considerations

The design considerations for the circuit of Fig. 2-3 are the same as those for Fig. 2-1 (Sec. 2-1.1) except for the effect of C_3 and R_C. The value of R_C should be chosen on the basis of voltage gain, even though current gain will be increased when voltage gain increases. R_C should be substantially smaller than R_E. Otherwise, there will be no advantage to the partially bypassed design. However, a smaller value for R_C will require a larger value for C_3, since the C_3 value is dependent upon the R_C value and the desired low-frequency cutoff point.

The value of C_3 can be found by:

$$\text{capacitance} = \frac{1}{6.2FR}$$

where capacitance is in farads, F is the low-frequency limit in hertz, and R is the value of R_C in ohms.

2-3.2. Design Example

Assume that the circuit of Fig. 2-3 is to be used in place of the Fig. 2-1 circuit and that the desired voltage gain is 25. Selection of component values, supply voltages, operating point, etc., is the same for both circuits. Therefore, the only difference in design is selection of values for C_3 and R_C. The value of

R_C should be the value of R_L, divided by the desired voltage gain, or 2000 ÷ 25 = 80 Ω.

The low-frequency limit of 30 Hz requires a C_3 capacitance value of $1 ÷ (6.2 × 30 × 80)$, or 67 μF. The voltage value of C_3 should be 1.5 times the maximum voltage involved: $1.0 × 1.5 = 1.5$ V (3 V for practical design).

2-4. Direct-Coupled Voltage Amplifier

Figure 2-4 is the working schematic of a two-stage, direct-coupled amplifier. When stable voltage gains greater than about 20 are required and it is not practical to bypass the emitter resistor of a single stage, two or more transistor amplifiers can be used. The output of one transistor is fed to the input of a second transistor. Generally, no more than three transistors should ever be required, with two transistors adequate to do most jobs of voltage amplifica- tion. The stages can be coupled by means of capacitors. For example, two (or three) basic stages, such as shown in Fig. 2-1, could be connected together. The value of the coupling capacitors would be dependent upon the base resistors (R_B) of the following stages. This creates a low-frequency-limit problem, as previously described.

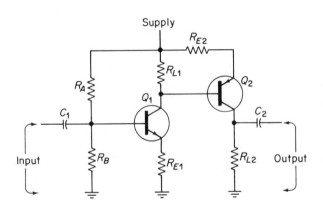

Input impedance $\approx R_B$

Output impedance $\approx R_{L2}$

Q_1 voltage gain $\approx \dfrac{R_{L1}}{R_{E1}}$

Q_2 voltage gain $\approx \dfrac{R_{L2}}{R_{E2}}$

Overall voltage gain $\approx 70\%$ of combined Q_1 and Q_2 voltage gains

$R_L > 5\,R_E$

$R_L \approx 10\,R_E$

$R_B \approx 10\,R_E$

$R_B < 20\,R_E$

Q_2 collector voltage $\approx 0.5 \times$ supply voltage (as adjusted by R_A)

Fig. 2-4. Basic direct-coupled voltage amplifier.

The stages can also be direct-coupled as shown in Fig. 2-4. This eliminates the interstage coupling capacitor, as well as R_A and R_B for the second stage. Note that such direct-coupled pairs (often known as *compounds*) use complementary *NPN* and *PNP* transistors so that bias voltages will be convenient.

Two- and three-stage compounds are available as *hybrid circuits*. [Such hybrid circuits consist of resistors, capacitors, diodes, and transistors, all contained in a single hermetically sealed package. Hybrid circuits are similar to integrated circuits (ICs) except that ICs usually are complete functioning circuits, such as a complete audio amplifier, operational amplifier, etc.]

It is generally easier to design with hybrid circuits than with individual components, since impedance relationships, Q-point, etc., have been calculated by the circuit manufacturer. Also, the data sheets supplied with the hybrid circuits provide information regarding source voltage, gain, impedances, etc. The data-sheet information can be followed to adapt the hybrid circuit for a specific application. However, in some cases it is necessary to select values of components external to the hybrid-circuit package. For this reason, and since it may be necessary to design a multistage amplifier (with individual components) for some special application, the following design considerations and examples are provided.

2-4.1. Design Considerations

Note that the circuit elements for transistor Q_1 in Fig. 2-4 are essentially the same as for the circuit of Fig. 2-1. Also note that the same circuit arrangement is used for transistor Q_2, except that R_A and R_B are omitted (as is the coupling capacitor between stages).

The design considerations for the circuit of Fig. 2-4 are the same as those for Fig. 2-1 (Sec. 2-1.1) except for the following:

The input impedance of the complete circuit is approximately equal to R_B. The output impedance is set by R_{L2}.

The value of C_1 is dependent upon the low-frequency limit and the value of R_B. The value of C_2 is dependent upon the low-frequency limit and the value of input resistance of the following stage (or the load).

2-4.2. Design Example

Assume that the circuit of Fig. 2-4 is to be used as a voltage amplifier. The desired output is 7 V (peak to peak) with a 500-Ω impedance. The input is 100 mV, also from a 500-Ω impedance. (This requires a voltage gain of 70.) The low-frequency limit is 30 Hz, with a high-frequency limit of 100 kHz. Minimum distortion is desired. (The circuit should not be overdriven.) A source of 20 V is specified. The circuit will not be battery-operated. A 2N3568 silicon *NPN* is to be used for the input; a 2N3638 *PNP* is to be used for the output.

Supply voltage and operating point. The supply voltage, 20 V, is specified. Since this is approximately 3 times the desired output of 7 V, there should be no distortion.

First establish the Q-point for the output stage. The collector voltage should be approximately one-half the source, or $20 \div 2 = 10$ V, at the Q-point.

Load resistance and collector current. The value of R_{L2} is specified as 500 Ω. Use the nearest standard value of 510 Ω. With a 10-V drop across R_{L2}, the Q_2 collector current is $10 \div 510$, or 19.6 mA (rounded off to 20 mA).

Emitter resistance, current, and voltage. When two stages are direct-coupled in the stabilized circuit of Fig. 2-4, the overall voltage gain is about 70 per cent of the combined gains of each stage. Since the required voltage gain is 70, the combined gains should be 100 (or a gain of 10 for each stage). To provide a voltage gain of 10 in the output stage, the value of R_{E2} should be one-tenth of R_{L2}, or $510 \div 10 = 51$ Ω. The current through R_{E2} will be the collector current of 20 mA, plus the base current. Assuming a circuit gain of 10, the base current is $20 \div 10$, or 2 mA. The combined currents through R_{E2} are $20 + 2$, or 22 mA. This will produce a drop of 1.12 V across R_{E2} (rounded off to 1 V).

Output base voltage and input collector voltage. The base of Q_2 should be 0.5 V from the emitter voltage. Since Q_2 is *PNP*, the base should be more negative (or less positive) than the emitter. The emitter of Q_2 is at $+19$ V (20 V $-$ 1 V). Therefore, the base of Q_2 should be $+18.5$ V. This sets the collector voltage for Q_1 at the Q-point.

Input stage resistances and current. The value of R_B (input impedance) is specified as 500 Ω. Use the nearest standard value of 510 Ω. As shown in Fig. 2-4, R_B should not be greater than 20 times, or less than 5 times, R_{E1}. An R_B/R_{E1} ratio of 10 is nominal for reasonable stability. Therefore, R_{E1} should be $510 \div 10$, or 51 Ω. With the R_{E1} value established, and a stage voltage gain of 10 desired, the value of R_{L1} should be 51×10, or 510 Ω.

With a 1.5-V drop across R_{L1}, the collector current is $1.5 \div 510$, or 2.94 mA (rounded off to 3 mA). The current through R_{E1} will be the collector current of 3 mA, plus the base current. Assuming a gain of 10, the base current is $3 \div 10$, or 0.3 mA. The combined currents through R_{E1} are $3 + 0.3$ mA, plus the 2-mA base current of Q_2, or $3 + 0.3 + 2 = 5.3$ mA. This produces a drop of about 270 mV across R_{E1}.

The base voltage of Q_1 is about 0.5 V higher than the emitter voltage, or $0.5 + 0.270 = 0.770$ V. With a 0.770-V drop across R_B, the current through R_B is $0.770 \div 510$, or about 1.5 mA.

The value of R_A should be sufficient to drop the 20-V source to 19.230 V, so that the base is 0.770 V above ground. The current through R_A is the current through R_B of 1.5 mA, plus the base current of 0.3 mA, or 1.5 + 0.3 = 1.8 mA. The resistance required to produce a 19.230-V drop with 1.8 mA is 19.230 ÷ 1.8, or about 10.7 kΩ. Use an 11-kΩ standard resistance as the trial value. Adjust R_A for the desired collector Q-point voltage at Q_2 rather than Q_1. [As described in a later section, the final adjustment of R_A will be made for distortion-free 7-V output signal at the collector of Q_2 (with a 100-mV input signal applied to Q_1).]

Input and output capacitors. Even though there is direct coupling between the stages, coupling capacitors are necessary at the input and output, unless the external device is provided with capacitors for isolation of the direct current. The value of C_1 forms a high-pass filter with R_B. The high limit of 100 kHz can be ignored. The low-frequency limit of 30 Hz requires a capacitance value of 1 ÷ (3.2 × 30 × 510), or 20 μF (approximately), for an approximate 1-dB drop. The value of C_2 is found in the same manner, except the resistance value R must be the load resistance. Since the load and input impedances are the same, the capacitance value of C_2 will be the same as C_1. The voltage values of C_1 and C_2 should be 1.5 times the maximum voltage involved, or 20 × 1.5 = 30 V.

Transistor selection. In this case, the transistors were specified in the design example. However, it is always wise to check basic transistor characteristics against the proposed circuit parameters *before* making any connections.

The maximum collector voltages for the 2N3568 and 2N3638 are 80 and 25 V, respectively, both well above the 20-V source.

The powers dissipated for the 2N3568 and 2N3638 are 55 and 200 mW, respectively, both well below the 300-mW data-sheet limit.

Minimum betas for the 2N3568 and 2N3638 are 40 and 20, respectively, both well above the required 10.

2-5. Transformerless Multistage Audio Amplifiers

In addition to the direct-coupled amplifier of Sec. 2-4, there are a number of transformerless multistage circuits used to provide voltage, current, or power amplification at audio frequencies. The circuits for the most important of these are the Darlington pair configurations (compounds) (Figs. 2-5 through 2-7), the phase inverter or splitter (Fig. 2-8), the emitter-coupled amplifier (Fig. 2-9), the differential amplifier (Fig. 2-10), the transformerless series-output amplifier (Fig. 2-11), the quasi-complementary amplifier (Fig. 2-12), and the full-complementary amplifier (Fig. 2-13). Each of these circuits, as well as various combinations of the circuits, is available in IC or hybrid packages.

Because such circuits are readily available in great variety at low cost, no detailed circuit-design procedures will be given here. Instead, we will concentrate on how these circuits can be combined to perform a specific function. A classic example of this is the combination of circuits necessary for a full-complementary audio amplifier. The audio industry has generally agreed on this configuration for high-fidelity circuits where transformers are not wanted. Such a circuit can provide up to 100 W of power into a loudspeaker, without the weight of transformers or the frequency limitations of coupling capacitors.

It should be noted that the complementary circuits found in ICs are for low-power applications, usually in the order of a few watts maximum. Where higher power is required, the circuit must be designed using individual components.

For these reasons, we shall describe the design procedures for transformerless circuits (using individual components) that will produce up to 100 W of power output. As is described in the design procedures, power output is dependent upon selection of component values and source voltages. The circuits can be used for any power output (not exceeding 100 W).

Before going into the design considerations and examples, the basic theory for the circuits shown in Figs. 2-5 through 2-13 will be discussed. Although it is assumed that the reader is familiar with amplifier circuit theory, you may not understand *why* a particular circuit is used to meet a given design problem.

2-5.1. Darlington Compounds

Figure 2-5 shows the basic Darlington circuit (known as the *Darlington compound*), together with two practical versions of the circuit. As shown, the Darlington compound is an emitter follower (or common collector) driving a second emitter follower. An emitter follower provides no voltage gain but can provide considerable power gain.

The main reason for using a Darlington compound (especially in audio work) is to produce high current (and power) gain. For example, Darlington compounds are often used as audio drivers to raise the power of a signal from a voltage amplifier to a level suitable to drive a final power amplifier. Darlingtons are also used as a substitute for a driver section (or to eliminate the need for a separate driver).

Darlingtons as basic common collectors (***emitter followers***). When the Darlington is used as a common collector, as shown in Fig. 2-5, the output impedance is approximately equal to the load resistance R_L. The input impedance is approximately equal to beta$^2 \times R_L$. The current gain is approximately equal to the average beta of the two transistors, squared. However, in most common collectors, power gain is of primary concern. That is, the

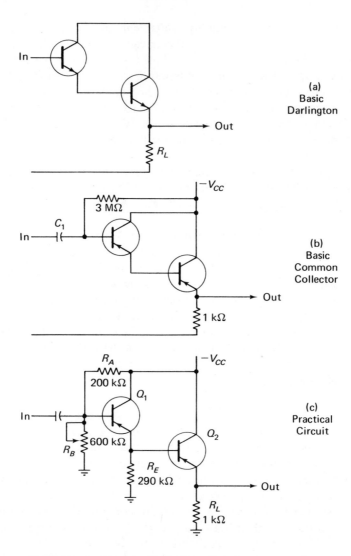

Fig. 2-5. Basic and practical Darlington compounds.

designer is interested in how much signal power can be increased across a given output load.

As an example, assume that the value of R_L (in Fig. 2-5c) is 1 kΩ and that the average beta is on the order of 15. This results in an input impedance of about 225 kΩ ($15^2 \times 1000$) and an output impedance of 1 kΩ. Now assume that a 2.5-V signal is applied at the input and an output of 2 V appears across

R_L. The input power is $2.5^2/225 \text{ k}\Omega \approx 0.028 \text{ mW}$. The output power is $2^2/1 \text{ k}\Omega \approx 4 \text{ mW}$. The power gain is $4/0.028 \approx 140$.

Darlingtons as basic common-emitter amplifiers. Darlington compounds can be used as common-emitter amplifiers to provide voltage gain. This is done by adding a collector resistor to any of the circuits in Fig. 2-5, and taking the output from the collector rather than the emitter. In effect, Q_1 then becomes a common-collector driving Q_2, which appears as a common-emitter amplifier. The entire circuit then appears as a common-emitter amplifier and can be used to replace a single transistor. Such an arrangement is often used where high voltage gain is desired.

A more practical method of using a Darlington as a common-emitter amplifier is to eliminate R_B and R_E (Fig. 2-5c), ground the emitter of Q_2, and transfer R_L to the collector of Q_2. Such an arrangement is shown in Fig. 2-6. The circuit of Fig. 2-6 is stabilized by the collector feedback through R_A, which holds both collectors at a potential somewhat less than 0.5 V from the base of Q_1. Note that both collectors are at the same voltage, and that this voltage is approximately equal to two base–emitter voltage drops (or about 1.5 V for two silicon transistors).

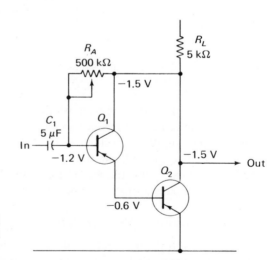

Fig. 2-6. Darlington compound with collector feedback and common-emitter output.

With the circuit of Fig. 2-6, the current gain is approximately equal to the ratio of R_A/R_L. Both the input and output voltage swings are somewhat limited in the Fig. 2-6 circuit. The input is biased at approximately 1 to 1.5 V. However, voltage gains of 100 (or more) are possible, since input impedance (or resistance) is approximately equal to the ratio of R_A/current gain, or equal to R_L. (With input and output impedances approximately equal to R_L, the voltage gain follows the current gain.)

Multistage Darlingtons. Darlington compounds need not be limited to two transistors. Three (and even four) transistors can be used in the Darlington circuit. A classic example of this is the General Electric circuit of Fig. 2-7 (which is available in hybrid form, in a TO-5 style package). This circuit is essentially a common-collector and common-emitter Darlington, followed by a common-emitter amplifier.

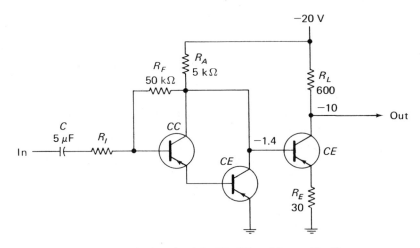

Fig. 2-7. General Electric *CC–CE–CE* multistage Darlington compound.

With R_I out of the circuit, both the input and output impedances are set by R_L. (In practice, the input impedance is slightly higher than R_L, typically on the order of 700 to 800 Ω.) With R_I removed, the voltage gain is about 1000. When R_I is used, the input impedance is approximately equal to R_I, and the voltage gain is reduced accordingly. For example, if R_I is 10 kΩ, the 1000 voltage gain drops to about 50.

2-5.2. Phase Inverter or Splitter

Figure 2-8 is the basic schematic of a phase inverter or splitter. As shown, the circuit is essentially a single-stage amplifier, with an output taken from the emitter as well as the collector. The values of R_L and R_E are usually equal, or near equal, so that there will be no voltage gain. The collector output is 180° out of phase with the input and, therefore, 180° out of phase with the emitter.

The main reason for using a phase inverter or splitter in audio work is to produce two 180° out-of-phase signals to drive a push-pull output stage. (A phase inverter is the transformerless version of the interstage transformer with an untapped primary and a center-tapped secondary.)

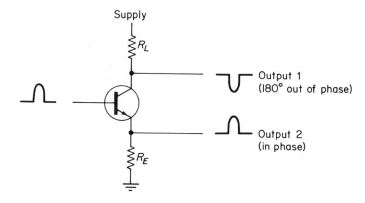

Fig. 2-8. Basic phase inverter or splitter.

2-5.3. Emitter-Coupled Amplifier

Figure 2-9 is the basic schematic of an emitter-coupled amplifier. This circuit is similar to the phase inverter in that two 180° out-of-phase outputs can be taken from the one input. Unlike the single-stage phase inverter, the emitter-coupled amplifier provides high gain. Therefore, an emitter-coupled amplifier can be used in a design where a low-voltage input must be amplified to drive a push-pull output stage.

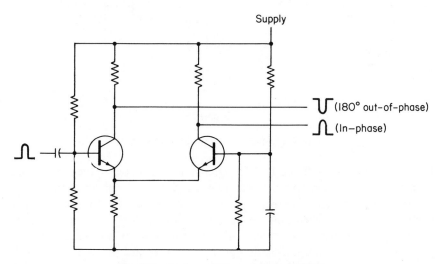

Fig. 2-9. Basic emitter-coupled amplifier.

2-5.4. Differential Amplifier

Figure 2-10 is the basic schematic of a differential amplifier. The differential amplifier is similar to the emitter-coupled amplifier, except that the

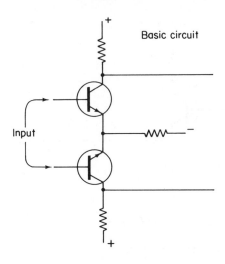

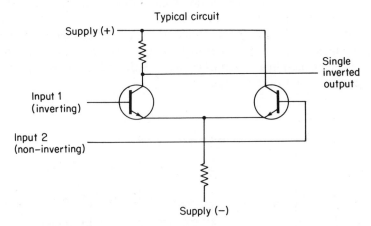

Fig. 2-10. Basic differential amplifier.

two output signals are the result of a *signal difference* between the two inputs. An output is produced only when there is a difference in signals at the input. Differential amplifiers are particularly useful (especially in oscilloscope, electronic meter, and recording instrument amplifiers) because signals common to both inputs (known as *common-mode signals*) are eliminated, or greatly reduced. A common-mode signal (like power-line pickup) drives both bases in phase with equal amplitude ac voltages, and the amplifier behaves as though the transistors were in parallel. The emitter resistor introduces emitter feedback, which reduces the common-mode signal gain without reducing the differential signal gain.

As described in Chapter 3, a differential amplifier is used as the input circuit of an operational amplifier. The ability of a differential amplifier to prevent conversion of a common-mode signal into a difference signal is expressed by its common-mode rejection ratio.

2-5.5. Transformerless Series Output Amplifier

Figure 2-11 shows two typical transformerless series-output amplifiers used in audio design. One configuration requires two power supplies but omits the coupling capacitor to the load. This configuration provides better low-frequency response (since there is no capacitor), but can be inconvenient because of the two power supplies. The configuration with a single power supply has reduced low-frequency response since the coupling capacitor forms a high-pass filter with the load resistance.

Either configuration of series output has two drawbacks. A phase inverter is required to drive the series-output stage, even if gain is not required. Also, an additional driver stage may be necessary to bring the power up from the output of a voltage amplifier to a level required by the series-output stage.

These problems are overcome by means of a *complementary* circuit, either quasi-complementary or full-complementary.

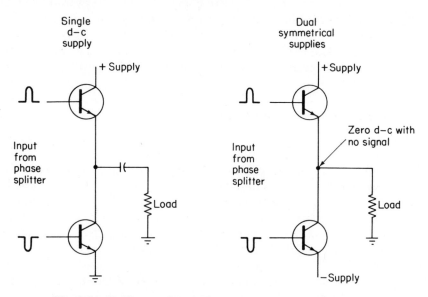

Fig. 2-11. Basic transformerless series output amplifiers.

2-5.6. Quasi-Complementary Amplifier

Figure 2-12 is the schematic of a quasi-complementary output. This circuit consists of a Darlington compound using *NPN* transistors and a direct-coupled compound using an input *PNP* and an output *NPN*. Both base

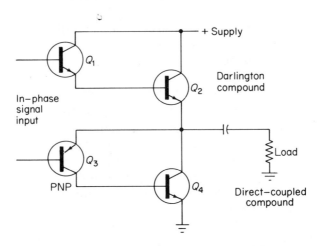

Fig. 2-12. Basic quasi-complementary amplifier.

signals can be in phase (although they are often at different voltage levels) so that a phase inverter is not needed.

For example, if the input is positive-going, Q_1 will be forward-biased, as is Q_2. A positive-going input at Q_3 will reverse bias Q_3 since it is *PNP*. This will produce a negative-going output from Q_3 to Q_4, which is *NPN*, and reverse bias Q_4. When the signal at the bases of Q_1 and Q_3 is negative-going, the condition will reverse (Q_3 will be reverse-biased; Q_4 will be forward-biased).

2-5.7. Full-Complementary Amplifier

Figure 2-13 shows two versions of the full-complementary output. Either version has an advantage over the quasi-complementary in that both halves of the circuit are identical. This makes it easier to match both halves (for positive and negative signals) to minimize distortion that could be caused by uneven amplification of the signal.

One complementary circuit uses two Darlington compounds. This circuit is also known as a *dual-Darlington* output, and is used where power gain is needed. The other circuit uses two direct-coupled compounds, and is used where voltage gain is most needed.

Phase inversion is not required for either circuit. A positive-going input will forward-bias Q_1 and Q_2, and reverse-bias Q_3 and Q_4. A negative-going input will produce opposite results.

As in the case of series-output circuits (Sec. 2-5.5) the load-coupling capacitor can be omitted if two power supplies are used (one positive and one negative). The tradeoff between the inconvenience of two power supplies versus improved low-frequency response must be decided by circuit requirements.

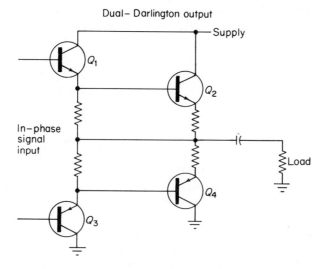

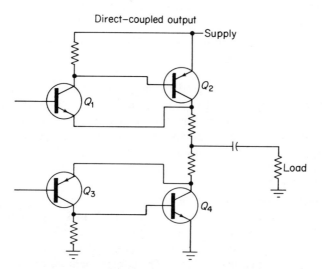

Fig. 2-13. Basic full-complementary amplifier.

As a rule of thumb, a 2000-μF capacitor into a 4-Ω load (such as a 4-Ω loudspeaker) will produce an approximate 3-dB drop at 20 Hz (a 10-V output will drop to 7 V).

2-5.8. Secondary Breakdown and Short-Circuit Protection

One problem with any of the output circuits is that one (or more) of the collectors will operate without a load resistance. In effect, the output load forms the load for the collector. Although the working load does present a load similar to that of a collector resistor, there is one major difference. The working load in any of the output circuits is not in series with the collector, as is the load resistor in a basic class A amplifier. In the simple class A amplifiers (Secs. 2-1 through 2-4), the collector load resistor reduces the collector voltage when heavy current is flowing, and raises the voltage only when a light current is present. If there should be a short in the output, heavy current may flow but the collector voltage will drop.

In the various series- and complementary-output circuits, a short circuit of the load will produce heavy current, without reducing the collector voltage. This can result in the destruction of the transistors. Therefore, some short-circuit protection should be considered for any output configuration involving circuits without collector resistors. (This includes most class B and AB low-frequency amplifiers, as well as class C RF circuits, which are discussed in later sections and chapters.)

The destructive condition is usually known as *secondary breakdown*, or *second breakdown*, and results from a sudden channeling of collector current into a localized area of the transistor. Secondary breakdown is usually prevented by limiting the collector current–voltage product.

2-6. Examples of Transformerless Direct-Coupled Audio Amplifiers

The following examples are used to illustrate the problems involved with transformerless amplifiers. Two circuits are discussed. Both circuits were developed by Motorola using their complementary plastic transistors.

A four-transistor circuit, capable of delivering 2 to 5 W of audio power with an approximate 0.25-V input signal, is shown in Fig. 2-14. A six-transistor circuit, delivering between 7 and 35 W with 0.1 to 0.45-V input signals, is shown in Fig. 2-15.

The most significant difference between the circuits of Figs. 2-14 and 2-15 is the output configuration. The circuit of Fig. 2-14 is essentially the series-output amplifier (with single power supply) described in Sec. 2-5.5. The output configuration of the six-transistor circuit (Fig. 2-15) is a composite complementary pair described in Sec. 2-5.7. The complementary pairs serve as extra driver transistors to supply the higher drive current necessary for higher power.

Four-transistor circuit. As shown in Fig. 2-14, the output circuit is formed by Q_3 and Q_4. To ensure maximum output signal swing, the dc voltage at the emitters of Q_3 and Q_4 must be one-half of V_{CC}. To prevent clipping on either half-cycle of the output signal, the dc voltage at the Q_3-Q_4 emitters must "track" with any variations in V_{CC}. The dc feedback arrangement meets these requirements, in that the amplifier is essentially a unity-gain voltage follower for direct current.

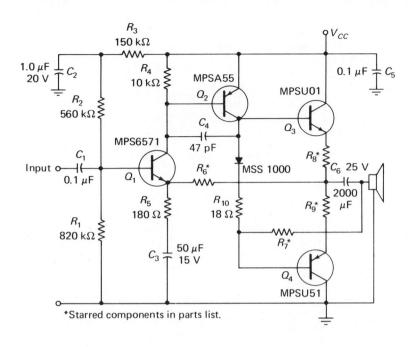

Parts List

	3 W	5 W
V_{CC}	18 V	22 V
R_6	3.9 kΩ	4.7 kΩ
R_7	470 Ω	560 Ω
R_8	0.75 Ω	0.47 Ω
R_9	0.75 Ω	0.47 Ω

*See text.

Amplifier Output		
	3 W	5 W
Idle current, nominal	25 mA	25 mA
Current drain at rated power output	290 mA	370 mA
Nominal sensitivity for rated power output	0.25 V	0.25 V
THD at rated maximum power output, 50 Hz to 15 kHz, nominal	2%	2%
THD at rated power output, nominal	0.25 %	0.25%
Input impedance, typical	300 kΩ	300 kΩ
Maximum power output at 5% THD, without current-limiting diodes	4.5 W	6.7 W
Maximum power output at 5% THD with current-limiting diodes*	4.0 W	6.0 W

Fig. 2-14. Four-transistor transformerless audio amplifier. Courtesy Motorola.

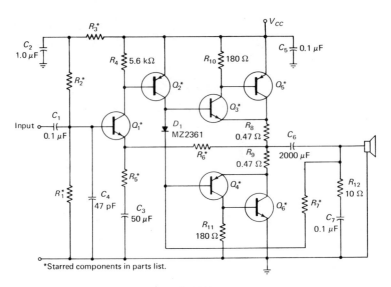

*Starred components in parts list.

Parts List

	7 W	10 W	15 W	20 W	25 W	30 W	35 W
V_{CC}	26 V	30 V	35 V	43 V	46 V	50 V	54 V
R_1	2.7 MΩ	2.7 MΩ	560 kΩ	560 kΩ	220 kΩ	220 kΩ	220 kΩ
R_2	1.2 MΩ	1.0 MΩ	330 kΩ	330 kΩ	150 kΩ	150 kΩ	150 kΩ
R_3	390 kΩ	820 kΩ	120 kΩ	120 kΩ	47 kΩ	47 kΩ	47 kΩ
R_5	100 Ω	82 Ω	100 Ω	75 Ω	220 Ω	270 Ω	270 Ω
R_6	8.2 kΩ	8.2 kΩ	10 kΩ	10 kΩ	10 kΩ	10 kΩ	10 kΩ
R_7	3.9 kΩ	3.9 kΩ	4.7 kΩ	5.6 kΩ	5.6 kΩ	6.8 kΩ	8.2 kΩ
Q_1	MPS6571	MPS6571	MPS6571	2N5088	2N5088	2N5088	2N5088
Q_2	2N5087	2N5087	2N5087	2N5087	MPSA56	MPSA56	MPSA56
Q_3	MPSA05	MPSA05	MPSU05	MPSU05	MPSU05	MPSU05	MPSU05
Q_4	MPSA55	MPSA55	MPSU55	MPSU55	MPSU55	MPSU55	MPSU55
Q_5	MJE371	MJE371	MJE105	MJE105	MJE2901	MJE2901	MJE2901
Q_6	MJE521	MJE521	MJE205	MJE205	MJE2801	MJE2801	MJE2801

	Amplifier Output (W)						
	7	10	15	20	25	30	35
Idle current, nominal (mA)	25	25	25	25	25	25	25
Current drain at rated output power (mA)	420	500	580	700	750	850	950
Sensitivity at rated output power (V)	0.1	0.1	0.1	0.1	0.45	0.4	0.45
THD at rated output power, typical (%)	2	2	1	1	0.5	0.5	0.5
THD at 1/2 rated output power, typical (%)	0.5	0.5	0.25	0.25	0.1	0.1	0.1
Input impedance, typical (kΩ)	800	700	200	200	90	90	90

Fig. 2-15. Six-transistor transformerless audio amplifier. Courtesy Motorola.

76

The voltage at the base of Q_1 is approximately one-half V_{CC} and is set by the divider composed of resistors R_1, R_2, and R_3. The Q_1 emitter voltage "follows" the base voltage, so that the dc output voltage is approximately equal to the emitter voltage. The Q_1 base is actually slightly greater than one-half of V_{CC} (since R_1 is larger than R_2 and R_3). This compensates for the small drop across resistor R_6 and the $V_{BE(on)}$ of Q_1.

The ac gain of the Fig. 2-14 circuit is set by the ratio of R_6 to R_5:

$$A_V = \frac{R_5 + R_6}{R_5} = \frac{R_6}{R_5} \qquad \text{when } R_6 \gg R_5$$

Capacitor C_3 allows the bottom of R_5 to be at signal ground, and provides dc isolation for the emitter of Q_1.

Elimination of *crossover distortion* is desirable, especially at low listening levels. (Crossover distortion is discussed further in Sec. 2-8.) The collector current of Q_2 (through the diode and R_{10}) produces a dc voltage that slightly forward-biases the base–emitter junctions of output transistors Q_3 and Q_4. With the output transistors forward-biased, the amplifier operates in class A for small signal outputs, and in class B for large output swings. (In class A operation, amplifier collector current flows at all times; in class B, collector current flows in the presence of an input signal.) Keep in mind that Q_3 and Q_4 are complementary (*PNP* and *NPN*). One transistor is conducting (collector current flowing) with the other transistor cut off (or near cutoff) on each half-cycle.

Objectionable power-supply hum is filtered from the input circuit by resistor R_3 and capacitor C_2. Capacitor C_5 protects the V_{CC} line against oscillations that could occur when high-current transients are present.

High-frequency oscillations may occur with varying source and load impedances. Capacitor C_4 provides a high-frequency rolloff at approximately 30 kHz to prevent such oscillations. (Rolloff is discussed further in Sec. 2-12.)

Six-transistor circuit. As shown in Fig. 2-15, transistors Q_3 and Q_5 form the composite *NPN* transistor, and Q_4 and Q_6 form the composite *PNP* transistor. Driver transistors Q_3 and Q_4 operate as emitter followers to establish the output voltage; at the same time they are common-emitter amplifiers supplying drive current to output transistors Q_5 and Q_6.

Both ac and dc feedback is applied to the composite pairs by resistors R_8 and R_9. To prevent thermal runaway, which can occur with an increase in collector–base leakage current due to high temperature, resistors R_{10} and R_{11} are placed between the base and emitter of the output transistors. The resistance of R_{10} and R_{11} is sufficiently low to prevent forward biasing of the output transistors by temperature-induced leakage currents.

High-frequency oscillations are suppressed by capacitor C_4. The network composed of resistor R_{12} and capacitor C_7 prevents oscillations that arise due to the increased impedance of the loudspeaker at high frequencies. The functions of the remaining circuits in Fig. 2-15 are identical to those of Fig. 2-14.

Keeping direct current out of the loudspeaker. The circuits of Figs. 2-14 and 2-15 use the output capacitor as a bootstrap to provide driving current to the *PNP* side of the negative half-cycle. This method requires the collector current of the predriver transistor Q_2 to flow through the loudspeaker. Some designers consider direct current in the loudspeaker as objectionable. One objection is that there can be a loud "pop" in the loudspeaker when power is first applied. Another objection is that a fixed direct current in the loudspeaker can change the acoustic characteristics, resulting in poor reproduction of sound. Some designers consider that a constant direct current tends to burn out the loadspeaker windings.

If the bootstrapping method of Figs. 2-14 and 2-15 is considered as objectionable, for whatever reason, the circuit can be modified as shown in Fig. 2-16. The capacitor C_{BS} in conjunction with the top resistor labeled $\frac{1}{2}R_7$ supplies the required drive current needed to ensure a full negative signal swing.

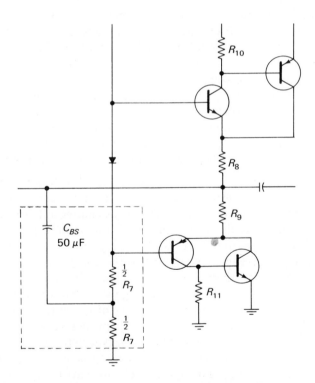

Fig. 2-16. Bootstrapping R_7 with a separate capacitor to keep direct current from the loudspeaker. Courtesy Motorola.

Overload protection. The amplifiers of Figs. 2-14 and 2-15 are not generally intended for use in sealed systems, so provisions for overload protection is desirable. One method is to add two diodes, D_2 and D_3, to the output circuit, as shown in Fig. 2-17. This method uses the forward voltage drop of diodes D_1, D_2, and D_3 to provide a clamp at the base of the output tran-

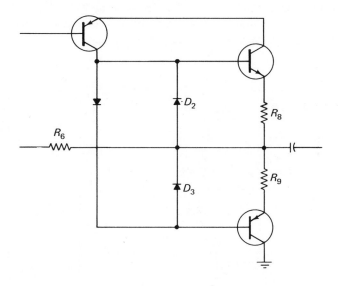

Amplifier Version (W)	Short-Circuit Current (A)	Diode D1	Diodes D2 and D3	R8 (Ω)	R9 (Ω)
35	2.1	MZ2361	MZ2361	0.47	0.39
30	2.0	MZ2361	MZ2361	0.47	0.39
25	1.6	MZ2361	MSS1000	0.47	0.39
20	1.55	MZ2361	MSS1000	0.47	0.39
15	1.35	MZ2361	MSS1000	0.47	0.47
10	1.3	MZ2361	MSS1000	0.47	0.47
7.0	1.25	MZ2361	MSS1000	0.47	0.47
5.0	1.0	MSS1000	MSS1000	0.47	0.47
3.0	0.8	MSS1000	MSS1000	0.47	0.47

Fig. 2-17. Connection of diodes D_2 and D_3 for overload protection. The diodes and circuit values are shown in the table. Courtesy Motorola.

sistors (Fig. 2-14) or the drivers (Fig. 2-15), which limits the maximum current in the output transistors.

Figure 2-18a shows the equivalent circuit of the clamp on the positive half-cycle, and Fig. 2-18b shows the action on the negative half-cycle. The chart in Fig. 2-17 shows the dc level at which the various amplifiers limit under short-circuit conditions, along with parts values and diode types used in the overload protection circuits. The 20- and 35-W versions use a 0.39-Ω resistor for R_9, rather than the 0.47-Ω, to prevent clipping, which can occur on the negative half-cycle with the larger resistor.

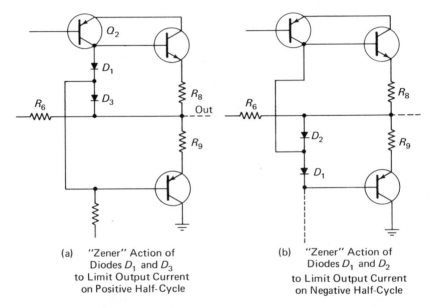

(a)   "Zener" Action of Diodes D_1 and D_3 to Limit Output Current on Positive Half-Cycle

(b)   "Zener" Action of Diodes D_1 and D_2 to Limit Output Current on Negative Half-Cycle

Fig. 2-18. Use of diodes for overload protection. Courtesy Motorola.

The overload network protects the amplifier against accidental short circuits in the output terminals and against an overload due to connecting extra loudspeakers. The problem of overloads in audio amplifiers is discussed further in Sec. 2-7.

Specifying transistors. Let us examine several sections in an amplifier circuit to illustrate one approach to the problem of specifying transistors. The basic circuit of Fig. 2-15 with a 25-W 8-Ω power rating is referred to in the following discussion.

In a power amplifier the logical place to begin design is the output terminal. For 25 W into an 8-Ω load, the peak-to-peak output voltage, or signal voltage swing, is given by:

$$V_{PP} = 2\sqrt{R(2)(P_{\text{rms}})}$$
$$= 2\sqrt{8(2)(25)}$$
$$= 40 \text{ V}$$

where V_{PP} is the peak-to-peak output voltage, R is the output load resistance, and P_{rms} is the rms output power.

The power-supply voltage must be

$$V_{CC} = V_{PP} + 2I_{PK}R_E + V_{CE}(Q_5) + V_{CE}(Q_6)$$

I_{PK} is found from the equation

$$I_{PK} = \frac{\frac{1}{2}V_{PP}}{\text{load resistance}} = \frac{\frac{1}{2}(40)}{8} = 2.5 \text{ A}$$

Figure 2-15 shows R_E (R_8 or R_9) to be 0.47 Ω, and the recommended power transistors are the MJE2801 and MJE2901. The data sheets for these devices lists a minimum h_{fe} of 25 at 3 A, and V_{CE} of 2 V for the output units. These numbers do not change significantly at 2.5 A, so the V_{CE} of Q_5 and Q_6 can be taken as 2 V. Note that the $V_{CE(\text{sat})}$ measurement found on many data sheets is meaningless in this application because the output transistors are operated as large-signal class B emitter followers, and do not saturate.

Using these figures to find V_{CC},

$$V_{CC} = 40 + 2(2.5)(0.47) + 2 + 2 \approx 46 \text{ V}$$

Adding 10 per cent for a high line voltage, the maximum V_{CC} is about 50 V. The following are minimum specifications for the output transistors:

$$BV_{CEO} = 50 \text{ V min at 200 mA}$$

$$h_{fe} = 30 \text{ min at 2.5 A and } V_{CE} \text{ of 2 V}$$

$$V_{BE(\text{on})} = 1.2 \text{ V max at 3 A and } V_{CE} \text{ of 2 V}$$

$$I_{CBO} = 0.1 \text{ mA max at 25°C and } V_{CB} \text{ of 50 V}$$

$$I_{CBO} = 2 \text{ mA max at 150°C and } V_{CB} \text{ of 50 V}$$

The MJE2801 and MJE2901 meet these requirements satisfactorily.

The collector current of the driver transistors is the base current of the output units. The peak output current is 2.5 A, and h_{fe} of the output units is 30, minimum. The maximum collector current of each driver (Q_3 or Q_4) is then

$$I_{\text{max}} = \frac{I_{\text{load (pk)}}}{h_{fe} \text{ min } (Q_5 \text{ or } Q_6)} = \frac{2.5}{30} = 83 \text{ mA}$$

The recommended drivers for the 25-W amplifiers are the MPS-U05 (*NPN*) and MPS-U55 (*PNP*). The data sheet has a minimum h_{fe} specification of 50 at collector currents of 10 mA and 100 mA, and BV_{CBO} is 50 V.

Modification of circuits. The designer may wish to increase or decrease certain specifications. For example, it may be desirable to use a transistor with a 55-V BV_{CEO}, or perhaps a minimum h_{fe} of 30 or 40 for the driver and 20 for the outputs. In doing this, the overall effects must be considered. For instance, if the h_{fe} specifications on the output and driver units are relaxed, the designer must ensure sufficient collector current in the predriver (Q_3) to provide full drive to the output stages. The predriver is then operated at a higher collector current, which increases the base current, making additional compensation in the preamplifier (Q_1) necessary. Such compensation can take the form of a higher h_{fe} preamplifier, or a decrease in the impedance of the base-biasing resistors of the preamplifier. Another obvious solution is to specify a higher h_{fe} for the predriver, thus making a change in the preamplifier unnecessary.

If you make intelligent use of specifications, you can safely make numerous changes in the circuits of Figs. 2-14 and 2-15. For example, input impedances can be lowered or raised, power levels can be changed, or the circuits can be used to drive different load impedances. It is only necessary to decide exactly what the circuit requires; then the exact device can be specified. For example, if the circuit is to operate into a 16-Ω loudspeaker, the value of I_{PK} is changed from 2.5 to 1.25 A, reducing the V_{CC} to about 43 V. In turn, this reduces the required BV_{CEO} for the output transistors and reduces the maximum collector current I_{max} of the driver transistors.

Power supplies. When the amplifier circuits are used in consumer applications, the associated power supply does not need to be complex. The circuit of Fig. 2-19 is a typical example. Capacitor C_1 is usually 1000 to 4000 μF. Resistor R_1 and capacitor C_2 are selected to provide voltage V_2, which may be necessary for operation of other circuitry, such as preamplifiers, AM or

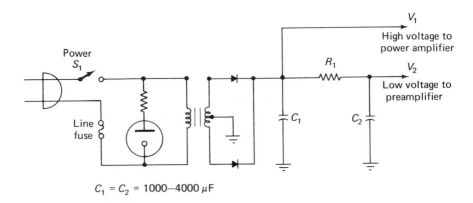

$C_1 = C_2 = 1000\text{–}4000\ \mu\text{F}$

Fig. 2-19. Typical power supply for transformerless amplifiers. Courtesy Motorola.

FM tuners, multiplex decoders, and so on. Power-supply circuits are discussed further in Chapter 6.

Heat sinks. The output transistors of all circuits shown in Figs. 2-14 and 2-15 require heat sinks. In addition, the driver transistors of the 15- to 35-W versions should be provided with heat sinks. These heat sinks should allow the dissipation capability of the drivers to be increased to about 10 per cent of the rated output power of the amplifier, which protects the devices under short-circuit conditions.

Discontinued parts. Note that the solid-state devices listed here may be discontinued from manufacture some time in the future. The manufacturer will then produce substitute devices with equivalent or superior characteristics, and will recommend these new devices in place of the existing parts. In that event, the circuits and design examples described here can still be used with these new devices, because the circuits allow for considerable variation in device characteristics. Also, the examples are sufficiently detailed that the designer can calculate any differences required. This same condition also applies to the circuits and examples described in Sec. 2-7.

2-7. Complementary Power Amplifier with Short-Circuit Protection

Figure 2-20 is the working schematic for a full complementary amplifier with short-circuit protection. Note that two power supplies are required, thus omitting the loudspeaker coupling capacitor.

Also note that some of the components are assigned values on the schematic. These are trial values for *all power output levels*. Other components are assigned reference designations. The values for these components depends upon the desired output power. Proper selection of these values, based on the design considerations and examples of this section, will provide power output at any level up to 100 W.

Tables 2-1, 2-2, and 2-3 list transistor types, resistance values and power supply voltages, and heat-sink requirements for power outputs of 35, 50, 60, 75, and 100 W. Other power outputs could be produced by means of intermediate values. However, the tabulated outputs should provide the designer with sufficient choice for most applications. The values tabulated are the nearest standard parts available that will meet or exceed the minimum specifications required for the particular amplifier.

The following is a brief *circuit description* to show the relationship of circuit function to design problems. Transistors Q_1, Q_2, Q_4, Q_6, Q_7, Q_8, Q_9, and Q_{10}, along with their associated components, comprise the standard full-complementary circuit. Transistors Q_1 and Q_2 are used in a differential amplifier configuration which, when used with the dual-power supply, pro-

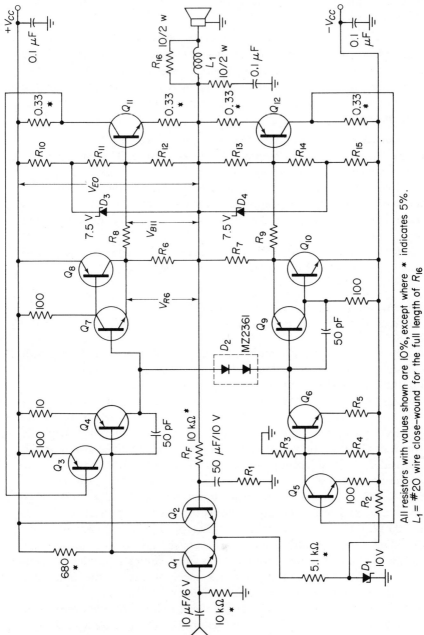

Fig. 2-20. Full-complementary amplifier with short-circuit protection. Courtesy Motorola.

All resistors with values shown are 10%, except where ∗ indicates 5%.
L_1 = #20 wire close-wound for the full length of R_{16}

84

Table 2-1. SEMICONDUCTOR COMPLEMENT FOR TRANSFORMERLESS AMPLIFIER

Output Power (watts rms)	Load Impedance (ohms)	Output Transistors		Driver Transistors		Predriver Transistors		Differential Amplifier Transistors	
		NPN Q_{10}	PNP Q_8	NPN Q_7	PNP Q_9	NPN Q_6	PNP Q_4	Single Channel	Dual Channel
35	4	MJ 2840	MJ 2940	MP SU 05	MP SU 55	MP SA 05	MP SA 55	MD 8001	MCF 8000
	8	MJE 2801	MJE 2901	MP SU 05	MP SU 55	MP SA 06	MP SA 56	MD 8001	MCF 8000
50	4	2N 5302	2N 4399	MP SU 05	MP SU 55	MP SA 06	MP SA 56	MD 8001	MCF 8000
	8	MJ 2841	MJ 2941	MP SU 06	MP SU 56	MP SA 06	MP SA 56	MD 8002	MCF 8001
60	4	2N 5302	2N 4399	MP SU 06	MP SU 56	MP SA 06	MP SA 56	MD 8001	MCF 8000
	8	MJ 2841	MJ 2941	MP SU 06	MP SU 56	MP SA 06	MP SA 56	MD 8002	MCF 8001
75	4	MJ 802	MJ 4502	MP SU 06	MP SU 56	MP SA 06	MP SA 56	MD 8001	MCF 8000
	8	MJ 802	MJ 4502	MM 3007	2N 5679	MM 3007	MM 4007	MD 8003	MCF 8002
100	4	MJ 802	MJ 4502	MP SU 06	MP SU 56	MP SU 06	MP SU 56	MD 8002	MCF 8001
	8	MJ 802	MJ 4502	MM 3007	2N 5679	MM 3007	MM 4007	MD 8003	MCF 8002

Q_{11} = MPSL01 Q_5 = MPSA20 Q_{12} = MPSL51 Q_3 = MPSA70
D_1 = MZ500–16   or MZ92–20* D_2 = MZ2361 D_4 and D_4 = IN5236B or
MZ 92–16A*

*For a low-cost Zener diode, an emitter–base junction of a silicon transistor can be substituted. A transistor similar to the MPS6512 can be used for the 7.5-V Zener. A transistor similar to the Motorola BC 317 can be used for the 10-V Zener dioder.

vides a convenient means for setting the dc voltage level to be direct-coupled to the loudspeaker. Note that the differential amplifier transistors and their associated components are part of a sealed package, as described in a later paragraph.

Table 2-2. RESISTOR VALUES* AND SOURCE VOLTAGES FOR TRANSFORMERLESS AMPLIFIER

Output Power (watts rms)	Load Impedance (ohms)	R_1 5%	R_2 10%	R_3 5%	R_4 5%	R_5 5%	R_6, R_7 5%	R_8, R_9 10%	R_{10}, R_{15} 5%	R_{11}, R_{14} 5%	R_{12}, R_{13} 5%	V_{CC} (volts)
35	4	820	2.7 kΩ	18 kΩ	1.2 kΩ	120	0.39	390	2.7 kΩ	1.5 kΩ	470	±21
	8	560	3.9 kΩ	22 kΩ	1.2 kΩ	180	0.47*	240	3.0 kΩ	1.2 kΩ	470	±27
50	4	680	3.3 kΩ	22 kΩ	1.2 kΩ	100	0.33	360	3.3 kΩ	1.5 kΩ	470	±25
	8	470	4.7 kΩ	27 kΩ	1.2 kΩ	150	0.43*	270	3.9 kΩ	1.2 kΩ	470	±32
60	4	620	3.9 kΩ	22 kΩ	1.2 kΩ	120	0.33	430	3.9 kΩ	1.5 kΩ	470	±27
	8	430	5.6 kΩ	33 kΩ	1.2 kΩ	120	0.39	300	4.7 kΩ	1.2 kΩ	470	±36
75	4	560	4.7 kΩ	27 kΩ	1.2 kΩ	91	0.33	620	5.6 kΩ	1.8 kΩ	470	±30
	8	390	6.8 kΩ	33 kΩ	1.2 kΩ	150	0.39	390	6.8 kΩ	1.5 kΩ	470	±40
100	4	470	5.6 kΩ	33 kΩ	1.2 kΩ	68	0.39	1.0 kΩ	8.2 kΩ	2.2 kΩ	470	±34
	8	330	8.2 kΩ	39 kΩ	1.2 kΩ	100	0.39	510	9.1 kΩ	1.8 kΩ	470	±45

All resistors are ½ W except R_6 and R_7, which are 5 W, and those marked, which are 2 W. All resistances not indicated to be in kilohms are in ohms.

Table 2-3. Transistor Heat-Sink Requirements for Transformerless Amplifier*

Output Power (watts rms)	Load Impedance (ohms)	Output Transistor Heat Sink (θ_{CA})† (°C/W)	Driver Transistor Heat Sink (θ_{CA})‡ (°C/W)
35	4	4.2	None
	8	2.4	None
50	4	3.0	60
	8	2.4	60
60	4	2.5	60
	8	2.0	60
75	4	1.6	35
	8	1.6	70
100	4	1.0	20
	8	1.0	50

*Heat-sink information is based on minimum required for safe operation under shunted load at 50°C ambient temperature.
†All output transistors are in TO–3 packages, except MJE2801/2901 (35 W/8 Ω) which are in the Case 90 Thermopad (Motorola trademark) plastic package.
‡All driver transistors are in plastic Uniwatt (Motorola trademark) packages, except those marked*, which are metal-cased TO–5.

Resistor R_F provides 100 per cent dc feedback from the output line to the input (at the base of Q_2), resulting in excellent dc stability. (This is a form of *overall feedback*, or *loop feedback*, since it involves more than one stage, as described in Sec. 2-1.) The resistance ratio of R_F to R_1 determines the *closed-loop* ac voltage gain of the amplifier. (The term "closed-loop gain" refers to the gain with feedback. Open-loop gain is gain without feedback.)

Transistor Q_4 functions as a high-gain, common-emitter driver. Since the output transistors (Q_7, Q_8, Q_9, Q_{10}) function as emitter followers (no voltage gain), Q_4 must be capable of handling the full-load voltage swing.

Transistor Q_6 serves as a constant-current source for the dc bias current, which flows through Q_4 and the bias diode D_2. (Transistor Q_6 eliminates the need for the large "bootstrap" capacitor found in some complementary audio amplifier designs.)

Transistors Q_7 and Q_8 form a compound pair which functions as an emitter follower with high current gain, and have unity voltage gain, for the positive portion of the output signal. Transistors Q_9 and Q_{10} function similarly for the negative portion of the output signal.

Zener diode D_1 is used to set the dc voltage through the differential amplifier and provide ac hum rejection from the negative power supply.

Transistors Q_3, Q_5, Q_{11}, and Q_{12} and semiconductors D_3 and D_4, along with their associated resistors, comprise the *short-circuit protection network*.

Resistors R_8, R_{10}, R_{11}, and R_{12} form a voltage-summing network. The voltage appearing at the base of transistor Q_{11} is thus determined by the collector current of Q_8 flowing through R_6 and the voltage appearing from the source $(+V_{CC})$ to the output. This summing network, since it detects both the voltage and current of Q_8, effectively senses the peak-power dissipation occurring in Q_8. At a predetermined power level in Q_8, the summing network can be chosen so that transistor Q_{11} conducts sufficiently to turn on transistor Q_3. Transistor Q_3 then steals the drive current from the base of transistor Q_4, and thus limits the power dissipated in Q_8. Diode D_3 is used to prevent Q_{11} from turning on, under normal load conditions, when the output signal swings negative. Resistors R_9, R_{13}, R_{14}, and R_{15}, along with transistors Q_{12} and Q_5 and diode D_4, similarly limit the power dissipation occurring in the output transistor Q_{10}.

2-7.1. Design Considerations

High-fidelity loudspeaker systems can appear capacitive or inductive as well as resistive. The current and voltage appearing in the amplifier will thus be out of phase when the load appears reactive. A 60° phase shift is not uncommon. At a 60° phase shift, half the source voltage and the full load current can appear simultaneously at the output transistor, or the full source voltage and one-half the full load current can appear, depending upon whether the load is capacitive or inductive.

Such a condition (simultaneous large voltage and current) could result in secondary breakdown (Sec. 2-5.8), just as if it were caused by a short circuit in the load. The short-circuit protection arrangement will prevent transistor damage (within limits) no matter what the cause. However, the protection circuit must be designed so that it will not interfere with normal operation, or with loads caused by the normal (up to 60°) phase shift at the output.

The *minimum peak-power level* to which the short-circuit dissipation can be limited is the product of peak current and voltage appearing at the output transistor under the worst-case allowable phase shift. This means that if the normal phase shift is 60°, the short-circuit power dissipation will be determined by:

$$P_{PD(\text{short circuit})} = \frac{V_{CC} \times I_{\text{peak}}}{2}$$

where P_{PD} is the peak dissipation for *each output transistor* (also the total *average power dissipation* for the amplifier).

The average power dissipation of each transistor is:

$$P_{AD(\text{short circuit})} = \frac{\frac{1}{2}V_{CC} \times I_{\text{peak}}}{2}$$

The worst-case average power dissipation in driver transistors Q_7 and Q_9 is the power dissipation expressed as P_{AD} divided by the current gain of the output transistor.

Because of the nature of the short-circuit protection network, the safe operating requirements (based on an overload of 1-s minimum) for the output transistor occurs at V_{CC} and is the same as the peak dissipation expressed as P_{PD}.

The maximum thermal resistance (or the minimum power-dissipation rating) required for each output transistor is found by:

$$\theta_{JC(max)} = \frac{T_{J(max)} - T_A - \theta_{CA} \times P_{AD}}{P_{AD}}$$

where $T_{J(max)}$ is the maximum junction temperature rating of the device, T_A is the maximum ambient temperature, and θ_{CA} is the thermal resistance of the heat sink, including a mica insulating washer, if used.

The minimum power-dissipation rating of the transistor is found by:

$$P_{DM} = \frac{T_{J(max)}}{\theta_{JC(max)}}$$

Refer to Sec. 1-4 for a further discussion of thermal resistance, heat sinks, and temperature design problems.

The following *performance characteristics* should be studied relative to design considerations for a specific application of the amplifier. These performance characteristics are typical for all the power output levels tabulated in Tables 2-1 and 2-2.

Output power. The amplifier will deliver its full-rated (rms) output power into the nominal load impedance providing the power supply has adequate regulation. (Refer to Chapter 6 for design data on regulated power supplies.) Figure 2-21 shows the power output versus load impedance.

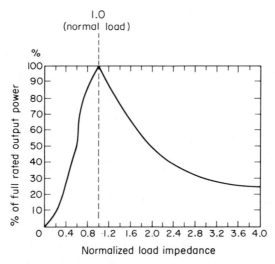

Fig. 2-21. Output power versus load impedance. Courtesy Motorola.

Input sensitivity. 1 V rms into 10 kΩ for full-rated output power.

Frequency response. Less than 3-dB rolloff from 10 Hz to 100 kHz, referenced to 1 kHz.

Power bandwidth. Full-rated output power ±0.5 dB from 2 Hz to 20 kHz (see Fig. 2-22).

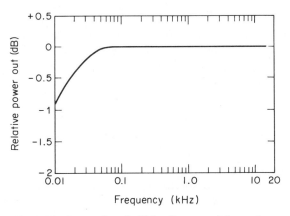

Fig. 2-22. Power bandwidth. Courtesy Motorola.

Total harmonic distortion. Less than 0.2 per cent at any power level between 100 mW and full-rated output, and at any frequency between 20 Hz and 20 kHz (see Fig. 2-23).

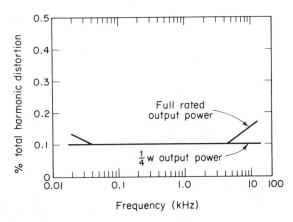

Fig. 2-23. Total harmonic distortion versus frequency. Courtesy Motorola.

Intermodulation distortion. Less than 0.2 per cent at any power level from 100 mW to full-rated output (60 Hz and 7 kHz mixed 4:1).

Square-wave response. See Fig. 2-24.

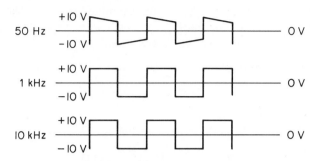

Fig. 2-24. Square-wave response. Courtesy Motorola.

Short-circuit power dissipation in each output transistor. See Fig. 2-25.

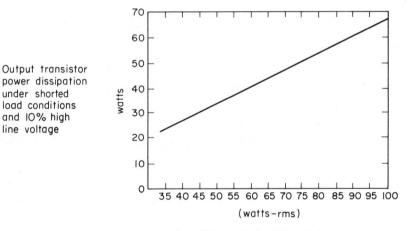

Fig. 2-25. Transistor power-dissipation curve. Courtesy Motorola.

2-7.2. Design Example

Assume that the circuit of Fig. 2-20 is to be used as a high-fidelity amplifier. The output must be 60 W rms into an 8-Ω loudspeaker or system, with normal operation allowed into ±60° reactive load. The amplifier must operate safely at 50°C ambient temperature with the output shorted. Distortion (both harmonic and intermodulation) must be less than 0.2 per cent. An input of no more than 1 V rms (1.4 V peak) must be required to produce the full 60-W output into the 8-Ω load.

Resistor R_1. The value of R_1 is approximately equal to peak input voltage $\div$ peak-load voltage $\times$ R_F. The value of R_F is arbitrarily fixed at 10 kΩ. The peak input voltage is specified in the examples as 1.4 V. The peak-load voltage is determined by first finding the rms-load voltage, then multiplying the result by 1.4. The rms-load voltage is found by:

$$\text{rms-load voltage} = \sqrt{\text{power output} \times \text{load impedance}}$$
$$= \sqrt{60 \times 8}$$
$$= 22 \text{ V}$$

Thus, 22×1.4, or 31 V, is the peak-load voltage. Therefore,

$$R_1 = \frac{1.4}{31} \times 10 \text{ k}\Omega = 450 \ \Omega$$

Use the next lowest standard value: 430 Ω.

Source voltage (V_{CC}). The source voltage (both positive and negative) is approximately equal to the peak-load voltage (31 V) + the voltage across R_6 + saturation and voltage-drop losses. The sum of the R_6 drop (approximately 1.5 to 2 V) and an arbitrary 3-V saturation and voltage-drop loss is equal to 5 V. Then, 31 V + 5 V = 36 V (source).

Resistor R_2. The maximum value of R_2 is approximately equal to:

$$\frac{\text{source voltage} - \text{voltage across } D_1}{\text{differential amplifier bias current} + D_1 \text{ current}}$$

The differential amplifier bias current is arbitrarily set at 2 mA. An additional 2 mA is required for proper regulation by D_1. This produces 4 mA through R_2. Therefore,

$$R_{2(\text{max})} = \frac{36 - 10}{0.004} = 6.5 \ \Omega$$

Use the next lowest 10 per cent standard value, or 5.6 kΩ.

Resistors R_3, R_4, and R_5. The voltage at the base of Q_6, with respect to the *negative* supply voltage, should be kept under 2 V to prevent premature clipping of the negative portion of the output signal. The base voltage of Q_6 is determined by the dividers R_3 and R_4. With the base-voltage set at an arbitrary 1.3 V,

$$\frac{R_4}{R_3 + R_4} \times V_{CC} = 1.3 \ V$$

Letting $R_4 = 1.2$ kΩ (assume 1 mA through bias network R_3 and R_4),

$$\frac{1.2 \text{ k}\Omega}{R_3 + 1.2 \text{ k}\Omega} \times 36 = 1.3 \text{ V} \qquad \text{then } R_3 = 33 \text{ k}\Omega \text{ (approx.)}$$

Using standard values of 1.2 kΩ and 33 kΩ for R_4 and R_3, respectively, the base voltage of Q_6 will be

$$Q_6 = \frac{1.2 \text{ k}\Omega}{33 \text{ k}\Omega + 1.2 \text{ k}\Omega} \times 36 = 1.26 \text{ V}$$

The value of resistor R_5 is:

$$R_5 = \frac{Q_6 \text{ base voltage} - Q_6 \text{ base-emitter voltage}}{\text{maximum dc bias current of } Q_4 + 1 \text{ mA (for resistor tolerances)}}$$

The Q_6 base voltage is 1.26 V; the typical base-emitter voltage (from the data sheet) is 0.75. The maximum dc bias for Q_4 is found by:

$$\text{maximum dc bias for } Q_4 = \frac{\text{peak load current}}{h_{FE(\text{min})}(Q_9) \times h_{FE(\text{min})}(Q_{10})}$$

The peak-load current is found by:

$$\frac{\text{peak-load voltage}}{\text{load impedance}} = \frac{31}{8} = 3.9 \text{ A}$$

Therefore,

$$R_{5(\text{max})} = \frac{1.26 - 0.75}{(3.9 \text{ A}/1000) + 1 \text{ mA}} = 125 \, \Omega$$

Use the nearest standard value for R_5: 120 Ω. (The value of 1000 is taken from the data-sheet h_{FE} values of Q_9 and Q_{10}.)

Resistors R_6 and R_7. The voltage appearing across R_6 and R_7, under maximum load conditions, should be between 1.5 V and 2 V. Using the 1.5-V figure, then

$$R_6 = R_7 = \frac{1.5}{3.9} = 0.384 \, \Omega \qquad \text{(peak-load current)}$$

Let $R_6 = R_7 = 0.39 \, \Omega$.

Resistors R_8, R_{10}, R_{11}, and R_{12}. The voltage at the base of Q_{11} must be approximately 1.4 V for Q_3 to conduct and provide the short-circuit protection. This voltage is determined by the voltage–current product at Q_8 (as shown by the voltage across R_6) and the network of R_8, R_{10}, R_{11}, and R_{12}.

The maximum current–voltage product appearing at Q_8 under *normal load* conditions occurs at the $\pm 60°$ phase-shift limit of the reactive load. For a 60° phase shift, the following equations can be derived for turning Q_3 on during *shorted output*:

$$V_{B11} = \tfrac{1}{2}K_1 V_{R6(\text{max})} + K_2 V_{EO(\text{max})} \qquad \text{(A)}$$

$$= K_1 V_{R6(\text{max})} + \tfrac{1}{2}K_2 V_{EO(\text{max})} \qquad \text{(B)}$$

(See Fig. 2-20 for V_{R6}, V_{B11}, and V_{EO}.)

where

$$K_1 = \frac{R_{12}}{R_8 + R_{12}}$$

$$K_2 = \frac{\text{equiv. resistance of } R_{11} \text{ and } R_8 \text{ in parallel}}{(\text{equiv. resistance of } R_{11} + R_8 \text{ in parallel}) + R_{11} + R_{10}} \qquad \text{(C)}$$

Solving (A) and (B) simultaneously yields

$$K_1 V_{R_6(\text{max})} = K_2 V_{EO(\text{max})}$$

since $V_{B11} = 1.4$ V, and substituting (C) in equation (A) or (B) yields

$$K_1 V_{R_6(\text{max})} = K_2 V_{EO(\text{max})} = 0.933 \text{ V} \qquad \text{(D)}$$

Now under shorted output,

$$V_{EO(\text{max})} = V_{CC} = 36 \text{ V}$$

$$V_{R_6(\text{max})} = I_{\text{peak}} R_E = 3.9 \text{ A} \times 0.39 \text{ A} = 1.53 \text{ V}$$

Therefore,

$$\frac{R_{12}}{R_8 + R_{12}} \times 1.53 = 0.933 \qquad \text{[from equation (D)]} \qquad \text{(E)}$$

Let $R_{12} = 470 \ \Omega$, a convenient value; then solving (E) yields

$$R_8 = 300 \ \Omega$$

Let

$$R'_{12} = \frac{R_{12} \times R_8}{R_{12} + R_8} = 183 \ \Omega$$

Then

$$K_2 V_{EO(\text{max})} = \frac{R'_{12}}{R'_{12} + R_{11} + R_{10}} \times V_{EO(\text{max})} = 0.933 \qquad \text{(F)}$$

Also,

$$\frac{R'_{12}}{R'_{12} + R_{11}} \times V_{D3} = 0.933$$

in order for the Zener diode to conduct. Let $V_{D3} = 7.5$ V, a convenient value for an economical Zener diode. For equation (F),

$$\frac{183}{183 + 1.2 \text{ k}\Omega + R_{10}} \times 36 = 0.933 \qquad \text{so } R_{10} \approx 5.7 \text{ k}\Omega$$

This can be found as follows:

$$0.933 \div 36 = 0.026$$

$$183 \div 1383 + R_{10} = 0.026$$

$$183 \div 0.026 = 1383 + R_{10}$$

$$7028 = 1383 + R_{10}$$

$R_{10} = 7038 - 1383 = 5655 \ \Omega$ (rounded off to 5.7 kΩ). To allow V_{D3} to turn on at a higher line voltage (say 40 V), reduce R_{10} to 4.7 kΩ.

Thus, the values for the resistors in the short-circuit network are:

$$R_6 = R_7 = 0.39 \text{ k}\Omega$$

$$R_8 = R_9 = 300 \ \Omega$$

$$R_{10} = R_{15} = 4.7 \text{ k}\Omega$$

$$R_{11} = R_{14} = 1.2 \text{ k}\Omega$$

$$R_{12} = R_{13} = 470 \ \Omega$$

Output transistors. Transistors Q_8 and Q_{10} should have a breakdown $BV_{CEO} \geqq 80$ V at $I_C = 200$ mA (allowing for 10 per cent high-source voltage, $36 + 36 + 10$ per cent $= 80$). The $h_{FE} = 20$ minimum at $I_C = 4.0$ A, and $V_{CE} = 2.0$ V.

Motorola MJ2841 and MJ2941 meet these specifications. Their power-dissipation rating is 150 W at 25°C, with a maximum junction temperature of 200°C. Therefore, the thermal resistance from junction to case is

$$\theta_{JC} = \frac{200° - 25°}{150 \text{ W}} = 1.17°\text{C/W}$$

From previous calculations, the power dissipation occurring in *each transistor* during a shorted load condition is

$$P_D = \frac{\frac{1}{2}V_{CC} \times 3.9 \text{ A}}{150} = 35.1 \text{ W}$$

Allowing for a 30 per cent increase in P_D due to high source voltage, component tolerances, etc., $P_{D(\text{max})}$ is 46 W. This requires a heat sink for both Q_8 and Q_{10}. Also, the amplifier must operate at 50°C. Therefore, to find the required sink-to-ambient thermal resistance of the heat sink (including any washers),

$$\theta_{SA} = \frac{T_J - T_A - P_{D(\text{max})} \times \theta_{JC}}{P_{D(\text{max})}}$$

$$= \frac{200 - 50 - (46 \times 1.17)}{46}$$

$$= 2.0°\text{C/W}$$

Driver transistors. Transistors Q_7 and Q_9 should have a breakdown $BV_{CEO} \geqq 80$ V at $I_C = 10$ mA. The $h_{FE} = 50$ minimum is at $I_C = 100$ mA.

Motorola MPS-U06 and MPS-U56 meet these specifications. Their power-dissipation rating is 5 W at 25°C, with a maximum junction temperature of 135°C. Therefore, the thermal resistance from junction to case is

$$\theta_{JC} = \frac{135 - 25}{5 \text{ W}} = 22°\text{C/W}$$

The maximum power dissipation (or P_D) in each driver transistor is

$$\frac{\text{max } P_D \text{ in each output transistor (or 46 W)}}{h_{FE} \text{ of output transistor (under short circuit)}}$$

The current of the output transistors under short-circuit conditions is

$$\frac{\text{short-circuit power (or 46 W)}}{\frac{1}{2} \text{ source voltage (or 20 V)}} = 2.3 \text{ A}$$

The 20 V allows for a high source voltage. The output transistor data sheets show an $h_{FE} \geq 45$ at 2.3 A. Therefore, the P_D of each driver is $46/45 = 1.03$ W.

This requires a heat sink for both Q_7 and Q_9. Also, the amplifier must operate at 50°C. Therefore, to find the required sink-to-ambient thermal resistance of the heat sink (including any washers),

$$\theta_{SA} = \frac{T_J - T_A - (P_{D(\text{driver})} \times \theta_{JC})}{P_{D(\text{driver})}} = \frac{135 - 50 - (1.03 \times 22)}{1.03}$$

$$= 60°\text{C/W}$$

Predriver transistors. Transistors Q_4 and Q_6 should have a breakdown $BV_{CEO} \geq 80$ V. The $h_{fe} = 75$ at $I_C = 10$ mA and the $V_{CE} = 1$ V.

Motorola MPS-A06 and MPS-A56 meet these requirements. Their power-dissipation rating is 500 mW at 25°C, with a maximum junction temperature of 135°C. Therefore, the thermal resistance from junction to case is

$$\theta_{JC} = \frac{135 - 25}{500 \text{ mW}} = 0.21°\text{C/mW}$$

The maximum power dissipation of the predriver transistors is

$$\frac{P_{D(\text{driver})}}{h_{fe(\text{driver})}} + P_{DC} \quad \text{or} \quad \frac{1030 \text{ mW}}{50} + (40 \text{ V} \times 5 \text{ mA}) = 220 \text{ mW}$$

where P_{DC} is the power dissipation of maximum collector voltage (40) and bias current in R_5 (5 mA from previous calculation).

The 220-mW maximum power dissipation is below the 500-mW rating at 25°C. The derated power dissipation of the transistors at 50°C should be $(135 - 50)/0.21 = 405$ mW (well above the 220 mW).

Differential amplifier. The differential-amplifier circuit of Q_1 and Q_2 can be obtained as a complete package (or hybrid circuit, Sec. 2-4). The Motorola MCF8001 is such a package. The collector resistance should be between 10 and 15 per cent of the external emitter resistance, and the BV_{CEO} should be ≥ 80 V (as are the other transistors in the amplifier). Figure 2-26 shows connections between the differential package and the amplifier circuit components. Note that the MCF8001 is a dual-channel device (such as required for a stereo system) but only one channel is needed for connection into the amplifier of Fig. 2-20.

Short-circuit protection transistors. Transistors Q_3, Q_5, Q_{11}, and Q_{12} operate at low-current levels and can be TO-92-type plastic transistors. Q_{11} and Q_{12} should have an $h_{FE} \geq 40$ at 2 mA, and a $BV_{CEO} \geq 80$ V. Motorola

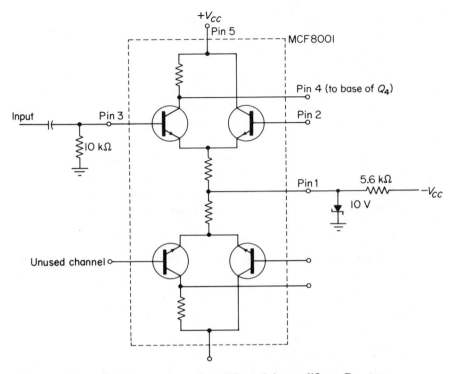

Fig. 2-26. Connections for differential amplifier. Courtesy Motorola.

MPS-L01 and MPS-L51 meet these specifications. Q_3 and Q_4 should have an $h_{FE} \geq 25$ at 1 mA, and a $BV_{CEO} \geq 10$ V. Motorola MPS-A20 and MPS-A70 meet these specifications.

2-8. Multistage Audio Amplifier with Transformer Coupling

Audio-amplifier stages can be coupled by means of transformers. As in the case of any design approach, transformers impose certain problems and have certain advantages. These can be traded off to meet a specific design need.

Inductive reactance. One major problem with transformers is the inductive reactance created by the transformer windings. Inductive reactance increases with frequency; at frequencies beyond about 20 kHz, the inductance reactance of iron core becomes so high that signals cannot pass or are greatly attenuated. For this reason, iron-core transformers are not used at higher frequencies. At low frequencies, the reactance drops to near zero, even with iron cores. Since transformers are placed across (or in shunt with) the amplifier circuits, the transformer winding acts as a short at low frequencies. If an amplifier must operate at very low frequencies (below about 20 Hz), transformers are not recommended.

Size and weight. For a transformer to handle any large amounts of power or current, the winding wire must be large. Also, the iron core must be large. Both of these add up to bulk weight. Except in certain cases, the added weight of transformers defeats the purpose of compact, lightweight, transistorized equipment.

Short-circuit burnout. Another problem with transformers is the danger of a short-circuit output, resulting in excessive, simultaneous voltage and current in the transistor collector. There is very little voltage drop across a transformer winding in a transistor collector circuit (compared with the drop across the load in an *RC* or direct-coupled amplifier). If a short circuit in the output (say due to a short across the load) causes heavy current flow, the transistor can be damaged. The destructive condition is known as *secondary breakdown*, or *second breakdown*. Secondary breakdown can be prevented (or minimized) by limiting the collector current–voltage product.

Another overload problem is presented when the load is a loudspeaker. Loudspeaker systems can appear *capacitive* or *inductive* as well as resistive. The current and voltage appearing in the amplifier will thus be out of phase when the load appears reactive. (With the normal in-phase condition, voltage is high when current is low, and vice versa.) A 60° phase shift is not uncommon for loudspeaker systems. At a 60° phase shift, the voltage is about one-half the supply when current is maximum. Thus, the output transistor receives full current and half voltage (or full voltage and half current), depending on whether the load is capacitive or inductive. Either way, the combined voltage and current can damage the transistor.

As a guideline, there will be no secondary breakdown if the output transistor can dissipate *twice* the power output of the amplifier (for single-ended transformer amplifiers). If the amplifier is push-pull, each output transistor must be capable of dissipating the full power output, if secondary breakdown is to be avoided.

Impedance matching. One of the major advantages to transformer coupling is the impedance-matching capability. The output impedance of a typical *RC* amplifier is on the order of several hundred (or thousand) ohms (generally set by the output collector resistor value). In audio systems (particularly for voice and music reproduction) this large output impedance must be matched to 4-, 8-, and 16-Ω loudspeaker systems. The severe mismatch results in power loss. With a transformer, the primary winding can be designed (or selected) to match the transistor circuit output impedance with the transformer secondary matching the loudspeaker (or other load) impedance.

Low supply voltage. Another major advantage of transformer coupling is caused by the low voltage drop across the transformer winding. Because of this low voltage drop, it is possible to operate a transformer-coupled amplifier with a much lower supply voltage than with an *RC* amplifier. As a guideline, the transformer-coupled amplifier can be operated at *one-half the supply voltage* required for a comparable *RC* amplifier.

Typical uses. Transformers are often used in the audio-amplifier sections of transistorized radio receivers and portable hi-fi systems. In these applications, very little power is required, so the transformers can be made compact and lightweight. Since there is no loss in impedance match and low voltage is required, the transformer-coupled circuits are ideal for battery operation. Transformers are also used in high-power, hi-fi/stereo systems and television audio sections, where the added weight is of little consequence.

Typical circuit. Figure 2-27 is the working schematic for a classic transformer-coupled audio amplifier. The circuit has a class *A* input or *driver stage*, and a class B push-pull output stage. The class A stage provides both voltage and power amplification as needed to raise the low input signal to a level suitable for the class B power output stage.

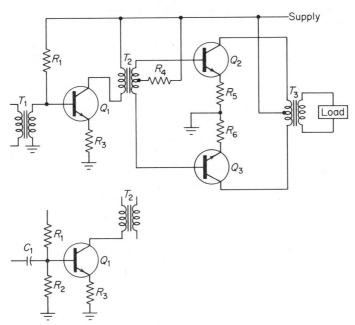

Alternate driver circuit with *RC* input

Fig. 2-27. Transformer-coupled audio amplifier. Courtesy Motorola.

The class A stage can be transformer-coupled or *RC*-coupled at the input as needed. Transformer coupling is used at the input, where a specific impedance-match problem must be considered in design. The class A input stage can be driven directly by the signal source or can be used with a preamplifier for very low-level signals. When required, a high-gain voltage amplifier (such as described in other sections of this chapter) is used as a *preamplifier*.

The push-pull output stage may be operated as a class B amplifier. That is, the transistors are cut off at the Q-point and draw collector current only in the presence of an input signal. Class *B* is the most efficient operating mode for audio amplifiers, since it draws the least amount of current (and no current where there is no signal). However, true class B operation can result in crossover distortion.

Crossover distortion. The effects of crossover distortion can be seen by comparing the input and output waveforms in Fig. 2-28. In true class B operation, the transistor remains cut off at very low signal inputs (because transistors have low current gain at cutoff) and turns on abruptly with a large signal. As shown in Fig. 2-28a, there is no conduction when the base–emitter voltage V_{BE} is below about 0.65 V (for a silicon transistor). During the instantaneous pause when one transistor stops conducting and the other starts conducting, the output waveform is distorted.

Distortion of the signal is not the only bad effect of this crossover distortion condition. The instantaneous cutoff of collector current can set up large voltage transients equal to several times the size of the supply voltage. This can cause the transistor to break down.

Crossover distortion can be minimized by operating the output stage as class AB (or somewhere between B and AB). That is, the transistors are forward-biased just enough for a small amount of collector current to flow at the Q-point. Some collector current is flowing at the lowest signal levels, and there is no abrupt change in current gain. The effects of this are shown in Fig. 2-28b. The combined collector currents result in a *composite curve* that is essentially linear at the crossover point. This produces an output that is a faithful reproduction of the input, at least as far as the crossover point is concerned. Of course, class AB is less efficient than class B, since more current must be used.

Some designers use an alternative method to minimize crossover distortion. This technique involves putting diodes in series with the collector or emitter leads of the push-pull transistors. Because the voltage must reach a certain value (typically 0.65 V for silicon diodes) before the diode will conduct, the collector current curve is rounded (not sharp) at the crossover point.

2-8.1. Design Considerations

The design considerations for transformer-coupled amplifiers are somewhat different from those of direct-coupled and resistance-coupled systems.

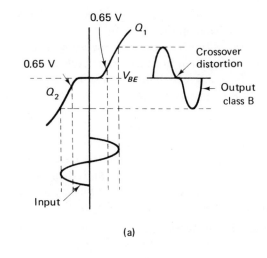

(a)

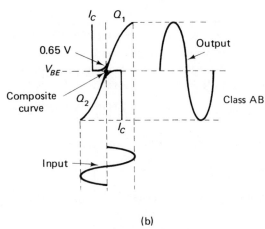

(b)

Fig. 2-28. Effects of crossover distortion, and how it is eliminated by forward bias of the base–emitter junction.

Collector efficiency and class of operation. The efficiency of an amplifier is determined by the ratio of collector input to output power. That is, an amplifier with 70 per cent efficiency will produce 7-W output for 10-W input (with power input being considered as collector source voltage multiplied by total collector current).

Typically, class B amplifiers can be considered as 70 to 80 per cent efficient (for design purposes). Class A amplifiers are typically in the 35 to 40 per cent efficiency range, with class AB amplifiers showing a 50 to 60 per cent efficiency.

In all cases, any design that produces an increase in collector current at the Q-point will produce correspondingly lower efficiency. This results in a tradeoff between efficiency and distortion.

The efficiency produced by a class of operation also affects the heat-sink requirements. Any design that produces more collector current at the Q-point requires a greater heat-sink capability. As a rule of thumb, *a class A amplifier requires double the heat-sink capability of a class B amplifier, all other factors being equal.*

In practical design, push-push amplifiers should be designed for true class B operation (transistors at or near cutoff at the Q-point), and then tested for distortion. If distortion is severe, increase the base–emitter forward bias so that some current flows at the Q-point, and then check distortion under identical test conditions. This approach is usually more realistic than trying to design an amplifier for a given class of operation.

Output power. The true output power of an audio amplifier can be determined by measuring the voltage across the load, then solving the equation

$$\text{power output} = \frac{\text{voltage}^2}{\text{load impedance}}$$

The output of a transformer-coupled amplifier is also related to the collector current (produced by the signal) and the primary impedance of the transformer. In a push-pull amplifier, the relationship is:

$$\text{power output} = \frac{\text{current}^2 \times \text{primary impedance}}{8}$$

In the single-ended amplifier, the relationship is:

$$\text{power output} = \frac{\text{current}^2 \times \text{primary impedance}}{2}$$

These relationships provide a basis for design of transformer-coupled amplifiers.

Transformer selection. Audio transformers are listed by primary impedance, secondary impedance, and power output capability. From a practical design standpoint, it is not always possible to find an off-the-shelf transformer with exact primary/secondary relationships at a given power rating. Most manufacturers will produce transformers with exact impedance relationships on a special order basis. However, this is usually not practical, except for special applications. Instead, the transformer should be selected for *exact secondary impedance*, and the nearest value of primary impedance, within the given power rating.

The following rules can be applied to selection of the transformer shown in Fig. 2-27.

Push-pull output transformer T_3. The rated power capability of transformer T_3 should be 1.1 times the desired power. The secondary impedance should match the load impedance (loudspeaker, etc.) into which the amplifier must operate. The primary impedance should be determined by maximum collector current passing through the primary windings, and the total collector voltage swing. The relationship is:

$$\text{primary impedance} = \frac{4(\text{supply voltage} - \text{minimum voltage})}{\text{maximum current}}$$

Minimum voltage can be determined by reference to the collector voltage–current curves for the transistors. As shown in Fig. 2-29, the minimum voltage point should be selected so that it is just to the right of the curved portion of the characteristics (where the curves start to straighten out). If curves are not available for the transistors, use an arbitrary 2 V for the minimum voltage. This will be satisfactory for most power transistors.

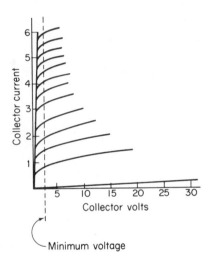

Fig. 2-29. Locating minimum voltage point on power-transistor curves.

Maximum current can be determined by the total collector voltage swing and the desired power output. The relationship is:

$$\text{maximum current} = \frac{2.1 \times \text{power output}}{(\text{supply voltage} - \text{minimum vo.tage})}$$

Push-pull input transformer T_2. Note that transformer T_2 is the input transformer for the push-pull output stage, and the output transformer for the single-ended driver. Therefore, the secondary impedance should be chosen to match the signal input impedance of the push-pull stage, while the primary of T_2 should match the driver output.

The rated power capability of T_2 should be equal to the input power of the driver stage (which is generally about 3 times the output to the push-pull stage).

The *total secondary impedance* of T_2 should be 4 times the signal input impedance of Q_2 and Q_3. This input impedance is found by dividing signal voltage by signal current.

The signal voltage and current are dependent upon the desired amount of collector signal current. In turn, the collector signal current is dependent upon the desired power output (from Q_2 and Q_3) and the primary impedance of T_3 (total primary impedance). The relationship is:

$$\text{collector signal current} = \sqrt{\frac{8 \times \text{power output}}{\text{total primary impedance}}}$$

With the required collector signal current established, the input signal voltage and current for Q_2 and Q_3 can be found by reference to the *transfer characteristic curves*. As discussed in Chapter 1, transfer curves are usually provided on power transistor data sheets. Typical transfer characteristics for power transistors are shown in Fig. 2-30. The transfer characteristics show the required base–emitter voltage (or signal voltage) and base current (or signal current) for a given collector current. However, an additional factor must be considered in establishing signal voltage. The base–emitter voltages shown in Fig. 2-30 can be considered as the signal voltages only

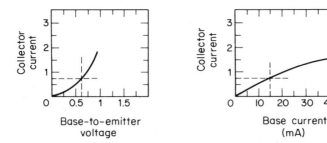

Fig. 2-30. Transfer curves for power transistors.

when the emitters of Q_2 and Q_3 are connected directly to ground. If emitter resistors (R_5 and R_6) are used to provide feedback stabilization, the voltage developed across the emitter resistors must be added to the base–emitter voltage to find a true signal voltage.

The values of R_5 and R_6 are chosen to provide a voltage drop approximately equal to the base–emitter voltage, when the collector signal current is passing through the emitter resistors. Since the base–emitter voltage of a silicon power transistor is typically less than 1 V, the values of R_5 and R_6 can arbitrarily be set to provide a drop of between 0.5 and 1 V with normal collector signal current.

The *primary impedance* of T_2 should match the output of Q_1. The relationship is:

$$\text{primary impedance} = \frac{\text{supply voltage} - \text{minimum voltage}}{\text{collector current at Q-point}}$$

However, for practical design, it is generally easier to select the transformer on the basis of secondary impedance and power rating, and then adjust the Q_1 output to match a primary impedance available with an off-the-shelf transformer. Note that a large primary impedance requires a high supply voltage, and vice versa. In turn, a low supply voltage is related to a larger collector current. Therefore, it may be necessary to trade off among supply voltage, available primary impedances (for off-the-shelf transformers with correct secondary impedance and power rating), and collector current.

Once the primary impedance of T_2 is established, the collector signal current for Q_1 can be determined. The relationship is:

$$\text{collector signal current} = \frac{2 \times \text{power output}}{\text{primary impedance}}$$

The power output from Q_1 must equal the required power input for Q_2 and Q_3. This power input is determined by:

$$\text{power input} = \text{signal voltage} \times \text{signal current}$$

Input transformer T_1. As shown in Fig. 2-27, transformer T_1 can be omitted if the resistance-coupled circuit is used. Generally, the transformer-coupled circuit is used if it is of particular importance to match the impedance of the signal source to the amplifier. When used, the primary impedance of T_1 should equal that of the signal source. The secondary of T_1 should match the signal input impedance of Q_1. The signal input impedance is found by:

$$\text{signal input impedance} = \frac{\text{signal voltage}}{\text{signal current}}$$

The signal (input) voltage and current are dependent upon the desired amount of collector signal current for Q_1 (previously established in calculating the secondary impedance of T_2). With the required collector signal current established, the input-signal voltage and current for Q_1 can be found by reference to the transfer characteristic curves. If emitter resistor R_3 is used to provide feedback stabilization, the voltage developed across the emitter resistor must be added to the base–emitter voltage to find a true signal voltage. The value of R_3 can be selected to provide a drop of between 0.5 and 1 V with normal collector signal current (as discussed for emitter resistors R_5 and R_6).

If transfer characteristic curves are not available for Q_1 (as is sometimes the case for low-power transistors), it is still possible to find the approximate signal voltage and current required to produce the necessary output. Signal voltage can be approximated by assuming that the base–emitter drop is 0.5 V. This must be added to any drop across emitter resistor R_3 (arbitrarily chosen to be between 0.5 and 1 V). Signal current can be approximated by dividing the desired collector signal current by beta of Q_1. Of course, since beta is variable, the input signal current can only be a rough approximation.

Source voltage. The source voltage required for Q_1 is approximately double that required for Q_2 and Q_3. However, since all three transistors will be operated from the same source, choose a source voltage that will match the requirements of Q_1. As a rule of thumb:

3 to 9 V should be used for power outputs up to 2 W.
6 to 15 V should be used for power outputs up to 20 W.
15 to 50 V should be used for power outputs up to 50 W.

These rules apply to the output of Q_1. Transistors Q_2 and Q_3 will require approximately one-half the voltage. For example, if Q_1 were to deliver 20 W (by itself), 6 to 15 V would be required. If Q_2 and Q_3 were to deliver the same 20 W, the voltage required would run between 3 and 7.5 V. Keep in mind that a higher source voltage will permit lower currents (and vice versa), all other factors being equal.

Transistor selection. Both frequency limit and power dissipation must be considered in selecting transistors for the circuit. In any audio system, the transistors should have an f_{ae} (Chapter 1) higher that 20 kHz, and preferably higher than 100 kHz. The power dissipation of Q_1 should be at least 3 times the required output of Q_1. The power dissipation of Q_2 and Q_3 should be between 1.3 and 1.5 times the required output of the amplifier circuit (at T_3) If any power dissipation exceeds about 1 W, heat sinks will be required. The required power dissipation ratings for the heat sinks should be calculated as described in Chapter 1.

Base-bias resistances. Resistors R_1, R_2, and R_4 serve to drop the source voltage to a level required at the transistor bases. The drop across R_4 should (in theory) bias Q_2 and Q_3 at cutoff. That is, no base current should flow except in the presence of a signal. This is not practical since the complete absence of base current would result in no drop across R_4. Also, such a bias condition would probably result in crossover distortion. As a starting point for Q_2 and Q_3 bias values, assume that a small residual base current is going to flow in each transistor. The value of this residual current should be set arbitrarily at one-tenth of the normal signal current (previously calculated). The combined residual currents of Q_2 and Q_3 will flow through R_4. The value of R_4 is then calculated by:

$$\text{resistance } (R_4) = \frac{\text{source voltage}}{2 \times \text{residual current}}$$

If transformer T_1 is used, resistor R_2 is omitted. Under these conditions, the drop across R_1 should bias Q_1 at the operating point. The base current of Q_1 flows through R_1 and produces the drop. The base current used to calculate the Q_1 input impedance can be used as a starting point for the calculation of R_1 value. The relationship is:

$$\text{resistance } (R_1) = \frac{\text{source voltage}}{\text{base current}}$$

If transformer T_1 is not used, resistor R_2 is added. The value of R_2 will determine the approximate input impedance of the amplifier circuit, and should be so selected. Generally, R_2 should be greater than 500 Ω, and less than 20 kΩ, for any audio circuit. When R_2 is used, the voltage drop from base to ground will also appear across R_2, resulting in some current flow through R_2. This current flow must be added to the base current in calculating the value of R_1.

Coupling capacitor. When transformer T_1 is used, capacitor C_1 is omitted. When used, C_1 forms a high-pass filter with R_2, producing some loss at the low-frequency end of the audio range. Using a 1-dB loss as the low-frequency cutoff point, the value of C_1 can be found by:

$$\text{capacitance} = \frac{1}{3.2FR}$$

where capacitance is in farads, F is the low-frequency limit in hertz, and R is resistance of R_2 in ohms.

2-8.2. Design Example

Assume that the circuit of Fig. 2-27 is to be used as an audio amplifier. The desired output is 10 W into an 8-Ω impedance. The signal is from a 500-Ω impedance source, and is approximately 1 V. The amplifier is to be transformer-coupled at the input and output. The low- and high-frequency limitations of transformer coupling can be ignored. However, crossover distortion should be minimized (or eliminated if possible). The maximum supply voltage available is 13.5 V.

Power and voltage amplification. The output voltage required to produce 10 W across an 8-Ω impedance is approximately 9 V ($E = \sqrt{PR}$). Since the input available is approximately 1 V, the total voltage gain is only 9. However, the input power is approximately 2 mW, assuming an input impedance of 500 Ω, and 1 V ($P = E^2 \div R$). This requires a total power gain of 5000 (10 W $\div$ 2 mW). Therefore, both class A Q_1 and class B Q_2 and Q_3 should be designed as power amplifiers.

Transformer T_3. The rated power capability of T_3 should be at least 10 W. The secondary impedance of T_3 should be 8 Ω, to match the 8-Ω output load. The primary impedance of T_3 is dependent upon a tradeoff among power output, supply voltage, and collector current for Q_2 and Q_3. The power output is fixed at 10 W, while the maximum supply voltage available is 13.5 V.

First, find minimum voltage for the collectors of Q_2 and Q_3. If curves such as those shown in Fig. 2-21 are available, select an arbitrary minimum voltage point just to the right of the curved portion. If curves are not available, use an arbitrary 2 V. Use 2 V in this case.

The maximum current required will then be $(2.1 \times 10)/(13.5 - 2)$, or approximately 1.83 A. With maximum current established, the absolute minimum impedance would $[4 \times (13.5 - 2)]/1.83$, or approximately 25 Ω. Now assume that the only off-the-shelf transformer (of suitable size and weight) with a 10-W capability and an 8-Ω secondary has a 100-Ω primary impedance. This value will produce a signal current less than the maximum. The collector current required to produce the desired power output (10 W) with the 100-Ω primary is $\sqrt{8 \times 10/100}$, or 0.9 A.

Transformer T_2. The total secondary impedance of T_2 should be 4 times the signal input impedance of Q_2 and Q_3, where the impedance is the signal voltage divided by the signal current.

Both signal voltage and current are dependent upon collector signal current. Using the value of 0.9 A as the required collector current for Q_2 and Q_3, the input signal voltage and current can be found by reference to the transfer curves. Assuming that the transfer curves are similar to those of Fig. 2-30, a base current of 15 mA and a base–emitter voltage of 0.6 V is required for a collector current of 0.9 A.

Next, assume that *stabilizing emitter resistors* R_5 and R_6 are going to be selected to provide a 0.9-V drop with the collector current of 0.9 A. This requires 1-Ω, 1-W resistors for R_5 and R_6.

The 0.9 V developed across R_5 and R_6 must be added to the 0.6-V base–emitter voltage to establish a signal voltage of 1.5 V. With this 1.5-V signal voltage and the 15-mA base (signal) current, the required power input to Q_2 and Q_3 is 1.5×15 mA = 22.5 mW.

Using the same values to find signal input impedance $1.5 \div 0.015 = 100$ Ω. Therefore, the *total secondary impedance* of T_2 should be 4 times 100, or 400 Ω.

Assume that an off-the-shelf interstage or driver transformer is found with a 400-Ω center-tapped secondary, a power capability of more than 22.5 mW, and a primary impedance of 40 Ω. With these values established for T_2, the next step is to calculate the power output and collector current for Q_1.

Transistor Q_1. The collector current for Q_1 is determined by required power output (which is the same as power input to Q_2 and Q_3, or 22.5 mW), and the primary impedance of T_2. The relationship is collector signal current $= 2 \times$ (power output) ÷ primary impedance, or 2×22.5 mW ÷ 40 = 1.1 mA.

Input transformer T_1. The primary impedance of T_1 should equal that of the signal source (500 Ω, as stated in the design problem). The secondary of T_1 should match the signal impedance of Q_1, where impedance equals signal voltage divided by signal current.

Both signal voltage and current are dependent upon collector signal current. Using the value of 1.1 mA as the required collector current for Q_1, the input signal voltage and current can be found by reference to the transfer curves. If stabilizing emitter resistor R_3 is used, the voltage drop across R_3 must be added to the base–emitter voltage of Q_1 to find the true signal voltage. Assume that R_3 is selected to provide a 0.5-V drop (typical) with the collector current of 1.1 mA. This requires $0.5 \div 1.1$ mA, or 454 Ω. (A 470-Ω resistor is the nearest standard.)

Now assume that transfer curves are not available for Q_1, but that beta is given as 10. Assume that the base–emitter drop is 0.5 V. This must be added to the drop across R_3 to produce a total input signal of 1 V. The input signal current is found by dividing the collector current by beta, or 1.1 mA $\div 10 = 0.11$ mA base current.

With an input signal of 1 V and an input current of 0.11 mA, the input impedance to Q_1 is $1 \div 0.00011$, or 9090 Ω. The secondary of T_1 should be selected for this value (9000 as a standard).

Transistors Q_1, Q_2, *and* Q_3. All three transistors must have an f_{ae} greater than 20 kHz. If f_{ae} is not given on the data sheet, the value can be calculated as described in Sec. 1–3. Usually, f_{ae} (or f_{ab}, or some similar term) will be found on all data sheets. However, in transistor catalog listings, the only frequency value shown (generally) is f_T. Since it is usually more convenient (and practical) to select transistors by means of catalogs rather than data sheets, it is necessary to have a simple means of approximating f_{ae} when h_{fe} and f_T are given. Since $f_T = f_{ae} \times h_{feo}$, then $f_{ae} = f_T \div h_{feo}$. Simply assume that $h_{feo} = h_{fe}$, and divide f_T by h_{fe} to find f_{ae} in a catalog listing. With the possible exception of certain very high-power transistors, practically any present-day transistor will have an f_{ae} greater than 20 kHz.

The power dissipation rating of Q_1 should be 3 times the required (Q_1) output, or 3×22.5 mW $= 67.5$ mW. The power dissipation of Q_2 and Q_3 should be between 1.3 and 1.5 times the required amplifier output, or $1.3 \times 10 = 13$ W, $1.5 \times 10 = 15$ W. Power dissipation of audio transistors is often listed in catalogs as P_T, or total power.

Transistor Q_1 can operate without a heat sink. Transistors Q_2 and Q_3 will require heat sinks. The °C/W rating for the heat sinks can be calculated as described in Sec. 1-4.

Some audio-power-transistor catalogs list two P_T dissipation values. One value is the free air rating (usually at 25°C), while the other value is at some specific case temperature (usually 100°C). The free air rating can be used without a heat sink. The case temperature rating implies the use of a heat sink to take advantage of the given power dissipation value. For example, assume that a transistor is rated for 15-W dissipation at a case temperature of 100°C, and the ambient temperature is 25°C. The required heat-sink rating would then be: 100°C − 25°C = 75°C; 75°C ÷ 15 W, or 5°C/W. That is, a 5°C/W heat sink would permit the transistor to dissipate the full 15 W, but keep the case temperature at or below 100°C.

Base-bias resistances. Resistors R_1 and R_4 serve to drop the supply voltage to the level required at the transistor bases. As discussed, resistor R_2 is not required when T_1 is used. The voltage drops across R_1 and R_4 are dependent upon the resistance values, and the base currents.

In theory, there will be no base current for Q_2 and Q_3, except in the presence of a signal. As discussed, this is not practical, and will probably result in crossover distortion. To minimize the possibility of crossover distortion, select the value of R_4 on the basis of a residual current in the absence of a signal. The residual current is arbitrarily set at one-tenth the value of the full signal current, or $0.1 \times 15\,\text{mA} = 1.5\,\text{mA}$. The combined residual currents of Q_2 and Q_3 flow through R_4. Therefore, the value of R_4 is the source voltage divided by 2 times the residual current, or $13.5 \div 2(1.5\,\text{mA}) = 4500\,\Omega$.

The value of R_1 should be of suitable resistance to bias Q_1 at a point where the input signal will not overdrive the amplifier. Using the values established thus far, the base of Q_1 is assumed to be at approximately 1 V (0.5 V across R_3 and 0.5-V emitter to base). This value may be too high for a no-signal operating point and should be reduced by bias from R_1. As a first trial value for R_1, divide source voltage by base current ($13.5 \div 0.11\,\text{mA} = 122{,}727\,\Omega$ $\times 10^3$). If this reduces the base to a level where the amplifier is overdriven by the negative peaks of the input signal, reduce the value of R_1 from the first trial value. As a second trial value, subtract the desired 1-V base voltage from the source voltage, then divide by the base current, or $13.5 - 1 = 12.5 \div 0.11\,\text{mA} = 113{,}637\,\Omega$.

2-9. Operating and Adjustment Controls

The most common operating controls for audio circuits used with music or voice reproduction equipment (hi-fi, stereo, public address, etc.) are the *volume* or *loudness* control, the *treble* control, and the *bass* control. The other most common audio control is the *gain* control (found on such circuits as the operational amplifier, servo control amplifier, etc.). The gain and volume controls are often confused, since they both affect output of the amplifier circuit. A true gain control sets the *gain of one stage* in the amplifier, thus setting the overall gain of the complete amplifier. A true volume control sets the *level of the signal* passing through the amplifier, without affecting the gain of any or all stages. A gain control is usually incorporated as part of a stage, while a volume control is usually found between stages, or at the input to the first stage.

2-9.1. Design Considerations for Volume Controls

As shown in Fig. 2-31, the basic volume control is a variable resistance or potentiometer, connected as a voltage divider. The voltage output (or signal level) is dependent upon the volume control setting.

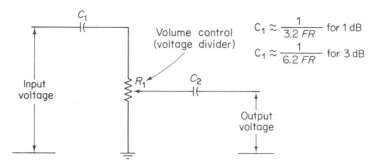

Fig. 2-31. Basic volume control.

If the audio circuit is to be used with voice or music, the volume control should be of the *audio taper* type, where the voltage output is not linear throughout the setting range. (The resistance element is not uniform.) This produces a nonlinear voltage output to compensate for the human ear's non-linear response to sound intensity. (The human ear has difficulty in hearing low-frequency sounds at low levels, and responds mainly to the high-frequency components.) Generally, the audio taper controls used with transistors are of the type where large changes are produced at the high-loss end.

If the audio circuit is not used with voice or music, the volume control should be of the linear type (unless there is some special circuit requirement). With such a control, the actual voltage or signal is directly proportional to the control setting.

No matter what type of volume control is used, it should be isolated from the circuit elements. If a volume control is incorporated as part of the circuit (such as the collector or base resistance), any change in volume setting can result in a change of impedance, gain, or bias. The simplest method for isolating a volume control is to use coupling capacitors as shown in Fig. 2-31. However, the capacitors create a low-frequency response problem. As in the case of coupling capacitors previously described, capacitor C_1 (of Fig. 2-31) forms a high-pass RC filter with volume control potentiometer R_1. Coupling capacitor C_2 forms another high-pass filter with the input resistance of the following stage. Using a 1-dB loss as the low-frequency cutoff point, the value of C_1 can be found by:

$$\text{capacitance} = \frac{1}{3.2FR}$$

where capacitance is in farads, F is the low-frequency limit in hertz, and R is resistance in ohms.

The volume control should be located at a low-signal-level point in the amplifier. The most common location for a volume control is at the amplifier input stage, or between the first and second stages.

When a volume control is located at the amplifier input, the control's resistance forms the input impedance (approximately). Volume controls are available in standard resistance values. Select the standard resistance value nearest the desired impedance.

When a volume control is located between stages, the resistance value should be selected to match the output impedance of the previous stage. Use the nearest standard resistance value. This will produce the least loss.

Very little current is required for a volume control that is isolated as shown in Fig. 2-31. Therefore, the power rating (in watts) required is quite low. Usually, 1 or 2 W is more than enough for any volume control used in transistor audio amplifier circuits. This makes it possible to use *noninductive* composition potentiometers. Do not use wirewound potentiometers for any audio application. The inductance produced by wirewound potentiometers can reduce frequency response of the circuit.

2-9.2. Design Example for Volume Control

Assume that the volume control of Fig. 2-31 is to be used between two resistance-coupled stages. The value of C_2 is already selected as part of the existing circuit. The values of C_1 and R_1 must be selected. The collector resistance of the previous stage is 2200 Ω. The supply voltage for the previous stage is 10 V. The desired low-frequency limit is 150 Hz. The amplifier will be used for voice and music amplification.

Resistance value. The output impedance of the previous stage can be considered as 2200 Ω (the collector resistance value). The nearest standard-value volume control is 2500 Ω. An audio-taper-type volume control should be used since voice and music amplification is involved. Any 2500-Ω, 1-W, composition, audio taper volume control should meet all requirements.

Capacitance value. The low-frequency limit of 150 Hz requires a capacitance value of $1/(3.2 \times 150 \times 2500) = 0.83$ μF, for an approximately 1-dB drop. Any standard value above 0.83 μF (say 1.0 μF) will be satisfactory. The voltage value of C_1 should be 1.5 times the maximum voltage (supply voltage) involved, or $10 \times 1.5 = 15$ V.

2-9.3. Volume Control using Attenuators and Pads

In some applications, the volume of audio signals is controlled by various attenuators and pads (such as T, L, O, and H pads). These attenuators and pads are made up of several interrelated resistances, all mechanically coupled to a common control shaft. Such attenuators and pads are commercially available. Therefore, no detailed design data will be given here. If further information on attenuators and pads is desired, refer to the author's *Handbook of Electronic Charts, Graphs, and Tables* (Prentice-Hall, Inc., Englewood Cliffs, N.J., 1970)

2-9.4. Design Considerations for Gain Controls

As shown in Fig. 2-32, the basic gain control is a variable resistance or potentiometer, serving as one resistance element in the amplifier circuit. Any of the three resistors (base, emitter, or collector) could be used as the gain control, since stage gain is related to each resistance value (all other factors remaining constant). However, the emitter resistance is the most logical choice for a gain control. If the collector resistance is variable, the output impedance of the stage will change as the gain setting is changed. A variable base resistance will produce a variable input impedance. A variable emitter resistance will have minimum effect on input or output impedance of the stage, but will directly affect both current and voltage gain.

With all other factors remaining constant, a decrease in emitter resistance will raise both current gain and voltage gain. An increase in emitter resistance will lower stage gain.

The resistance value of an emitter gain control should be chosen on the same basis as the emitter resistor (Sec. 2-1), except that the desired value should be the midpoint of the control range. For example, if a 500-Ω fixed

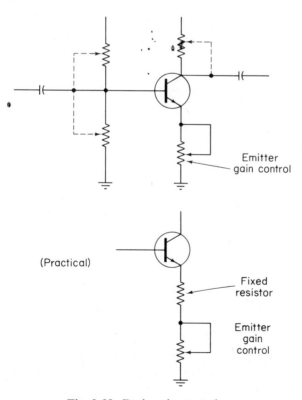

Fig. 2-32. Basic gain control.

resistor is normally used as the emitter resistance (or if 500 Ω is the calculated value for proper stage gain, bias stability, etc.), then the variable gain control should be 1000 Ω.

In practical applications, it is usually desirable to connect an emitter gain control in series with a fixed resistance. If the gain control is set to the minimum resistance value (0 Ω), there will still be some emitter resistance to provide gain stabilization and prevent thermal runaway. As a rule of thumb, the series resistance should be no less than one-twentieth of the collector resistor value. This will provide a maximum stage gain of 20.

If the gain control must provide for reduction of the stage voltage gain from some nominal point down to unity, the maximum value of the control should equal the collector resistance. If reduction to unity current gain is desired, the maximum value of the control should equal the input (base) resistance.

An audio taper potentiometer should not be used as a gain control, unless there is some special circuit requirement. However, the potentiometer used should be of the noninductive, composition type. The wattage rating of an emitter gain control should be the same as for an emitter resistor. Usually, 1 or 2 W is sufficient for any stage of a voltage amplifier. Use of a gain control in a power amplifier should be avoided. If a gain control must be used in power amplifiers, incorporate the control in the input stage where emitter current will be minimum.

2-9.5. Design Example for Gain Control

Assume that the practical circuit of Fig. 2-32 is to be provided with an emitter resistance gain control. The value of the collector resistance is 5000 Ω. The source voltage is 20 V. The calculated emitter resistance is 500 Ω, to provide a nominal stage voltage gain of 10. The gain control must provide a continuously variable gain from unity to 20.

Fixed series resistance value. To provide a maximum gain of 20, the series resistance value must be $\frac{1}{20}$ of the collector resistance, or 5000 $\times$ 1/20, or 250 Ω.

Variable resistance value. To provide a minimum gain of unity, the total emitter resistance (with variable control set to maximum resistance) must be equal to the collector resistance of 5000 Ω. Therefore, the variable resistance should be 5000 $-$ 250, or 4750 Ω.

2-9.6. Design Considerations for Treble Controls

A treble control provides a means of adjusting the high-frequency response of an audio amplifier. Such adjustment may be necessary because of variations in response of the human ear, or to correct frequency response of a particular recording. As shown in Fig. 2-33, a typical treble control circuit

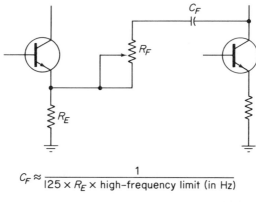

$$C_F \approx \frac{1}{125 \times R_E \times \text{high-frequency limit (in Hz)}}$$

$$R_F \approx R_E \times \text{maximum desired voltage gain}$$

Fig. 2-33. Basic treble control.

consists of a variable resistance or potentiometer and a fixed capacitor. These two components form a feedback circuit between two stages.

At any given frequency, the amount of feedback (and thus the frequency response) is set by adjustment of the potentiometer. As frequency increases, the capacitor reactance decreases. This also results in a change of feedback (and frequency response).

The voltage gain of two stages with feedback is approximately equal to the impedance of the feedback circuit divided by the source impedance. In this case, the source impedance is the emitter resistance (R_E) value. The feedback impedance (Z_F) is the vector sum of the potentiometer (R_F) resistance value and the capacitor (C_F) reactance value.

The voltage gain of the two stages can be set to any desired level for any given frequency by means of the feedback circuit. Of course, the closed-loop gain (gain with feedback) cannot be greater than that which could be produced by the two stages without feedback (open-loop gain). Although the feedback circuit has some effect on voltage gain at all frequencies, the greatest effect is on the high-frequency end of the audio range, as shown in Fig. 2-34.

The values of C_F and R_F are dependent upon desired gain, high-frequency limit, and source impedance. To find a trial value for C_F, use $\frac{1}{125} \times F \times R_E$, where F is the high-frequency limit of the audio amplifier. To find a trial value for R_F, use $R_E \times$ maximum desired voltage gain (for the two stages).

2-9.7. Design Example for Treble Control

Assume that the circuit of Fig. 2-33 is to provide a maximum voltage gain of 50, at 20 kHz, and that the emitter resistor is 50 Ω.

Feedback capacitance. The trial value for the feedback capacitance is $\frac{1}{125} \times 20,000 \times 50$, or 0.008 μF.

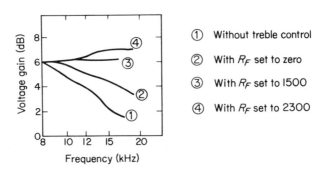

Fig. 2-34. Effect of treble control on voltage gain.

Feedback resistance. The trial value for the feedback resistance is 50 × 50, or 2500 Ω.

With R_F set to its full value, at a frequency of 20 kHz, the impedance of the feedback circuit is approximately 2700 Ω. This would provide a maximum voltage gain of 54. With R_F set to minimum, the impedance (at 20 kHz) is approximately 1000 Ω, providing a gain of about 20.

2-9.8. Design Considerations for Bass Control

A bass control provides a means of adjusting the low-frequency response of an audio amplifier. Such adjustment may be necessary because of variations in response of the human ear. As discussed, the human ear does not respond as well to low-frequency sounds at low levels as it does to high-frequency sounds at the same level. Also, coupling capacitors present high reactance to low-frequency signals. Both of these conditions require that the low-frequency signals be boosted (in relation to high-frequency signals).

As shown in Fig. 2-35, a typical bass control circuit consists of a variable resistance or potentiometer and a fixed inductance. These two components form a bypass circuit around the volume-control coupling capacitor. The

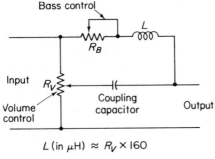

L (in μH) $\approx R_V \times 160$

$R_B \approx R_V \times 10$

Fig. 2-35. Basic bass control.

bass control could be placed elsewhere in the circuit, around another coupling capacitor. However, by placing the bass circuit at the volume control, it is possible to boost low-frequency signals more when the volume is low.

As the frequency is lowered, the inductance of L decreases, bypassing more of the signal around the coupling capacitor C. At any frequency, the amount of bypass (and thus the frequency response) is set by adjustment of the potentiometer R_B. Since the upper end of the volume control R_V is bypassed by L and R_B, the amount of bypass increases as the volume control is set to low level. Thus, the bypassed low-frequency signals will be stronger than the high-frequency signals, at low-volume levels.

The values of L and R_B are dependent upon the total resistance of the volume control. To find a trial value for L, use $R_V \times 160 \times L$ (in μH). A trial value for R_B is $R_V \times 10 = R_B$ (in ohms).

2-9.9. Design Example for Bass Control

Assume that the circuit of Fig. 2-35 is to be used with a 2500-Ω volume control.

Bypass inductance. The trial value for the bypass inductance is 2500 $\times$ 160, or 400,000 μH (400 mH).

Bypass resistance. The trial value for the bypass resistance is 2500 $\times$ 10, or 25,000 Ω.

2-10. Audio Filters

The most precise filters are composed of inductors and capacitors (LC filters). Except for special applications, the use of LC filters for audio frequencies below about 500 Hz is not recommended. The inductances required at low frequencies are quite large, thus making the inductances heavy and bulky. RC filters should be used below 500 Hz. The author's *Handbook of Electronic Charts, Graphs, and Tables* (Prentice-Hall, Inc., Englewood Cliffs, N.J., 1970) provides a complete description of LC-filter circuit design.

The simplest way of filtering audio frequency signals is to use resistance–capacitance (RC) filters. Low frequencies are removed by a series capacitor and shunt resistor (Fig. 2-36), while high frequencies are filtered by interchanging the resistor and capacitor (Fig. 2-37).

2-10.1. Design Considerations for RC Filters

Filter attenuation is rated in terms of decibel drop at a given frequency. Generally, RC filters are designed to produce a 3-dB drop (to 0.707 of input) at a selected cutoff frequency. The relationships for capacitance and resistance

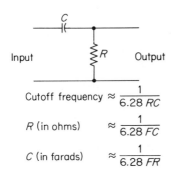

$$\text{Cutoff frequency} \approx \frac{1}{6.28\ RC}$$

$$R\ \text{(in ohms)} \approx \frac{1}{6.28\ FC}$$

$$C\ \text{(in farads)} \approx \frac{1}{6.28\ FR}$$

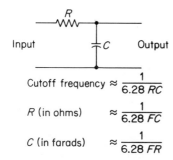

Fig. 2-36. Basic high-pass (low-cut) *RC* filter.

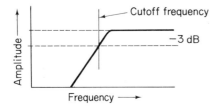

$$\text{Cutoff frequency} \approx \frac{1}{6.28\ RC}$$

$$R\ \text{(in ohms)} \approx \frac{1}{6.28\ FC}$$

$$C\ \text{(in farads)} \approx \frac{1}{6.28\ FR}$$

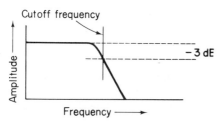

Fig. 2-37. Basic low-pass (high-cut) *RC* filter.

values versus cutoff frequency for *RC* filters with a 3-dB drop are: cutoff frequency = $1/6.28RC$; $R = 1/6.28FC$; $C = 1/6.28FR$, where *F*, or frequency, is in hertz, *C* is in farads, and *R* is in ohms.

Note that the high-pass filter of Fig. 2-36 (sometimes called a low-cut filter) is formed by the capacitance and resistance of an *RC* coupled amplifier.

As discussed in previous sections of this chapter, the value of a coupling capacitor is found by $1/3.2FR$. This is to produce a 1-dB drop. Generally, minimum drop is desired for a coupling capacitor (unless the circuit is being designed as a filter, or if a greater drop can be tolerated by design requirements).

Both the high-pass (low-cut) and low-pass (high-cut) filter circuits can be combined to provide a *bandpass RC* filter, as shown in Fig. 2-38. As a rule of thumb, the high-frequency limit (F_H) must be at least 10 times the low-frequency limit (F_L) for such a bandpass circuit to be effective. If the high- and low-frequency limits are not greater than 10:1, there will be considerable interaction between the combined circuits. Even with a 10:1 ratio, some interaction may occur. Therefore, the equations of Fig. 2-38 are approximate.

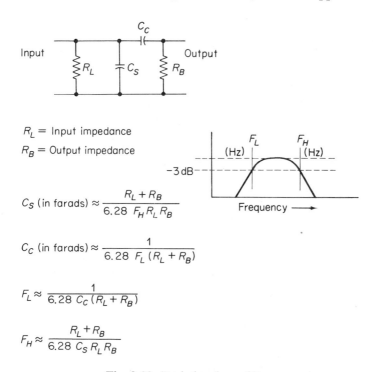

R_L = Input impedance

R_B = Output impedance

$$C_S \text{ (in farads)} \approx \frac{R_L + R_B}{6.28\ F_H R_L R_B}$$

$$C_C \text{ (in farads)} \approx \frac{1}{6.28\ F_L (R_L + R_B)}$$

$$F_L \approx \frac{1}{6.28\ C_C (R_L + R_B)}$$

$$F_H \approx \frac{R_L + R_B}{6.28\ C_S R_L R_B}$$

Fig. 2-38. Basic bandpass filter.

A single *RC* filter will provide a gradual transition from the passband to the cutoff region. If a rapid transition is necessary for design, two or more *RC* filter stages can be combined as shown in Fig. 2-39. The increase in attenuation at the cutoff frequency, and at frequencies above and below the cutoff point, for a two-stage high-pass filter are shown in Fig. 2-40. A similar

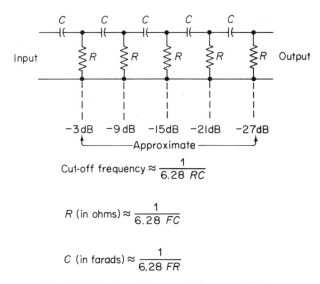

$$\text{Cut-off frequency} \approx \frac{1}{6.28\ RC}$$

$$R\ \text{(in ohms)} \approx \frac{1}{6.28\ FC}$$

$$C\ \text{(in farads)} \approx \frac{1}{6.28\ FR}$$

Fig. 2-39. Basic multistage high-pass RC filter.

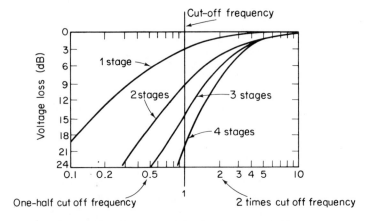

Fig. 2-40. Voltage-loss curve for high-pass RC filter stages.

curve for a two-stage low-pass filter is shown in Fig. 2-41. Any number of stages may be added to an RC filter. As a rule of thumb, each stage will increase the attenuation by 6 dB at the cutoff frequency.

In any RC filter, either the capacitor or the resistor value could be assumed (to find the other value for a given cutoff frequency). However, in practical design, the resistance value is usually assumed, since the resistor is chosen to meet other circuit requirements. For example, the resistance in a high-pass filter may also form the circuit's input or output impedance.

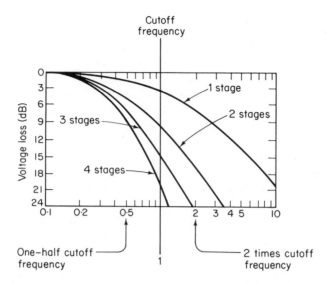

Fig. 2-41. Voltage-loss curve for low-pass *RC* filter stages.

2-10.2. Design Example for High-Pass (Low-Cut) Filter

Assume that the circuit of Fig. 2-36 is to provide a 3-dB drop at 15 Hz. The value of *R* must be 1000 Ω. The value of *C* is found by $1 \div 6.28 \times 1000 \times 15$, or 10 μF.

Now assume that the same basic circuit is to provide 33-dB drop at 15 kHz. The same values would be used, but additional stages would be added.

To find the number of additional stages, subtract the 3-dB drop (for the first stage) from 33 dB, then divide the remaining 30-dB drop by 6 dB, to find that five additional stages must be added. Thus, a total of six stages must be used.

2-10.3. Design Example for Low-Pass (High-Cut) Filter

Assume that the circuit of Fig. 2-37 is to provide a 3-dB drop at 300 Hz. The value of *R* must be 1000 Ω. The value of *C* is $1 \div 6.28 \times 1000 \times 300$, or 0.5 μF.

2-10.4. Design Example for Bandpass Filter

Assume that the circuit of Fig. 2-38 is to provide a bandpass between 300 Hz and 7 kHz. The signal must be down 3 dB at the high and low frequencies. The values of R_L and R_B must be 1000 Ω. The value of C_C is $1 \div [6.28 \times 300 \times (1000 + 1000)]$, or 0.26 μF. The value of C_S is $(1000 + 1000) \div (6.28 \times 7000 \times 1000 \times 1000)$, or 0.045 μF.

2-11. Audio-Circuit Test Procedures

The following sections describe test procedures for audio circuits. The procedures can be applied at any time during design. As a minimum, the tests should be made when the circuit is first completed in breadboard form. If the test results are not as desired, the component values should be changed as necessary to obtain the desired results. The circuit should be retested in final form (with all components soldered in place). This will show if there is any change in circuit characteristics due to the physical relocation of components. The test procedures include a series of notes regarding changes in component values on test results. This information is summarized at the end of the chapter.

The following sections do not describe every possible test to which audio circuits can be subjected. However, they do include all basic tests necessary for "typical" audio circuit operation.

2-12. Frequency Response of Audio Circuits

The frequency response of an audio amplifier, or filter, can be measured with an audio signal generator and a meter or oscilloscope. When a meter is used, the signal generator is tuned to various frequencies, and the resultant circuit output response is measured at each frequency. The results are then plotted in the form of a graph or *response curve*, such as shown in Fig. 2-42. The procedure is essentially the same when an oscilloscope is used to measure audio-circuit frequency response. However, an oscilloscope gives the added benefit of visual distortion analysis, as discussed in a later section.

2-12.1. Basic Frequency-Response Measurement

The basic frequency-response measurement procedure (with either meter or oscilloscope) is to apply a *constant-amplitude* signal while monitoring the circuit output. The input signal is varied in frequency (but not amplitude) across the entire operating range of the circuit. Any well-designed audio circuit should have a constant response from about 20 Hz to 20 kHz. With direct-coupled amplifiers, the response can be extended from a few Hz up to 100 kHz (and higher). The voltage output at various frequencies across the range is plotted on a graph similar to that shown in Fig. 2-42.

1. Connect the equipment as shown in Fig. 2-42. Set the generator, meter, and oscilloscope controls as necessary. It is assumed that the reader is thoroughly familiar with the use of electronic test equipment. The voltage divider shown in Fig. 2-42 is that described in Chapter 1 (Fig. 1-16).
2. Initially, set the generator output frequency to the low end of the frequency range. Then set the generator output amplitude to the desired input level.

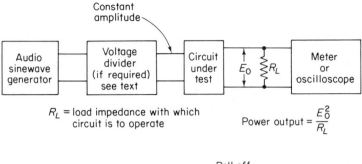

R_L = load impedance with which circuit is to operate

Power output = $\dfrac{E_0^2}{R_L}$

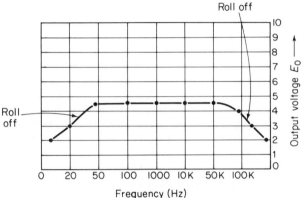

Fig. 2-42. Frequency-response test connections and typical response curve.

For example, the power-amplifier circuit of Fig. 2-20 requires 1-V-rms input to produce full power output.

3. In the absence of a realistic test input voltage, set the generator output to an arbitrary value. A simple method of finding a satisfactory input level is to monitor the circuit output (with the meter or oscilloscope) and increase the generator output at the circuit center frequency (or at 1 kHz) until the circuit is overdriven. This point will be indicated when further increases in generator output do not cause further increases in meter reading (or the output waveform peaks begin to flatten on the oscilloscope display). Set the generator output just *below* this point. Then return the meter or oscilloscope to monitor the generator voltage (at circuit input) and measure the voltage. Keep the generator at this voltage throughout the test.

4. If the circuit is provided with any operating or adjustment controls (volume, loudness, gain, treble, bass, etc.), set these controls to some arbitrary point when making the initial frequency response measurement. The response measurements can then be repeated at different control settings if desired.

5. Record the circuit output voltage on the graph. Without changing the generator output amplitude, increase the generator frequency by some fixed

amount, and record the new circuit output voltage. The amount of frequency increase between each measurement is an arbitrary matter. Use an increase of 10 Hz at the low end and high end (where rolloff occurs), and an increase of 100 Hz at the middle frequencies.

6. Repeat this process, checking and recording the circuit output voltage at each of the check points in order to obtain a frequency-response curve. With a typical audio amplifier circuit, the curve will resemble that of Fig. 2-42, with a flat portion across the middle frequencies and a rolloff at each end. A bandpass filter will have a similar response curve. High-pass and low-pass filters will produce curves with rolloff at one end only. (High-pass has a rolloff at the low end, and vice versa.)

7. After the initial frequency-response check, the effect of operating or adjustment controls should be checked. Volume, loudness, and gain controls should have the same effect all across the frequency range. Treble and bass control may also have some effect at all frequencies. However, a treble control should have the greatest effect at the high end, while a bass control will affect the low end most.

8. Note that generator output may vary with changes in frequency, a fact often overlooked in making a frequency-response test of any circuit. Even precision laboratory generators can vary in output with changes in frequency, thus resulting in considerable error. Therefore, it is recommended that the generator output be monitored after each change in frequency (some audio generators have a built-in output meter). Then, if necessary, the generator output amplitude can be reset to the correct value. Within extremes, it is more important that the generator output amplitude *remain constant* rather than at some specific value when making a frequency-response check.

2-12.2. Voltage-Gain Measurement

Voltage-gain measurement in an audio amplifier is made in the same way as frequency response. The ratio of output voltage to input voltage (at any given frequency, or across the entire frequency range) is the voltage gain. Since the input voltage (generator output) must be held constant for a frequency-response test, a voltage-gain curve should be identical to a frequency-response curve.

2-12.3. Power Output and Gain Measurement

The power output of an audio amplifier is found by noting the output voltage E_O across the load resistance R_L (Fig. 2-42), at any frequency, or across the entire frequency range. Power output is $E_O^2 \div R_L$.

To find power gain of an amplifier, it is necessary to find both the input and output power. Input power is found in the same way as output power, except that the impedance at the input must be known (or calculated). This is not always practical in some amplifiers, especially in designs where input impedance is dependent upon transistor gain. With input power known (or estimated), the power gain is the ratio of output power to input power.

Generally, a power gain is not required by design specifications. Instead, an *input sensitivity* specification is used. Input sensitivity specifications require a minimum power output with a given voltage input (such as 100-W output with 1-V-rms input).

2-12.4. Power-Bandwidth Measurement

Many audio-amplifier design specifications include a power-bandwidth factor. Such specifications require that the audio-amplifier deliver a given power output across a given frequency range. For example, the circuit of Fig. 2-20 will produce full power output up to 20 kHz (as shown in Fig. 2-22), even though the frequency response is flat up to 100 kHz. That is, voltage (without load) will remain constant up to 100 kHz, while power output (across a normal load) will remain constant up to 20 kHz.

2-12.5. Load-Sensitivity Measurement

An audio-amplifier circuit of any design will be sensitive to changes in load. This is especially true of power amplifiers but can also be the case with voltage amplifiers. An amplifier will produce maximum power when the output impedance of the amplifier is designed to be the same as the load impedance. This is shown by the curve of Fig. 2-21 (the load-sensitivity curve for the circuit of Fig. 2-20). If the load is twice the output circuit impedance (ratio of 2.0), the output power will be reduced to approximately 50 per cent. If the load is 40 per cent of the output impedance (ratio of 0.4), the output power will be reduced to approximately 25 per cent.

Any power amplifier circuit should be checked for load sensitivity during some stage of design. Such a test will often show up defects in design that would not easily be found with the usual frequency-response and power-output tests.

The circuit for load-sensitivity measurement is the same as for frequency response (Fig. 2-42), except that load resistance R_L should be variable. [Never use a wire-wound load resistance. The reactance can result in considerable error. If a non-wire-wound variable resistance of sufficient wattage rating is not available, use several fixed resistances (carbon or composition) arranged to produce the desired resistance values.]

Measure the power output at various load impedance/output impedance ratios. To make a comprehensive test of the circuit under design, repeat the load-sensitivity test across the entire frequency range.

2-12.6. Dynamic-Output-Impedance Measurement

The load-sensitivity measurement test just described can be reversed to find the dynamic output impedance of an amplifier circuit during design. The connections and procedure (Fig. 2-42) are the same, except that the load resistance R_L is varied until maximum output power is found. Power is removed and R_L is disconnected from the circuit. The dc resistance of R_L (measured with an ohmmeter) is equal to the dynamic output impedance. Of course, the value applies only at the frequency of measurement. The test should be repeated across the frequency range.

2-12.7. Dynamic-Input-Impedance Measurement

To find the dynamic input impedance of an amplifier circuit, use the circuit of Fig. 2-43. The test conditions should be identical to those for frequency response, power output, etc. That is, the same audio generator, operating load, meter or oscilloscope, and frequencies should be used.

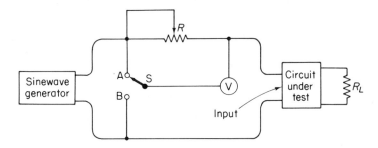

Fig. 2-43. Measuring dynamic input impedance.

The signal source is adjusted to the frequency (or frequencies) at which the circuit will be operated. Switch S is moved back and forth between position A and B, while resistance R is adjusted until the voltage reading is the same in both positions of the switch. Resistor R is then disconnected from the circuit, and the dc resistance of R is measured with an ohmmeter. The dc resistance of R is then equal to the dynamic impedance at the circuit input.

Accuracy of the impedance measurement is dependent upon the accuracy with which the dc resistance is measured. A noninductive resistance must be used. The impedance found by this method applies only to the frequency used during the test.

2-12.8. Audio-Amplifier Signal Tracing

An oscilloscope is the most logical instrument for checking design circuits, whether they are complete audio-amplifier systems, or a single stage. The oscilloscope will duplicate every function of an electronic voltmeter in troubleshooting, signal tracing, and performance-testing design circuits. In addition, the oscilloscope offers the advantage of a visual display for such common audio equipment conditions as distortion, hum, noise, ripple, and oscillation.

An oscilloscope is used in a manner similar to that of an electronic voltmeter when signal tracing audio circuits. A signal is introduced into the input by the signal generator. The amplitude and waveform of the input signal are measured on the oscilloscope. The oscilloscope probe is then moved to the input and output of each stage, in turn, until the final output is reached. The gain of each stage is measured as a voltage on the oscilloscope. In addition, it is possible to observe any change in waveform from that applied to the input. Thus, stage gain and distortion (if any) are established quickly with an oscilloscope.

2-12.9. Checking Distortion by Sine-Wave Analysis

The connections for audio circuit signal tracing with sine waves are shown in Fig. 2-44. The procedure for checking amplifier distortion by means of sine waves is essentially the same as that described in Sec. 2-12.8. The primary

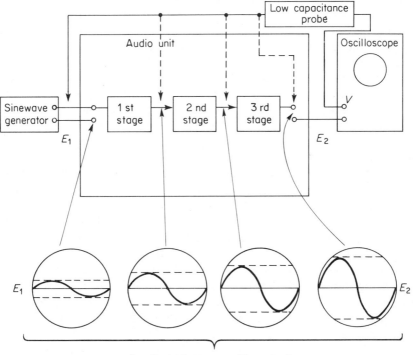

Amplitude increases with each stage
waveform remains substantially the same

Fig. 2-44. Basic audio signal tracing with an oscilloscope.

concern, however, is deviation of the amplifier (or stage) output waveform from the input waveform. If there is no change (except in amplitude), there is no distortion. If there is a change in the waveform, the nature of the change will often reveal the cause of distortion. For example, the presence of second or third harmonics will distort the fundamental.

In practice, analyzing sine waves to pinpoint design problems that produce distortion is a difficult job, requiring considerable experience. Unless the distortion is severe, it may pass unnoticed. Therefore, sine waves are best used where *harmonic distortion* or *intermodulation distortion* meters are combined with oscilloscopes for distortion analysis. If an oscilloscope is to be used alone, square waves provide the best basis for distortion analysis. (The reverse is true for frequency response and power measurements.)

2-12.10. Checking Distortion by Square-Wave Analysis

The procedure for checking distortion by means of square waves is essentially the same as for sine waves. Distortion analysis is more effective with square waves because of their high odd-harmonic content, and because it is easier to see a deviation from a straight line with sharp corners than from a curving line.

As in the case of sine-wave distortion testing, square waves are introduced into the circuit input, while the output is monitored on an oscilloscope (see Fig. 2-45). The primary concern is deviation of the amplifier (or stage) output waveform from the input waveform (which is also monitored on the oscilloscope). If the oscilloscope has the dual-trace feature, the input and output can be monitored simultaneously. If there is a change in waveform, the nature of the change will often reveal the cause of distortion. For example, a comparison of the square-wave response shown in Fig. 2-24 against the "typical" patterns of Fig. 2-45 indicates possible low-frequency phase shift. In any event, such a comparison would indicate some problem in the low-frequency response that could probably be corrected by a change in design values.

The third, fifth, seventh, and ninth harmonics of a clean square wave are emphasized. Therefore, if an amplifier passes a given audio frequency and produces a clean square-wave output, it is safe to assume that the frequency response is good up to at least *9 times* the fundamental frequency. For example, if the response at 10 kHz is as shown in Fig. 2-24, it is reasonable to assume that the response is the same at 100 kHz. This is convenient since not all audio generators will provide an output up to 100 kHz (required by many amplifier design specifications).

2-12.11. Harmonic-Distortion Measurement

No matter what amplifier circuit is used or how well the circuit is designed, there is always the possibility of odd or even harmonics being present with the fundamental. These harmonics combine with the fundamental and produce distortion, as is the case when any two signals are combined. The effects of second and third harmonic distortion are shown in Fig. 2-46.

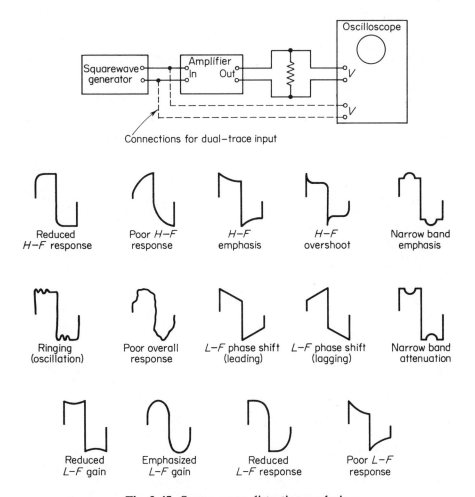

Fig. 2-45. Square-wave-distortion analysis.

Commercial harmonic distortion meters operate on the *fundamental suppression* principle. As shown in Fig. 2-46, a sine wave is applied to the amplifier input, and the output is measured on the oscilloscope. The output is then applied through a filter that suppresses the fundamental frequency. Any output from the filter is then the result of harmonics. This output can also be displayed on the oscilloscope. (Some commercial harmonic-distortion meters use a built-in meter instead of, or in addition to, an external oscilloscope.) When the oscilloscope is used, the frequency of the filter output signal can be checked to determine harmonic content. For example, if the input was 1 kHz and the output (after filtering) was 3 kHz, it would indicate third harmonic distortion.

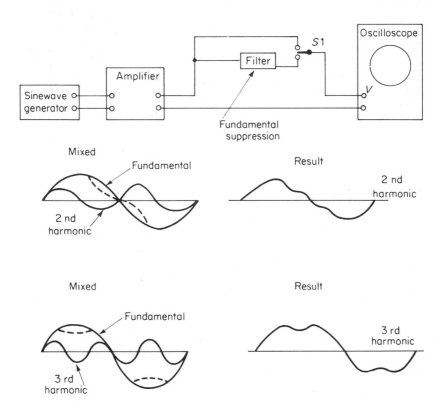

Fig. 2-46. Harmonic-distortion analysis.

The percentage of harmonic distortion can also be determined by this method. For example, if the output without filter was 100 mV, and with filter it was 3 mV, a 3 per cent harmonic distortion would be indicated.

In some commercial harmonic-distortion meters, the filter is tunable so that the amplifier can be tested over a wide range of fundamental frequencies. In other harmonic-distortion meters, the filter is fixed in frequency but can be detuned slightly to produce a sharp null.

When a design circuit is tested over a wide range of frequencies for harmonic distortion, and the results plotted on a graph similar to that of Fig. 2-23, the percentage is known as *total harmonic distortion* (THD). Note that the THD shown in Fig. 2-23 is less than 0.2 per cent. Also note that harmonic distortion can vary with frequency and power output.

2-12.12. Intermodulation-Distortion Measurement

When two signals of different frequency are mixed in any circuit, there is a possibility of the lower-frequency signal amplitude-modulating the higher-frequency signal. This produces a form of distortion known as *intermodulation distortion*.

Commercial intermodulation-distortion meters consist of a signal generator and high-pass filter as shown in Fig. 2-47. The signal-generator portion of
the meter produces a high-frequency signal (usually about 7 kHz) that is
modulated by a low-frequency signal (usually 60 Hz). The mixed signals are
applied to the circuit input. The amplifier output is connected through a high-
pass filter to the oscilloscope vertical channel. The high-pass filter removes
the low-frequency (60-Hz) signal. Therefore, the only signal appearing on the
oscilloscope vertical channel should be the high-frequency (7-kHz) signal. If
any 60-Hz signal is present on the display, it is being passed through as modulation on the 7-kHz signal.

Figure 2-47 also shows an intermodulation test circuit that can be fabricated in the shop or laboratory. Note that the high-pass filter is designed to

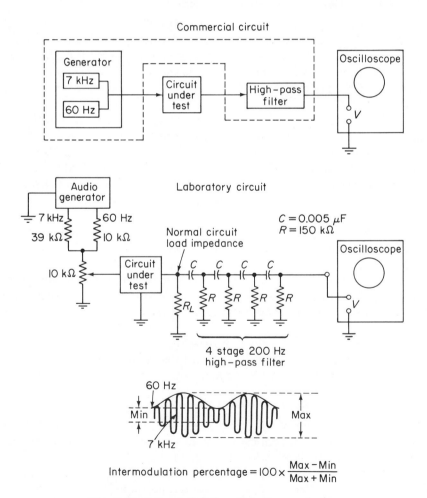

Fig. 2-47. Intermodulation-distortion analysis.

pass signals above about 200 Hz. The purpose of the 39-kΩ and 10-kΩ resistors is to set the 60-Hz signal at 4 times the 7-kHz signal. Most audio generators provide for a line-frequency output (60 Hz) that can be used as the low-frequency modulation source.

If the laboratory circuit of Fig. 2-47 is used instead of a commercial meter, set the generator line-frequency output to 2 V (if adjustable). Then set the generator audio output (7 kHz) to 2 V. If the line-frequency output is not adjustable, measure the value and then set the generator audio output to the same value.

The percentage of intermodulation distortion can be calculated using the equation of Fig. 2-47.

2-12.13. Background-Noise Measurement

If the vertical channel of an oscilloscope is sufficiently sensitive, an oscilloscope can be used to check and measure the background-noise level of a design circuit, as well as to check for the presence of hum, oscillation, etc. The oscilloscope vertical channel should be capable of a measurable deflection with about 1 mV (or less) since this is the background-noise level of many amplifiers.

The basic procedure consists of measuring amplifier output with the volume or gain control (if any) at maximum but without an input signal. The oscilloscope is superior to a voltmeter for noise-level measurement since the frequency and nature of the noise (or other signal) are displayed visually.

The basic connections for background-noise-level measurement are shown in Fig. 2-48. The oscilloscope gain or sensitivity control is increased until there is a noise or "hash" indication.

It is possible that a noise indication could be caused by pickup in the lead wires. If in doubt, disconnect the leads from the circuit but not from the oscilloscope.

If it is suspected that there is 60-Hz line hum present in the circuit output (picked up from the power supply or any other source), set the oscilloscope "sync" control to line. If a stationary signal pattern appears, it is due to line hum.

If a signal appears that is not at the line frequency, it can be due to oscillation of the circuit or to stray pickup. Short the amplifier input terminals. If the signal remains, it is probably oscillation of the circuit.

2-12.14. Phase-Shift and Feedback Measurements

In any circuit, there will be some phase shift between input and output signals. This is usually not critical for audio-amplifier circuits. One exception is in operational amplifiers, where feedback from output to input is used to control gain, etc. For that reason, the procedures for measurement of phase-shift and feedback levels in design circuits are postponed until Chapter 3.

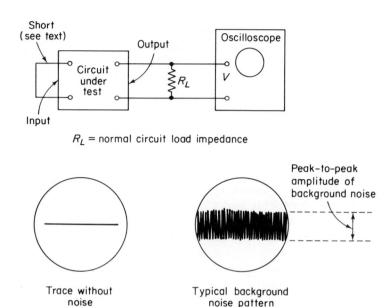

Fig. 2-48. Measuring circuit background noise and hum.

2-13. Analyzing Design Problems with Test Results

The following notes apply primarily to solving design problems (poor frequency response, lack of gain, etc.) in audio circuits. However, many of the procedures can be applied to other circuits.

When design circuits fail to perform properly (or as hoped they would perform) a *planned procedure* for isolating the problem is very helpful. Keep in mind that transistor-circuit troubleshooting is difficult at best. This is especially true when the circuit involves more than one stage, since the stages are interdependent.

A special problem arises in analyzing the failure of design circuits. The first requirement in logical troubleshooting is a thorough knowledge of the circuit's performance when operating normally. However, a failure in a trial circuit just designed can be the result of component failure or improper trial values for components. For example, an existing amplifier may show low gain, based on past performance. A newly designed amplifier circuit may show the same results, simply because it is the best gain possible with the selected trial components.

Start all trouble analysis of design circuits with the basic procedures of Sec. 1-7.4. Try to isolate trouble on a stage-by-stage basis. For example, if the circuit has two or more stages, and gain is low for the overall circuit, measure the gain for each stage. With trouble isolated to a particular stage, try to determine which half of the stage is at fault.

Any transistor stage has two halves, input and output. Generally, the input is base–emitter, with the emitter–collector acting as the output. Keep in mind that a defect in one half will affect the other half. An obvious example of this is where low input current (base) will produce low output current (collector). Sometimes less obvious is the case where output will affect input. For example, in a stage with an emitter resistor (for feedback stabilization), an open collector will appear to reduce the input impedance.

Circuits with loop feedback present a particular problem. A closed feedback loop causes all stages to respond as a unit, making it difficult to know which stage is at fault. This problem can be solved by opening the feedback loop. To do so, however, creates another problem, since the operating Q-point of one (or possibly all) stages will be disturbed.

Look for any transistor that is *full-on* (collector voltage very low) or *full-off* (collector voltage very high, probably near the supply voltage). Either of these conditions in a linear amplifier of any type is the result of component failure or improper design.

Design faults that appear under no-signal conditions (such as improper Q-point) are generally the result of improper bias relationships. Faults that appear only when a signal is applied can also be caused by poor bias relationships, but can also result from wrong component values.

A high-gain circuit is generally more difficult to bias than a low-gain circuit. Use the following procedure when it is difficult to find a good bias point (or Q-point) for a high-gain amplifier. Increase the input signal until the output waveform appears as a square wave (that is, overdrive the amplifier circuit; see Fig. 2-49). Then keep reducing the input signal, while adjusting the bias, until both positive and negative peaks are clipped by the same amount. If it is impossible to find any bias point where the signal peaks can be clipped symmetrically, a defect in design or components can be suspected. Unsymmetrical clipping can be caused by operating the transistor on a non-linear portion of its transfer curve or load line (improper bias), or by the fact that the transistor does not have a linear curve (or a very short linear curve). Try a different transistor. If the results are the same, change the circuit trial values (collector, emitter, and base resistances).

If it is impossible to obtain any Q-point, except very near full-on or full-off, this can be caused by excess *positive feedback*. While this condition is desirable in a multivibrator or oscillator (Chapter 5), it must be avoided in the design of linear amplifiers. As is discussed in Chapter 3, positive feedback can cause a linear amplifier to act like a flip-flop circuit or oscillator.

High-gain amplifiers should be carefully checked to find any tendency to oscillate or to exhibit abnormal noise (Sec. 2-12.13). High-gain amplifiers may also be very sensitive to supply-voltage changes. When a design circuit has

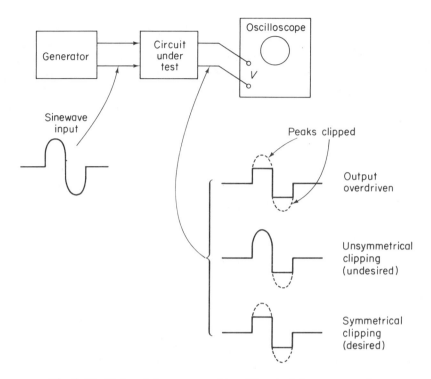

Fig. 2-49. Determining proper bias (Q-point) by means of oscilloscope displays.

been completed, it is often helpful to repeat all the basic test procedures (Sec. 2-12) with various supply voltages. Unless otherwise specified by design requirements, any circuit should perform equally well with a ± 10 per cent supply-voltage variation.

If oscillation occurs in any amplifier circuit (high or low gain), try moving the input and output leads. An amplifier may oscillate because input and output leads are close together. It may be necessary to physically relocate parts, or to shield parts, of an amplifier to prevent feedback that results in oscillation.

Low-frequency oscillation is often the result of poor supply-voltage filtering, or too many stages connected to the same supply-voltage point. Try isolating the stages with separate supply-voltage–filter capacitors.

The most common cause of *poor low-frequency response* is low capacitor values. The design procedures for this chapter provide for an approximate 1-dB loss at the low-frequency limit. If a greater loss can be tolerated, a lower capacitor (coupling capacitor or emitter-bypass capacitor) value can be used. If better low-frequency response is required by design specifications, increase the capacitor values.

The most common cause of *poor high-frequency response* is the input capacitance of transistors. As frequency increases, transistor input capacitance
decreases, changing the input impedance. This change in input impedance
usually results in decreased gain, all other factors remaining equal. Generally,
poor high-frequency response is not a problem over the audio range (up to
about 20 kHz) but can be a problem beyond about 100 kHz. The only practical solutions are to reduce stage gain, or change transistors.

3

OPERATIONAL AMPLIFIERS
WITH INTEGRATED CIRCUITS

The designation "operational amplifier" was originally adopted for a series of high performance dc amplifiers used in analog computers. These amplifiers were used to perform mathematical operations applicable to analog computation (summation, scaling, subtraction, integration, etc.). Today, the availability of inexpensive integrated-circuit (IC) amplifiers has made the *packaged operational amplifier* useful as a replacement for *any low-frequency amplifier*.

Because such packaged circuits are readily available in great variety at low cost, no detailed circuit-design procedures will be given here. Instead, we will concentrate on how to select external components to perform a given function, how to connect external power sources, how to mount the packages, and how to interpret IC data sheets. If further information on any phase of operational amplifiers is desired, refer to the author's *Manual for Operational Amplifier Users* (Reston Publishing Company, Inc., Reston, Va., 1976).

Most of the basic design information for a particular IC can be obtained from the data sheet. Likewise, a typical IC data sheet will describe a few specific applications for the IC. However, IC data sheets generally have two weak points. First, they do not show how the listed parameters relate to design problems. Second, they do not describe the great variety of applications for which a basic IC operational amplifier can be used.

In any event, it is always necessary to interpret IC data sheets. Each manufacturer has its own system of data sheets. It would be impractical to discuss all data sheets here. Instead, we will discuss typical information found on IC data sheets and see how this information affects simplified design.

3-1. Basic IC Operational Amplifier

IC operational amplifiers (op-amps) generally use several differential stages in cascade to provide both common-mode rejection (Sec. 2-5.4) and high gain. Therefore, they require both positive and negative power supplies. Since a differential amplifier has two inputs, it provides phase inversion for degenerative feedback and can be connected to provide in-phase or out-of-phase amplification. A conventional op-amp requires that the output be fed back to the input through a resistance or impedance. The output is fed back to the negative or inverting input so as to produce degenerative feedback (to provide the desired gain and frequency response). As in any amplifier, the signal shifts in phase as it passes from input to output. This phase shift is dependent upon frequency. When the phase shift approaches 180°, it adds to (or cancels out) the 180° feedback phase shift. Thus, the feedback is in phase with the input (or nearly so) and will cause the amplifier to oscillate. This condition of phase shift with increased frequency limits the bandwidth of an op-amp. The condition can be compensated for by the addition of a phase-shift network (usually an *RC* circuit, but sometimes a single capacitor).

3-1.1. Typical IC Op-Amp Circuit

The circuit diagram and the equivalent circuit (or symbol) as they appear on the data sheet of a typical IC op-amp are shown in Fig. 3-1. This Motorola circuit is a three-stage amplifier with the first stage a differential-in, differential-out amplifier designed for high gain, high common-mode rejection, and input overvoltage protection. [The input diodes prevent damage to the circuit should the input terminals be accidentally connected to the power-supply leads (or other undesired high voltage source).] The second stage is a differential-in, single-end-out amplifier with low gain and high common-mode rejection. Common-mode feedback is used from the second stage back to the first stage to further aid in the control of a common-mode input signal. Using these two differential-amplifier stages, and a common-mode feedback, a typical common-mode rejection of 110 dB is obtained. The third stage is a single-ended amplifier that provides high gain, voltage translation to a ground reference, output current drive capabilities, and output short-circuit protection.

3-1.2. Connecting a Power Source to an IC

Typically, op-amps require connection to both a positive and a negative power supply. This is because most op-amps use one or more differential amplifiers. When two power supplies are required, the supplies are usually equal or symmetrical (such as $+6$ V and -6 V, $+12$ V and -12 V, etc.). This is the case with the op-amp of Fig. 3-2, which normally operates with

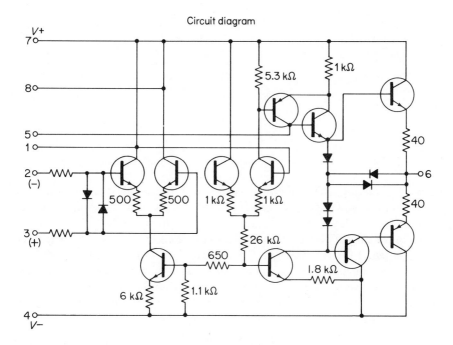

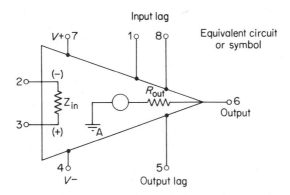

Fig. 3-1. Typical IC operational-amplifier circuit and symbol.

$+12$ V and -12 V. It is possible to operate an op-amp that normally requires two supplies from a single supply by means of special circuits (external to the op-amp). Such circuits are discussed at the end of this section. (Solid-state power supplies are discussed fully in Chapter 6.)

IC op-amp power connections. Unlike most discrete transistor circuits in which it is usual to label one power-supply lead positive and the other negative

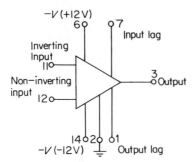

Fig. 3-2. Typical linear op-amp operating with symmetrical 12-V power supplies.

without specifying which (if either) is common to ground, it is necessary that all IC op-amp power-supply voltages be referenced to a common or ground (which may or may not be physical or equipment ground).

As in the case of discrete transistors, manufacturers do not agree on power-supply labeling for IC op-amps. For example, the circuit shown in Fig. 3-2 uses $+V$ to indicate the positive voltage and $-V$ to indicate the negative voltage. Another manufacturer might use the symbols V_{EE} and V_{CC} to represent negative and positive, respectively. As a result, the op-amp data sheet must be studied carefully before applying any power source.

Typical op-amp power-supply connections. Figure 3-3 shows typical power-supply connections for an op-amp. The protective diodes shown are recommended for any power-supply circuit in which the leads could be accidentally reversed. The diodes permit current flow only in the appropriate direction. The op-amp of Fig. 3-3 requires two power sources (of 12 V each) with the positive lead of one and the negative lead of the other tied to ground or common.

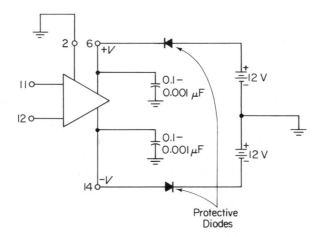

Fig. 3-3. Typical power-supply connections for op-amp.

The two capacitors shown in Fig. 3-3 provide for decoupling (signal bypass) of the power supply. Usually, disc ceramic capacitors are used. The capacitors should always be connected as close to the op-amp terminals as is practical, not at the power-supply terminals. This is to diminish the effects of lead inductance. It is particularly important to decouple each op-amp where two or more op-amps are sharing a common voltage supply. A guideline for op-amp power-supply decoupling capacitors is to use values between 0.1 and 0.001 μF.

In addition to the capacitor shown in Fig. 3-3, some op-amp layouts may require additional capacitors on the power lines. The main problem with op-amp power-supply connections is undesired oscillation due to feedback. Most modern op-amps are capable of producing high gain at high frequencies. If there is a feedback path through a common power-supply connection, oscillation will occur. In the case of IC op-amps, which are physically small, the input, output, and power-supply terminals are close, creating the ideal conditions for undesired feedback. To make the problem worse, most op-amps are capable of passing frequencies higher than those specified on the data sheet.

For example, an op-amp to be used in the audio range (say up to 20 kHz with a power gain of 20 dB) could possibly pass a 10-MHz signal without some slight gain. This higher-frequency signal could be a harmonic of signals in the normal operating range and, with sufficient gain, could feed back and produce undesired oscillation.

When laying out any op-amps, particularly IC op-amps in the breadboard or experimental stage, always consider the circuit to be of radio frequency (RF), even though the op-amp is not supposed to be capable of RF operation, and the circuit is not normally used with RF.

Grounding metal IC op-amp cases. The metal case of the IC op-amp shown in Fig. 3-3 is connected to terminal 2 and to no other point in the internal circuit. Thus, terminal 2 can and should be connected to equipment ground, as well as to the common or ground of the two power supplies.

The metal cases of some IC op-amps may be connected to a point in the internal circuit. If so, the case will be at the same voltage as the point of contact. For example, the case might be connected to pin 14 of the op-amp shown in Figs. 3-2 and 3-3. If so, the case will be below ground (or "hot") by 12 V. If the case is mounted directly on a metal chassis that is at ground, the op-amp and power supply will be damaged. Of course, not all IC op-amps have metal cases; likewise, not all metal cases are connected to the internal circuits. However, this point must be considered before using a particular IC op-amp.

Calculating current required for op-amps. The data sheets for op-amps usually specify a nominal operating voltage (and possibly a maximum operating voltage), as well as a "total device dissipation." These figures can be used to calculate the current required for a particular op-amp. Use the simple dc Ohm's law and divide the power by the voltage to find the current. However, certain points must be considered.

First, use the actual voltage applied to the op-amp. The actual voltage should be equal to the nominal operating voltage, but in no event higher than the maximum voltage.

Second, use the *total* device dissipation. The data sheet may also list other power dissipations, such as "device dissipation," which is defined as the dc power dissipated by the op-amp itself (with output at zero and no load). The other dissipation figures will always be smaller than the total power dissipation.

Power-supply tolerances. Typically, op-amps will operate satisfactorily with ±20 per cent power supplies. These tolerances apply to actual opera ، g voltage, not to maximum voltage limits. The currents (or power consumed) will vary proportionately.

Power-supply ripple and regulation are both important. Generally, solid-state power supplies with filtering and full feedback regulation are recommended, particularly for high-gain op-amps. Ideally, ripple (and all other noise) should be 1 per cent or less.

The fact that an op-amp generally requires two power supplies (due to the differential amplifiers) creates a particular problem with regard to offset. For example, if the V+ supply is 20 per cent high, and the V− supply is 20 per cent low, there will be an unbalance and offset, even though the op-amp circuits are perfectly balanced.

The effects of operating an op-amp beyond the voltage tolerance are essentially the same as those experienced when the op-amp is operated at temperature extremes. That is, a high power-supply voltage will cause the op-amp to "overperform," whereas low voltages will result in "underperformance." A low voltage will usually not result in damage to the op-amp, as is the case when operating the op-amp beyond the maximum rated voltage.

Single power-supply operation. An op-amp is generally designed to operate from symmetrical positive and negative power-supply voltages. This results in a high common-mode rejection capability as well as good low-frequency operation (typically a few hertz down to direct current). If the loss of very low-frequency operation can be tolerated, it is possible to operate op-amps from a single power supply, even though designed for dual supplies. Except for the low-frequency loss, the other op-amp characteristics should be unaffected.

The following notes describe a technique that can be used with most op-amps to permit operation from a single power supply, with a minimum of design compromise. The same maximum ratings that appear on the data sheet are applicable to the op-amp when operating from a single polarity power supply, and must be observed for normal operation.

The technique described here is generally referred to as the "split-Zener" method. The main concern in setting up for single supply operation is to *maintain the relative voltage levels*. With an op-amp designed for dual supply operation, there are three reference levels: $+V$, 0, and $-V$. For example, if the data sheet calls for ± 12-V supplies, the three reference levels are $+12$ V, 0 V, and -12 V.

For single supply operation, these same reference levels can be maintained by using $++V$, $+V$, and ground (that is, $+24$ V, $+12$ V, and 0 V), where $++V$ represents a voltage level double that of $+V$. This is illustrated in Fig. 3-4, where the op-amp is connected in the split-Zener mode. Note that there is no change in the relative voltage levels even though the various op-amp terminals are at different voltage levels (with reference to ground). Terminal 14 (normally connected to the -12-V supply) is at ground. Terminal 2 (normally ground or common) is set at one-half the total Zener voltage

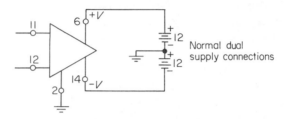

Single-supply connections

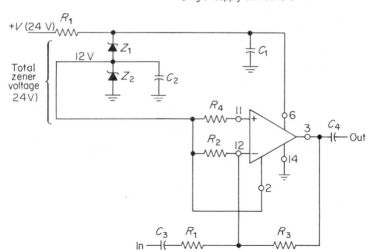

Fig. 3-4. Connections for single power-supply operation (with ground reference).

(+12 V). Terminal 6 (normally connected to the +12-V supply) is set at the full Zener voltage (+24 V).

With single supply, the differential input terminals (11 and 12), which are normally at ground in a dual supply system, must also be raised up one-half the Zener voltage (+12 V). Under these circumstances, the output terminal (3) is also at one-half the Zener voltage, plus or minus any offset (refer to Sec. 3-3).

To minimize offset errors due to unequal voltage drops caused by the input bias current across unequal resistances, it is recommended that the value of the input offset resistance R_4 be equal to the parallel combination of R_2 and R_3. (The problems of offset correction are discussed fully in Secs. 3-3 and 3-4.)

Any deviation between absolute Zener level will also contribute to an error in the op-amp output level. Typically, this is on the order of 50 to 100 μV per volt of deviation of Zener level. Except in rare cases, this deviation should be of little concern.

Note that the op-amp of Fig. 3-4 has a ground reference terminal (terminal 2). Not all op-amps have such terminals. Some op-amps have only +V and −V terminals or leads, even though the two levels are referenced to a common ground. That is, there is no physical ground terminal or lead on the op-amp.

Figure 3-5 shows the split-Zener connections for single supply operation with such op-amps. Here, the input terminals (A and B) are set at one-half the total Zener supply voltage: the −V terminal (D) is set at ground, and the +V terminal (F) is set at the full Zener voltage (+24 V).

Figures 3-4 and 3-5 both show a connection to positive power supplies. Negative power supplies can also be used. With a negative supply, the +V terminal is connected to ground and the −V terminal is connected to the total Zener supply (+24 V), with the input terminals and op-amp ground terminal (if any) connected to one-half the Zener supply. Of course, the polarity of the Zener diodes must be reversed (refer to Chapter 6).

Figures 3-4 and 3-5 both show a series resistance R_S for the Zener diodes. This is standard practice for Zener operation. (Refer to Sec. 6-2.) The approximate or trial value for R_S is found by:

$$\frac{(\text{maximum supply voltage} - \text{total Zener voltage})^2}{\text{safe power dissipation of Zeners}}$$

For example, assume that the total Zener voltage is 24 V (12 V for each Zener), that the supply voltage may go as high as 27 V, and that 2-W Zeners are used. Under these conditions,

$$\frac{(27 - 24)^2}{2} = 4.5 \, \Omega \qquad \text{for } R_S$$

From a design standpoint, operation of an op-amp with a single supply is essentially the same as with the conventional dual power supply. The following

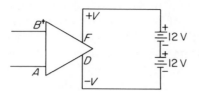

Normal dual-supply connections

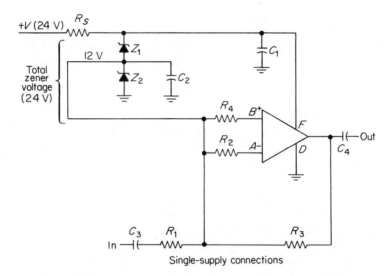

Single-supply connections

Fig. 3-5. Connections for single supply operation (without ground reference).

notes describe the basic differences in operational characteristics with both types of power supply.

The normal op-amp phase/frequency compensation methods are the same for both types of supply. (Phase/frequency compensation is described fully in Sec. 3-2.) The high-frequency limits are essentially the same. However, the low-frequency limit of an op-amp with a single supply is set by the values of capacitors C_3 and C_4. These capacitors are not required for dual supply operation. Capacitors C_3 and C_4 are required for single supply operation since both the input and output of the op-amp are at a voltage level equal to one-half the total Zener voltage (or 12 V, using our example). Thus, the op-amp cannot be used as a dc amplifier with the single supply system. In a dual supply system, the inputs and outputs are at 0 V.

The closed-loop gain (Sec. 3-2) is the same for both types of supply and is determined by the ratio R_3/R_1.

The values of decoupling capacitors C_1 and C_2 are essentially the same for both types of supply. However, it may be necessary to use slightly larger values with the single supply system, since the impedance of the Zeners is probably different from that of the power supply (without Zeners).

The value of R_2 should be between 50 and 100 kΩ for a typical op-amp. Values of R_2 much higher or lower than these limits can result in decreased gain or in an abnormal frequency response. From a practical design standpoint, choose trial values using the guidelines, and then run gain and frequency-response tests.

The value of R_4, the input offset resistance, is chosen to minimize offset error from impedance unbalance. As an approximate trial value, the resistance of R_4 should be equal to the parallel combination of R_2 and R_3. That is,

$$R_4 \approx \frac{R_2 R_3}{R_2 + R_3}$$

3-1.3. Typical IC Packages

Figure 3-6 shows some typical IC packages in common use. Packaging methods for ICs are in a constant state of development and change. Therefore, no detailed mounting techniques will be discussed here. In general, the ceramic packages are superior to the TO-5 style, since there is no problem of insulating the metal case. Ceramic packages are also superior to plastic packages, in that ceramic materials can operate at higher temperatures. However, ceramic packages are more expensive.

During the breadboard stage of design, any of the IC packages can be mounted in commercially available sockets. This will eliminate soldering and unsoldering the leads during design and test. Sockets manufactured by the Barnes Development Co. and the Sealectro Corp. are typical of the temporary mounts required during design.

3-1.4. Power Dissipation and Thermal Design Problems for ICs

The basic rules for ICs, regarding power dissipation and thermal considerations, are essentially the same as for the transistors as discussed in Sec. 1-4. The maximum allowable power dissipation (usually specified as P_d or P_D on IC data sheets) is a function of the maximum storage temperature T_S, the maximum ambient temperature T_A, and the thermal resistance from pellet to case θ_{PC}. The basic relationship is:

$$P_D = \frac{T_S - T_A}{\theta_{PC}}$$

All IC data sheets will not necessarily list all these parameters. It is quite common to list only the maximum power dissipation for a given ambient temperature, and then show a derating factor in terms of maximum power

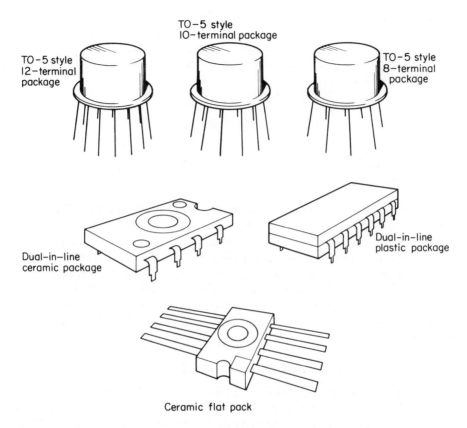

Fig. 3-6. Typical IC packages.

decrease for a given increase in temperature. For example, a typical IC might show a maximum power dissipation of 110 mW at 25°C, with a derating factor of 1 mW/°C. If such an IC were operated at 100°C, the maximum power dissipation would be $100 - 25$ or 75°C increase; $110 - 75$, or 35 mW.

In the absence of specific data-sheet information, the following typical temperature characteristics can be applied to the basic IC package types:

Ceramic flat pack:
> thermal resistance $= 140$°C/W
> maximum storage temperature $= 175$°C
> maximum ambient temperature $= 125$°C

TO-5 style package:
> thermal resistance $= 140$°C/W
> maximum storage temperature $= 200$°C
> maximum ambient temperature $= 125$°C

Dual in-line (ceramic):

> thermal resistance = 70°C/W
> maximum storage temperature = 175°C
> maximum ambient temperature = 125°C

Dual in-line (plastic):

> thermal resistance 150°C/W
> maximum storage temperature = 85°C
> maximum ambient temperature = 75°C

3-2. Design Considerations for Frequency Response and Gain

Most of the design problems for op-amps are the result of tradeoffs between gain and frequency response (or bandwidth). The *open-loop* (without feedback) gain and frequency response are characteristics of the basic op-amp circuit, but they can be modified with *phase compensation* networks. The *closed-loop* (with feedback) gain and frequency response are primarily dependent on *external feedback* components. Although op-amps are generally used in the closed-loop operating mode, the open-loop characteristics have considerable effect on operation, and must be considered when designing a feedback system for an op-amp.

3-2.1. Inverting and Noninverting Feedback

The two basic op-amp feedback systems, *inverting* feedback and *noninverting* feedback, are shown in Figs. 3-7 and 3-8, respectively. In both cases, the op-amp output is fed back to the inverting (or minus) input through an impedance Z_F to control frequency response and gain. In the inverting feedback system of Fig. 3-7, the input signal is also applied to the inverting input, resulting in an inverted output. In the noninverting system of Fig. 3-8, the input signal is applied to the noninverting input, producing a noninverted output (a positive input produces a positive output). The inverting feedback system of Fig. 3-7, in which a positive input produces a negative output, is the more commonly used.

The equations shown in Figs. 3-7 and 3-8 are classic guidelines. The equations do not take into account the fact that open-loop gain is not infinitely high and output impedance is not infinitely low. Thus, the equations contain built-in inaccuracies and must be used as *guides only*.

With both feedback configurations, the closed-loop gain, or the ratio V_{OUT}/V_{IN}, is approximately equal to the ratio Z_F/Z_R. Typically, Z_F and Z_R are fixed, composition (noninductive) resistors. Thus, if Z_F is a 100,000-Ω resistor, and Z_R is 1000 Ω, the resistance ratio is 100 : 1, and the approximately closed-loop gain is 100 (40 dB).

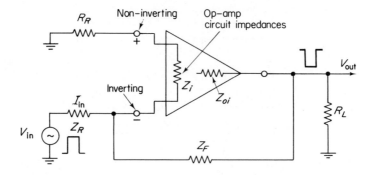

$$\frac{V_{out}}{V_{in}} = -\frac{Z_F}{Z_R} \text{ (Where open loop gain is infinite)}$$

$$\text{Loop gain} \approx \frac{\text{Open loop gain}}{\dfrac{Z_F}{Z_R}}$$

$$\text{Common mode rejection} \ll \frac{\text{Open loop gain} \times Z_i}{R_R}$$

$$\text{Common mode rejection} \gg \frac{R_R}{Z_i}$$

$$Z_{in} \approx \frac{Z_F}{\text{Open loop gain}} \quad Z_{in} \approx Z_R \text{ (Where open-loop is infinite)}$$

$$Z_{out} \approx \frac{Z_{oi}}{\text{Loop gain}}$$

Fig. 3-7. Inverting feedback op-amp (theoretical relationships). Courtesy RCA.

In theory, any voltage gain could be set by proper selection of Z_F and Z_R. There are obvious practical limits. For example, if the open-loop gain is less than 100, there is no Z_F/Z_R ratio that will produce a gain of 100. Likewise, even if the gain can be obtained, an improper Z_F/Z_R ratio could operate the

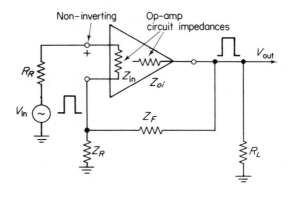

$$\frac{V_{out}}{V_{in}} = 1 + \frac{Z_F}{Z_R} \text{ (Where open-loop gain is infinite)}$$

$$\text{Loop gain} \approx \frac{\text{Open loop gain}}{1 + \dfrac{Z_F}{Z_R}}$$

$$\text{Common mode rejection} << \frac{\text{Open-loop gain} \times Z_i}{Z_i + R_R}$$

$$\text{Common mode rejection} >> \frac{Z_i + R_R}{Z_i}$$

$$Z_{in} \approx Z_i + \frac{\text{Open-loop gain} \times Z_i}{1 + \dfrac{Z_F}{Z_R}}$$

$$Z_{out} \approx \frac{Z_{oi}}{\text{Loop gain}}$$

Fig. 3-8. Noninverting feedback op-amp (theoretical relationships). Courtesy RCA.

op-amp in an unstable condition (as is discussed in later paragraphs of this section).

With both configurations, *loop gain* is defined as the ratio of open-loop gain to closed-loop gain. When loop gain is large, the inaccuracies in the equations of Figs. 3-7 and 3-8 decrease. For example, if open-loop gain is 1000 and closed-loop gain is 100, loop gain is 10. If closed-loop gain is reduced to 10, loop gain is increased to 100, and the equations are more accurate. However, as is discussed later, it is not always possible, nor is it always desirable, to operate with a large loop gain.

In the inverting configuration of Fig. 3-7, the input impedance of the system Z_{IN} (that appears to the input signal) is approximately equal to resistance Z_R. The output impedance of the system Z_{OUT} is approximately equal to the input impedance of the op-amp Z_I divided by the loop gain. The common-mode rejection is some value much greater than the ratio of R_R and Z_I. Thus, a large value of R_R is desirable for good common-mode rejection. However, it is not always possible to use large values of R_R.

In the noninverting configuration of Fig. 3-8, the system characteristics are similar to those for the inverting feedback, as shown by the equations. The major difference is that the op-amp input impedance Z_I has a greater effect in the case of noninverting (Fig. 3-8) operation.

3-2.2. Gain/Frequency Characteristics of a Theoretical Op-Amp

The gain/frequency relationships shown in Fig. 3-9 are based on a theoretical op-amp. That is, the open-loop gain remains flat as frequency increases and then begins to roll off at some particular frequency. The point at which the rolloff starts is sometimes referred to as a *pole*. The curve of Fig. 3-9 is known as a *one-pole plot* (and may also be known as a Bode plot, gain/bandwidth plot, or a frequency-response curve). At the first (and only) pole, the open-loop gain rolls off at 6 dB/octave, or 20 dB/decade. The term "6 dB/

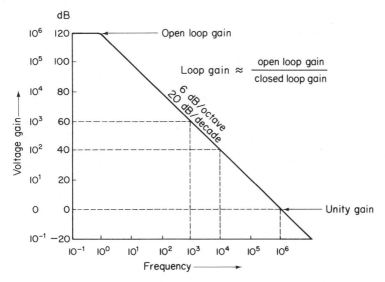

Fig. 3-9. Frequency-response curve of theoretical operational amplifier.

octave" means that the gain drops by 6 dB each time the frequency is doubled. This is the same as a 20-dB drop each time the frequency is increased by a factor of 10.

If the open-loop gain of an amplifier is as shown in Fig. 3-9, any stable closed-loop gain could be produced by the proper selection of feedback components, provided that the closed-loop gain is less than the open-loop gain. That only concern is a tradeoff between gain and frequency response.

For example, if a gain of 40 dB (10^2) is desired, a feedback resistance Z_F that is 10^2 times larger than the input resistance Z_R is used (such as a Z_F of 1000 and a Z_R of 10). The closed-loop gain is then flat to 10^4 Hz and rolls off at 6 dB/octave to unity gain (a gain of 1) at 10^6 Hz. If a 60-dB (10^3) gain is required instead, the feedback resistance Z_F is raised to 10^3 times the input resistance Z_R (Z_F of 10,000 and a Z_R of 10). This reduces the frequency response. Gain is flat to 10^3 Hz (instead of 10^4), followed by a rolloff of 6 dB/octave down to unity gain.

If should be noted that the gain/frequency response curves of practical op-amps rarely look like that shown in Fig. 3-9. A possible exception is an internally compensated IC op-amp, described later.

3-2.3. Gain/Frequency Characteristics of a Practical Op-Amp

The open-loop frequency-response curve of a practical op-amp more closely resembles that shown in Fig. 3-10. This curve is a three-pole plot. The first pole occurs at about 0.2 MHz, the second pole at 2 MHz, and the third pole at 20 MHz. (In a truly practical frequency-response curve there will be

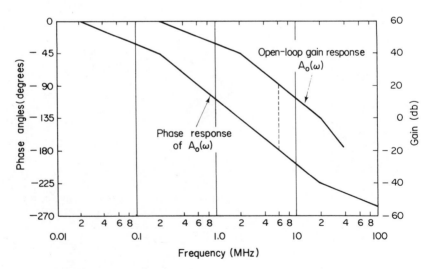

Fig. 3-10. Gain and phase response of an open-loop op-amp without phase compensation. Courtesy RCA.

no sharp breaks at the poles. Instead, the pole "corners" will be rounded and often difficult to distinguish. However, the curve of Fig. 3-10 is given here to illustrate certain frequency-response characteristics.)

In the curve of Fig. 3-10, gain is flat at 60 dB to the first pole, then rolls off to 40 dB at the second pole. Since there is a decade between the first and second poles (0.2 MHz to 2 MHz), the rolloff is 20 dB/decade, or 6 dB/octave. As frequency increases, rolloff continues from the second pole to the third pole, where gain drops from 40 dB to 0 dB. Thus, the rolloff is 40 dB/decade, or 12 dB/octave. At the third pole, gain drops below unity as frequency increases at a rate of 60 dB/decade, or 18 dB/octave.

Some op-amp data sheets provide a curve similar to that shown in Fig. 3-10. If the data sheets are not available, it is possible to test the op-amp under laboratory conditions, and draw an actual response curve (the frequency response and phase shift). Procedures for measurement of op-amp frequency response are similar to those for audio circuits (Sec. 2-12). Typical op-amp phase-shift measurements are described in Sec. 3-17.

3-2.4. Phase-Shift Problems

Note that the curve of Fig. 3-10 also shows the phase shift of an op-amp. As frequency increases, the phase shift between input and output signals of the op-amp increases. At frequencies up to about 0.02 MHz, the phase shift is zero. That is, the output signals are in phase with the noninverting input, and exactly 180° out of phase with the inverting input (which is generally used with op-amp feedback circuits, as discussed in Sec. 3-2.1). As frequency increases up to about 0.2 MHz, phase shift increases by about 45°. That is, the output signals are about 135° (180 − 45) output phase with the inverting input. At about 6 MHz, the output signals are in phase with the inverting input. Since the output is fed back to the inverting input through Z_F (Fig. 3-7), the input and output are in phase. If the output amplitude is large enough, the op-amp will oscillate.

Oscillation can occur if the output is shifted near 180° (and fed back to the inverting input). Even if oscillation does not occur, op-amp operation can become unstable. For example, the gain will not be flat. Usually, a *peaking condition* will occur, in which output remains flat up to a frequency near the 180° phase-shift point, and then gain will increase sharply to a peak. This is caused by the output signals being nearly in phase with input signals and reinforcing the input signals. At higher frequencies, gain drops off sharply and/or oscillation occurs. As a guideline, the op-amp should never be operated at a frequency at which phase shift is near 180° without compensation. In the op-amp of Fig. 3-10, the maximum uncompensated frequency would be about 5 MHz. This is the frequency at which open-loop gain is about +25 dB.

One guideline often mentioned in the op-amp literature is based on the fact that the 180° phase-shift point almost always occurs at a frequency at which the open-loop gain is in the 12 dB/octave slope. Thus, the guideline states that a closed-loop gain should be selected so that the unity gain is obtained at some frequency near the beginning of the 12 dB/octave slope (such as at 2 MHz in the op-amp of Fig. 3-10). However, this usually results in a very narrow bandwidth.

A more practical guide can be stated as follows: When a selected closed-loop gain is equal to or less than the open-loop gain at the 180° phase-shift point, the op-amp will be unstable. For example, if a closed-loop gain of 20 dB or less is selected, an op-amp with open-loop, uncompensated curves similar to Fig. 3-10 will be unstable.

To find the minimum closed-loop gain, simply note where the −180° phase angle intersects the phase-shift line. Then draw a vertical line up to cross the open-loop gain line. The closed-loop gain must be more than the open-loop gain at the frequency where the 180° phase shift occurs but less than the maximum open-loop gain. Using Fig. 3-10 as an example, the closed-loop gain must be greater than 20 dB but less than 60 dB.

Keep in mind that the guidelines discussed thus far apply to an uncompensated op-amp. With proper phase compensation, bandwidth (frequency response) and/or gain can be extended.

3-2.5. Op-Amp Phase-Compensation Methods

Op-amp design problems created by excessive phase shift can be solved by compensating techniques that alter response so that excessive phase shifts no longer occur in the desired frequency range. The following are the basic methods of phase compensation.

Closed-loop modification. The closed-loop gain of an op-amp can be altered by means of capacitors and/or inductances in the external feedback circuit (in place of fixed resistances). Capacitors and inductances change impedance with changes in frequency. This provides a different amount of feedback at different frequencies and changes the amount of phase shift in the feedback signal. The capacitors and inductances can be arranged to offset the undesired open-loop phase shift.

Phase-shift compensation by closed-loop modification is generally not used since the method can create impedance problems at both the high- and low-frequency limits of operation. However, closed-loop modification is used for applications in which the op-amp is part of a bandpass, band rejection, or peaking filter, as described in Secs. 3-8 and 3-9.

Input-impedance modification. The open-loop input impedance of an op-amp can be altered by means of resistors and capacitors connected at the op-amp input terminals. The impedance presented by the *RC* combination changes with frequency, thus altering the input impedance of the op-amp.

In turn, the change in input impedance (with frequency) changes the bandwidth and phase-shift characteristics of the op-amp. Such an arrangement causes the rolloff to start at a lower frequency than the normal open-loop response of the op-amp but produces a *stable rolloff* similar to the "ideal" curve in Fig. 3-9. With an op-amp properly compensated by an *RC* circuit at the input, the desired closed-loop gain can be produced by selection of external feedback resistors in the normal manner.

Phase-compensation techniques that alter the open-loop input impedance permit the introduction of a zero into the response. This zero can be designed to cancel one of the poles in the open-loop response. Typically, the first pole is canceled and the open-loop gain drops to zero at the second pole. That is, after modification, the response drops to zero at a frequency where the uncompensated response changes from 6 dB/octave to 12 dB/octave. This is shown in Fig. 3-11. In another, less frequently used input-modification design, the response drops to zero at the frequency where the uncompensated response changes from 12 dB/octave to 18 dB/octave. Both input-impedance-modification designs are discussed in Sec. 3-2.6.

Phase-lead compensation. The open-loop gain and phase-shift characteristics of an op-amp can be modified by means of a capacitor (or capacitors) connected to stages in the op-amp. Usually, the capacitors are connected between collectors in one of the high-gain differential stages. In other cases, the capacitors are connected from the collectors to ground. Generally, the capacitors are external to the op-amp, and are connected to internal stages by means of terminals provided on the package (such as the terminals shown in Fig. 3-1).

Phase-lead compensation requires a knowledge of the op-amp circuit characteristics. Usually, information for phase-lead compensation is provided on the op-amp data sheet. Typical phase-lead systems are described in Sec. 3-2.7.

Phase-lag compensation. The open-loop gain and phase-shift characteristics of an op-amp can be modified by means of a series capacitor and resistor connected to stages in the op-amp. There are two basic phase-lag compensation systems.

In one system, generally known as *RC rolloff, straight rolloff,* or *phase-lag rolloff compensation,* the open-loop response is altered by means of an *RC* network connected across a circuit component, such as across input or output of an op-amp gain stage.

In the other system, generally known as *Miller-effect rolloff* or *Miller-effect phase-lag compensation,* the open-loop response is altered by means of an *RC* network connected between the input and output of an inverting gain stage in the op-amp. The impedance of the compensating *RC* network then *appears* to be divided by the gain of that stage.

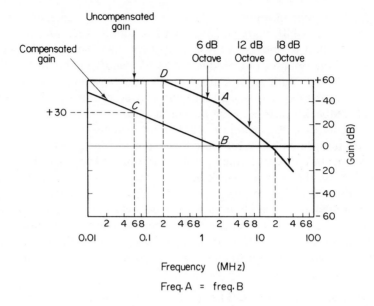

Frequency (MHz)

Freq. A = freq. B

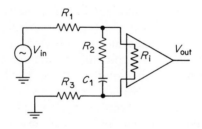

R_i = input impedance of op-amp

$R_1 = R_3$

$R_1 = R_3 = \left(\dfrac{\text{uncompensated gain (dB)}}{\text{compensated gain dB}}\right) R_i$

$R_2 = \dfrac{R_1 + R_3}{\left(\dfrac{\text{Freq.}D}{\text{Freq.}C} - 1\right)\left(1 + \dfrac{R_1 + R_3}{R_1}\right)}$

$C_1 = \dfrac{1}{6.28\ \text{Freq.}D \times R_2}$

$\text{Compensated gain} = \dfrac{\text{Uncompensated gain} \times R_i}{R_i + R_1 + R_3}$

$\text{Freq.}D = \dfrac{1}{6.28 \times R_2 \times C_1}$

Fig. 3-11. Phase compensation by modification of input impedance (early-rolloff method). Courtesy Motorola.

With either method, the rolloff starts at the corner frequency produced by the *RC* network. The Miller-effect rolloff technique requires a much smaller phase-compensating capacitor than that which must be used with the straight rolloff method. Also, the reduction in swing capability that is inherent in the straight rolloff is delayed significantly when the Miller-effect rolloff is used.

As with phase-lead, either method of phase-lag compensation requires a knowledge of the op-amp circuit characteristics. Usually, information for phase-lag compensation is provided on the op-amp data sheet, when such methods are recommended by the manufacturer. A typical straight rolloff system is described in Sec. 3-2.7. Miller-effect rolloff is discussed in Sec. 3-2.8.

How to select a phase-compensation method. A comprehensive op-amp data sheet will recommend one or more methods for phase compensation and will show the relative merits of each method. Usually, this is done by means of response curves for various values of the compensating network. Several examples are discussed in Secs. 3-2.6 through 3-2.8.

The recommended phase-compensation methods and values should be used in all cases. Proper phase compensation of an op-amp is at best a difficult, trial-and-error job. By using the data-sheet values it is possible to take advantage of the manufacturer's test results on production quantities of a given op-amp.

If the data sheet is not available or if the data sheet does not show the desired information, it is still possible to design a phase-compensation network using rule-of-thumb equations. The first step in phase compensation (when not following the data sheet) is to test the op-amp for open-loop frequency response and phase shift (as described in Secs. 2-12 and 3-17). Then draw a response curve similar to that in Fig. 3-10. On the basis of actual open-loop response and the information in Secs. 3-2.5 through 3-2.8, select trial values for the phase-compensating network. Then repeat the frequency-response and phase-shift test. If the response is not as desired, change the values as necessary.

Each method of phase compensation has its advantages and disadvantages. The main advantage of open-loop input impedance modification (Sec. 3-2.6) is that it can be done without data-sheet information (or with limited data). The only op-amp characteristic required is input impedance. This is almost always available in data-sheet or catalog form. If not, input impedance can be found by a simple test discussed in Sec. 3-17.

Phase lead and phase lag are the most widely accepted techniques for op-amps (particularly IC op-amps). These methods have an advantage (over input modification) in that the phase-compensation network is completely isolated from the feedback network. In the case of input modification, resistance in the phase-compensation network forms part of the feedback network.

Phase-lead and phase-lag compensation have certain disadvantages. A careful inspection of the information in Secs. 3-2.7 through 3-2.9 will show that it is necessary to know certain internal characteristics of the op-amp before an accurate prediction of the compensated frequency response can be found. In the case of phase-lead compensation (Sec. 3-2.7), the compensated response is entirely dependent upon value of the capacitors and must be found by actual test. In the case of phase-lag compensation (Secs. 3-2.8 and 3-2.9), the values for R and C of the compensation network are based on the uncompensated open-loop frequency at which gain changes from a 6 dB/ octave drop to a 12 dB/octave drop. This can be found by test of the uncompensated op-amp. However, to predict the frequency at which the compensated response starts to roll off (or the gain after compensation) requires a knowledge of internal-stage transconductance (or gain) and stage load. This information is usually not available and cannot be found by simple test.

To sum up, if the data sheet is available, use the recommended phase-compensation method. It will probably be phase lead or phase lag. If no phase-compensation information is available, use input-impedance modifications.

3-2.6. Phase Compensation by Modification of Input Impedance

There are two accepted methods for phase compensation using input-impedance modification. The first method, shown in Fig. 3-11, is the most widely used since it provides a straight rolloff similar to the ideal curve of Fig. 3-9. Once the input circuit is modified, conventional feedback can be used to select any point along the rolloff. That is, any combination of gain and frequency can be produced as described in 3-2.2. The main disadvantage to the method of Fig. 3-11 is early rolloff. Thus, if high gain is required, the bandwidth will be very narrow.

The method shown in Fig. 3-12 is used only where bandwidth is of greatest importance, and gain can be sacrificed. As shown, rolloff does not start until the breakpoint between 6 dB/octave and 12 dB/octave is reached, and gain is flat up to that point (no peaking condition). However, the method of Fig. 3-12 usually results in a little gain across the operating frequency range.

Early-rolloff method. Assume that the method of Fig. 3-11 is to be used with an op-amp having the characteristics shown in Fig. 3-10. That is, the uncompensated, open-loop gain is 60 dB, the 6 dB/octave rolloff starts at about 0.2 MHz, the 12 dB/octave starts at 2 MHz, and the 18 dB/octave starts at 20 MHz (which is also the point at which the open-loop gain drops to zero).

The first step in using the method shown in Fig. 3-11 is to note the frequency at which the uncompensated rolloff changes from 6 dB to 12 dB (point A of Fig. 3-11). The compensated rolloff should be zero (unity gain, point B) at the same frequency.

Draw a line up to the left from point B that *increases* at 6 dB/octave. For example, with point B at 2 MHz, the line should intersect 0.2 MHz as it

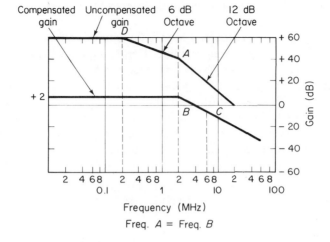

Freq. *A* = Freq. *B*

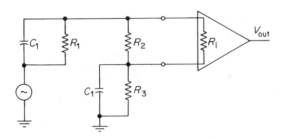

R_i = input impedance
    of op—amp

$$R_1 = R_3 = \frac{\left(\dfrac{1}{6.28 \times \text{Freq. } D}\right)}{C_1}$$

C_1 = See text

$$R_2 = \frac{\text{compensated gain} \times 2R_1}{\text{uncompensated gain}}$$

$$\text{Freq. } C \approx \frac{\left(\dfrac{2}{C_1 R_2}\right)}{6.28}$$

Fig. 3-12. Phase compensation by modification of input impedance (extended bandwidth method). Courtesy Motorola.

crosses the 20-dB gain point, and should intersect 0.02 MHz as it crosses the 40-dB point.

Any combination of compensated gain and rolloff starting frequency (point C) can be selected along the line. For example, if the rolloff starts at 0.2 MHz, the gain is about 20 dB, and vice versa.

Assume that the circuit of Fig. 3-11 is used to produce a compensated gain of 30 dB, with rolloff starting at about 0.06 MHz and dropping to zero (unity gain) at 2 MHz. The typical input impedance R_t is 10 kΩ. (Uncompensated gain, similar to that of Fig. 3-10, and typical input impedance can be found by referring to the data sheet or by actual test.)

Using the compensated-gain equation of Fig. 3-11, the relationship is:

$$30 \text{ dB} = \frac{(60 \text{ dB})(10,000)}{10,000 + R_1 + R_3}$$

Therefore,

$$R_1 + R_3 = \left(\frac{60}{30} - 1\right) \times 10,000$$

$$= 1 \times 10,000$$

$$= 10,000$$

$$\text{If } R_1 = R_3, \qquad R_1 = R_3 = 5000$$

Using the equation of Fig. 3-11, the value of R_2 is:

$$R_2 = \frac{10,000}{[(0.2/0.06) - 1][1 + (10,000/10,000)]}$$

$$= \frac{10,000}{2.3 \times 2}$$

$$= 2100 \text{ Ω (nearest standard value)}$$

The value of C_1 is:

$$C_1 = \frac{1}{2100(6.28)(0.2 \text{ MHz})} \approx 0.0004 \text{ μF} \quad \text{(nearest standard value)}$$

If the circuit of Fig. 3-11 shows an instability in the open-loop or closed-loop condition, try increasing the values of R_1 and R_3 (to reduce gain); then select new values for R_2 and C_1.

Extended-bandwidth method. Assume that the method of Fig. 3-12 is to be used with an op-amp having the characteristics shown in Fig. 3-10.

The first step in using the method shown in Fig. 3-12 is to note the frequency at which the uncompensated rolloff changes from 6 dB to 12 dB (point A of Fig. 3-12, 2 MHz). The compensated rolloff (point B) should start at the same frequency.

Assume that the circuit of Fig. 3-12 is used to produce a flat, compensated gain of 2 dB, with rolloff starting at 2 MHz. Find the approximate values for

R_1 through R_3 and C_1. Also, find the approximate maximum operating frequency (point C).

Assume a convenient value for each capacitor C_1, say 0.001 μF. Using this value, and the frequency (point D) at which the uncompensated 6-dB rolloff starts, find the value of R_1 and R_3 as follows:

$$R_1 = R_3 = \frac{1/(6.28 \times 0.2^6)}{0.001^{-6}} \approx \frac{0.8^{-6}}{0.001^{-6}} \approx 800 \, \Omega$$

Using a desired compensated gain of 2 dB, an uncompensated gain of 60 dB, and a value of 1600 for $2R_1$, the value of R_2 is:

$$R_2 = \frac{2 \times 1600}{60} \approx 53 \, \Omega$$

Using 0.001 μF for C_1 and 53 Ω for R_2, the approximate maximum operating frequency (point C) is:

$$\text{frequency C} = \frac{2/(0.001^{-6} \times 53)}{6.28} \approx 6 \, \text{MHz}$$

3-2.7. Phase-Lead-Compensation Examples

As discussed, phase-lead compensation requires the addition of a capacitor (or capacitors) to the basic op-amp circuit. Generally, the capacitors are external. However, some IC op-amps include an internal capacitor to provide fixed phase-lead compensation.

Phase-lead compensation requires a knowledge of internal op-amp circuit characteristics. As a result, the manufacturer's data sheet or similar information must be used. The alternative method is to use typical values for the compensating capacitor and test the results. Of course, this is time consuming and may not prove satisfactory, even after tedious testing.

The following are some examples of manufacturers' data on phase-lead compensation.

Figure 3-13 shows typical phase-lead compensation characteristics for an IC op-amp. The two external compensating capacitors tabulated in Fig. 3-13 are connected from the collectors of the first differential amplifier in the op-amp to ground by means of terminals on the IC.

The dashed lines in Fig. 3-13 illustrate the use of the curves for design of a 60-dB amplifier. First, the intersection of the various gain/frequency curves is followed out along the 60-dB line (horizontally) to the curve for a capacitor value of 0.001 μF. The intersection occurs at approximately 230 kHz. This means that if a 0.001-μF phase-lead capacitor is used, the op-amp response should be flat at 60 dB (within a 3-dB range) up to a frequency of 230 kHz. At higher frequencies, the op-amp output drops off. Thus, any gain up to 60 dB can be selected by proper choice of feedback and input resistances.

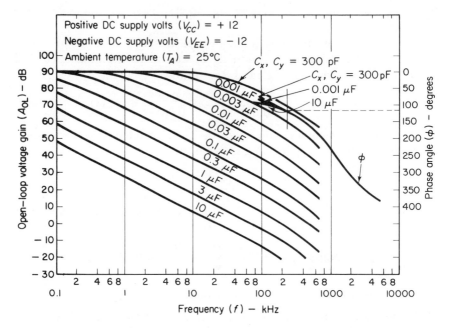

Fig. 3-13. Phase-compensation characteristics for CA3033 or CA3033A. Courtesy RCA.

Next, follow the 230-kHz line (vertically) until it intersects the phase curve. The intersection occurs at approximately 118°. This means that if a 0.001-μF capacitor is used, and the op-amp is operated at 230 kHz, the phase shift will be 118°. Thus, there is a 62° *phase margin* (180° − 118°) between input and output (assuming that the input signal is applied to the inverting input in the usual manner). A 62° phase margin should provide very stable operation.

Now assume that it is desired to operate at a higher frequency but still to provide the 60-dB gain. Follow the 60-dB line (horizontally) to intersect with the 300-pF curve (300 pF is the smallest recommended capacitor). The intersection occurs at approximately 600 kHz. However, the 600-kHz line intersects the phase curve at about 175°, resulting in a phase margin of about 5° (180° − 175°). This will probably produce unstable operation. Thus, if 60-dB gain must be obtained, the capacitor value must be larger than 300 pF.

The curves of Fig. 3-13 show that, for a given gain, a larger value of phase-lead capacitance reduces frequency capability, and vice versa. However, a reduction in frequency increases stability. For a given frequency of operation, capacitor size has no effect on stability, only on gain.

Figure 3-14 shows the characteristics of an IC *op-amp with built-in phase-lead capacitance*. Note that this rolloff approaches that of the "ideal" op-amp

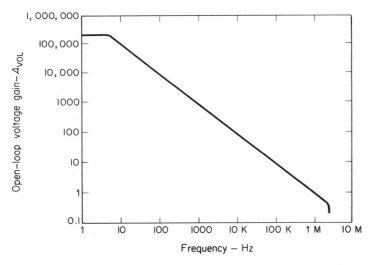

Fig. 3-14. Open-loop gain versus frequency characteristics of op-amp with internal phase compensation.

shown in Fig. 3-9. Any closed-loop gain (less than open-loop gain) can be selected by feedback resistances in the normal manner. Likewise, any combination of gain/frequency response can be selected. For example, if it is desired to start the rolloff at 10^4 Hz, choose a feedback resistance $Z_F = 10,000$ (and $Z_R = 100$). The closed-loop gain is then flat to 10^4 Hz and rolls off at 6 dB/octave to unity gain at 10^6 Hz.

Keep in mind that the gain/frequency characteristics of the op-amp in Fig. 3-14 *cannot be changed.* Thus, if a different gain/frequency response is required (say 10^2 gain at 10^5 Hz), the op-amp is unsuitable.

Figure 3-15 shows a somewhat different system of phase-lead graphs as they appear on the manufacturer's data. These graphs show both the uncompensated and compensated characteristics on the same graph. The curves marked A are the uncompensated open-loop voltage gain (A_{VOL}) and phase shift. The B curves show compensated gain and phase shift.

When uncompensated, the gain starts the 6-dB/octave rolloff at about 0.3 MHz, the 12-dB/octave rolloff at about 5 MHz, and the 18-dB/octave rolloff at about 20 MHz. The phase shift exceeds 180° at about 4.5 MHz. Thus, the absolute maximum uncompensated operating frequency is about 4 MHz.

With compensation, 6-dB/octave rolloff starts at about 0.002 MHz. Unity gain occurs at about 4.5 MHz. The compensated phase shift never exceeds about 150°, which occurs at about 10 MHz. Thus, stable operation is possible up to 10 MHz, even though there is no gain beyond about 4.5 MHz. As with

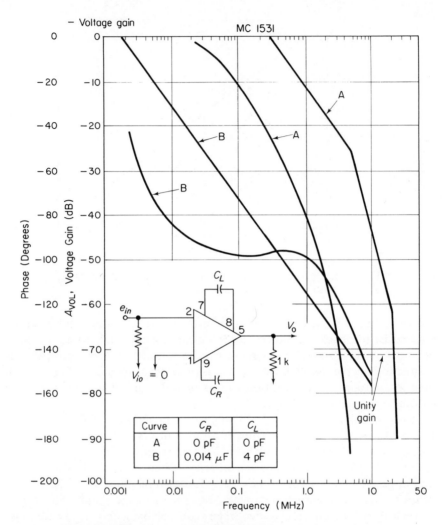

Fig. 3-15. Op-amp characteristics with and without phase-lead compensation. Courtesy Motorola.

the other examples, any combination of gain and frequency can be selected along the compensated A_{VOL} line by proper selection of feedback and input resistances.

3-2.8. Conventional Phase-Lag-Compensation Examples

As shown in Fig. 3-16, conventional phase-lag compensation requires the addition of a capacitor and resistor to the basic op-amp circuit. This RC network is connected across a circuit component (such as across a stage out-

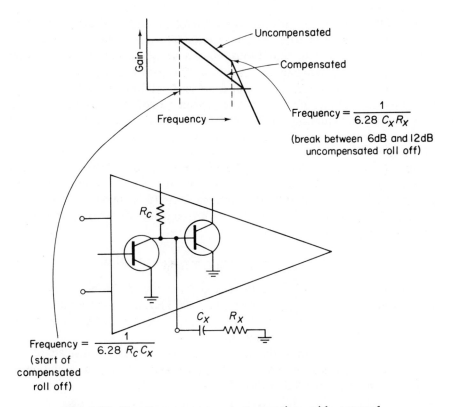

Fig. 3-16. Frequency-response compensation with external capacitor and resistor (RC rolloff or phase-lag compensation).

put or input, or across the op-amp output). Conventional phase-lag compensation is external to the op-amp circuit. Many IC op-amps have terminals provided for connection of external phase-compensation circuits to the internal circuit.

As in the case of phase-lead compensation, the manufacturer's information must be used for phase-lag compensation. The only alternative is to use the equations of Fig. 3-16 and test the results. This is not recommended unless the manufacturer's information is not available. Always use the published information, at least as a first trial value.

As shown in Fig. 3-16, the values of the external phase-compensating resistor and capacitor are dependent upon the frequency at which the uncompensated gain changes from 6 dB/octave to 12 dB/octave (generally known as the second pole of the uncompensated gain curve). Thus, it is relatively simple to find values for R and C if the uncompensated gain characteristics are known (or can be found by test). However, note that the frequency at which the compensated rolloff will start (the start of the compensated 6-dB/

octave rolloff) is dependent upon the value of the external capacitor C and internal characteristics of the op-amp circuit. Thus, even though satisfactory values of R and C can be found to produce a straight rolloff, there is no way of determining the frequency at which rolloff will start (using the equations).

If conventional phase-lag rolloff is to be used, and the values must be found by test (no manufacturer's data), assume a convenient value of R and find C using the equations of Fig. 3-16. Then test the compensated op-amp. If the compensated rolloff starts at too low a frequency, decrease the value of C and find a new corresponding value for R.

As an example, assume that the uncompensated rolloff changes from 6 dB/octave to 12 dB/octave at 10 MHz, and it is desired to have the compensated rolloff start at 300 kHz. Assume that R_X is 1000 Ω (a convenient value). Using the equation of Fig. 3-16, a frequency of 10 MHz, and an R_X of 1000 Ω, the value for C_X is:

$$C_X = \frac{1}{6.28 \times 10 \text{ MHz} \times 1000} \approx 16 \text{ pF}$$

Test the op-amp with C_X at 16 pF and R_X at 1000 Ω. If the compensated rolloff starts at some frequency lower than 300 kHz, increase the value of C_X and find a new value for R_X. If the rolloff starts above 300 kHz, decrease the value of C_X and use another value for R_X.

For example, if compensated rolloff starts at 100 kHz instead of the desired 300 kHz, increase the value of C_X to 30 pF. Using the equation of Fig. 3-16, the corresponding value of R_X is:

$$R_X = \frac{1}{6.28 \times 10 \text{ MHz} \times 30 \text{ pF}} \approx 530 \text{ Ω}$$

Figures 3-17 and 3-18 show typical phase-lag-compensation characteristics for an op-amp. Figure 3-17 shows the uncompensated gain characteristics (curve A), the phase-lag-compensated characteristics (curve B), and phase-lead characteristics (curve C). Note that the uncompensated gain changes from a 6 dB/octave rolloff to a 12 dB/octave rolloff at about 10 MHz. Thus, 10 MHz is used in the equation to find the values of R and C. As shown, the recommended values are 18 pF and 820 Ω. Rolloff starts at some frequency below 0.1 MHz, and continues at 6 dB/octave down to unity gain at about 33 MHz, when the recommended RC combination is used.

Although the phase-lag capacitance of 18 pF shown in curve B of Fig. 3-17 is sufficient to provide stable operation (no oscillation) in a resistance feedback system down to unity gain, the capacitance is not sufficient to provide flat, closed-loop response (flat to within ± 1 dB) below 20 dB. That is, if the feedback network is chosen to provide something less than 20 dB closed-loop gain, operation will be stable, but the response will not be flat (there will be peaking or shifts in gain greater than ± 1 dB).

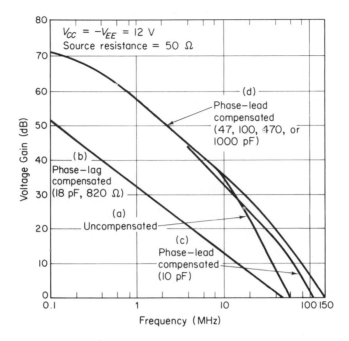

Fig. 3-17. Open-loop gain as a function of frequency for compensated and uncompensated op-amps. Courtesy RCA.

This condition is shown in Fig. 3-18. For example, if the closed-loop gain is set for 10 dB, the recommended capacitance value is 56 pF. For unity gain, the capacitance value must be increased to 68 pF. A corresponding value of R must be found in each case. For example, if 56 pF is used for 10-dB gain, R must be reduced to about 280 Ω. If C is made 68 pF for unity gain, R is reduced to about 230 Ω.

3-2.9. Miller-Effect Phase-Lag-Compensation Examples

As shown in Fig. 3-19, the problems in determining Miller-effect phase-lag values are essentially the same as for conventional phase lag. That is, the values of the compensating resistor and capacitor are dependent upon the frequency at which uncompensated rolloff changes from 6 to 12 dB. The frequency at which compensated rolloff starts is dependent upon compensating RC values and internal op-amp values. (This is also true for the frequency at which compensated rolloff reaches unity gain.)

Again, always use the manufacturer's information. If Miller-effect phase compensation values must be found without the manufacturer's data, assume

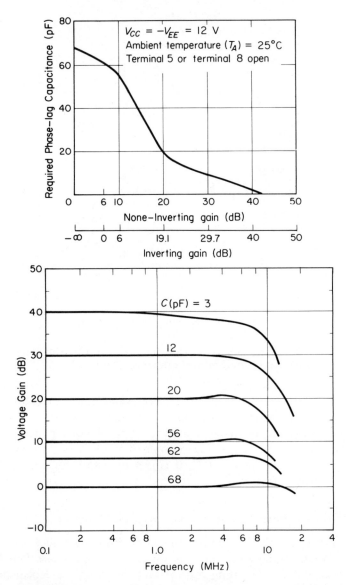

Fig. 3-18. Amount of phase-lag capacitance required to obtain a flat (± 1 dB) response and typical response characteristics. Courtesy RCA.

a convenient value for R and find C using the equations of Fig. 3-19. Decrease the value of C if compensated rolloff starts at too low a frequency, and vice versa.

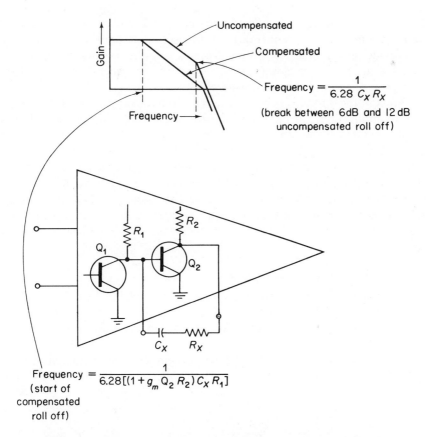

Fig. 3-19. Frequency-response compensation with external capacitor and resistor (Miller-effect rolloff).

3-3. Interpreting IC Data Sheets

The designer must have a thorough knowledge of op-amp characteristics to get the best possible results from the op-amp in any system. Op-amp characteristics provide a good basis for system design. However, commercial op-amps, either IC or discrete component, are generally designed for specific applications. Although some commercial op-amps are described as "general purpose," it is impossible to design an op-amp with truly universal characteristics.

For example, certain op-amps are designed to provide high-frequency gain at the expense of other performance characteristics. Other op-amps provide very high gain or high input impedance in low-frequency applications. IC op-amps, which are fabricated by the diffusion process, can be made suitable for comparator applications (where both halves of the differential amplifier circuits must be identical). Likewise, IC op-amps can be processed to provide high gain at low power-dissipation levels. For these reasons, any description of op-amp characteristics must be of a general nature, unless a specific application is being considered.

Most of the op-amp characteristics required for proper use of the op-amp in any application can be obtained from the manufacturer's data sheet or similar catalog information. There are some exceptions to this rule. For certain applications, it may be necessary to test the op-amp under simulated operating conditions. In using data-sheet information or test results, or both, it is always necessary to interpret the information. Each manufacturer has its own system of data sheets. It is impractical to discuss all data-sheet formats here. Instead, we shall discuss typical information found on op-amp data sheets, as well as test results, and see how this information affects the op-amp user.

3-3.1. Open-Loop Voltage Gain

The open-loop voltage gain (A_{VOL} or A_{OL}) is defined as the ratio of a change in output voltage to a change in input voltage at the input of the op-amp. Open-loop gain is always measured without feedback and usually without phase-shift compensation.

Open-loop gain is *frequency-dependent* (gain decreases with increased frequency). This is shown in Fig. 3-20. As shown, the gain is flat (within about

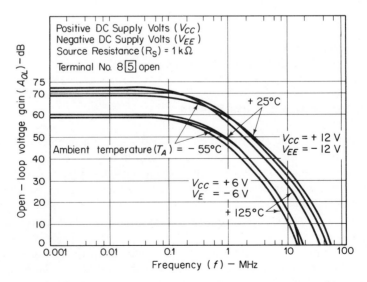

Fig. 3-20. Open-loop gain versus frequency for CA3008. Courtesy RCA.

±3 dB) up to frequencies of about 0.1 MHz. Then the gain rolls off to unity at frequencies above 10 MHz. The open-loop gain is also *temperature-dependent*, and dependent upon supply voltage, as shown in Fig. 3-20. Generally, gain increases with supply voltage. The effects of temperature on gain are different at different frequencies.

Ideally, open-loop gain should be infinitely high since the primary function of an op-amp is to amplify. In general, the higher the gain, the higher the accuracy of the op-amp transfer function (relationship of output to input). However, there are practical limits to gain magnitude and also levels at which an increase in magnitude buys little in the way of increased performance. The true significance of open-loop gain is many times misapplied in op-am operation, where in reality open-loop gain determines closed-loop accuracy limits rather than ultimate accuracy.

The numerical values of the open-loop gain (and the bandwidth) of an op-amp are of relatively little importance in themselves. The important requirement is that the open-loop gain must be greater than the closed-loop gain over the frequency of interest if an accurate transfer function is to be maintained. For example, if a 40-dB op-amp and a 60-dB op-amp are used in a 20-dB closed-loop-gain configuration, and the open-loop gain is decreased 50 per cent in each case (say due to component aging), the closed-loop gain of the 40-dB op-amp varies 9 percent and that of the 60-dB op-amp varies only 1 per cent.

The *frequency rolloff characteristics* are the prime determinants of op-amp frequency response. The greater the rate of rolloff prior to the intersection of the feedback ratio (closed-loop) frequency characteristics with the open-loop response (in the active region), the more difficult phase compensation of the op-amp becomes.

An 18-dB/octave rolloff is generally considered the maximum slope that can occur in the active region before proper phase compensation becomes extremely difficult or impossible to achieve. In addition, because op-amps have useful application down to and including unity gain, the active region of the op-amp may be considered as the entire-portion frequency characteristic above the 0-dB bandwidth. Thus, a well-designed op-amp should roll off at no greater than 18 dB/octave until below unity gain.

As discussed, open-loop gain can be modified by several compensation methods. A typical op-amp data sheet will show the results of such compensation, usually by means of graphs such as the one shown in Fig. 3-15.

After compensation is applied, the op-amp can be connected in the closed-loop configuration. The voltage gain under closed-loop conditions is dependent upon external components (the ratio of feedback resistance to input resistance). Thus, closed-loop gain is usually not listed as such on op-amp data sheets. However, the data sheet may show some typical gain curves with various ratios of feedback (Fig. 3-21 is an example). If available, such

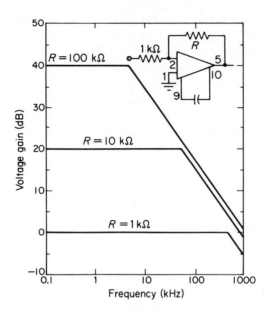

Fig. 3-21. Closed-loop voltage gain of IC versus frequency.

curves can be used directly to select values of feedback components (as well as phase-compensation components).

3-3.2. Phase Shift

Figure 3-22 shows the open-loop phase-shift curve of a typical op-amp. (Sometimes, such curves are included on the open-loop frequency-response graph.) Because a closed-loop gain of unity allows the highest frequency

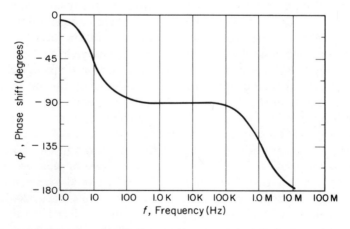

Fig. 3-22. Open-loop phase shift for MC1556. Courtesy Motorola.

response for the loop gain, the closed-loop unity gain frequency is considered the worst case for phase shift.

One figure of merit commonly used in evaluating the stability of an op-amp is *phase margin*. As discussed in Sec. 3-2, oscillations can be sustained if the total phase shift around the loop (from input to output, and back to input) can reach 360° before the total gain around the loop drops below unity (as the frequency is increased). Because an op-amp is normally used in the inverting mode, 180° of phase shift is available to begin with. Additional phase shift is developed by the op-amp due to internal circuit conditions.

Phase margin represents the difference between 180° and the phase shift introduced by the op-amp at the frequency at which loop gain is unity. A value of 45° phase margin is considered quite conservative to provide a guard against production variations, temperature effects, and other stray factors. This means that the op-amp should not be operated at a frequency at which the phase shift exceeds 135° (180° − 45°). However, it is possible to operate op-amps at frequencies where the phase shift is kept within the 160° to 170° region. Of course, the ultimate stability of an op-amp must be established by tests.

3-3.3. Bandwidth, Slew Rate, and Output Characteristics

The bandwidth, slew rate, output voltage swing, output current, and output power of an op-amp are all interrelated. These characteristics are frequency-dependent and depend upon phase compensation. The characteristics are also temperature- and power-supply-dependent, but to a lesser extent. Before discussing the interrelationship, let us define each of the characteristics.

Bandwidth for an op-amp is usually expressed in terms of open-loop operation. The common term is BW_{OL} at 3 dB, such as shown in Fig. 3-23. For example, a BW_{OL} of 800 kHz indicates that the open-loop gain of the op-amp will drop to a value of 3 dB below the flat or low-frequency level at a frequency of 800 kHz.

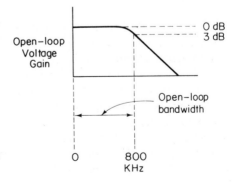

Fig. 3-23. Bandwidth and open-loop gain relationships.

Frequency range is sometimes used in place of open-loop bandwidth. The frequency range of an op-amp is often listed as "useful frequency range" (such as dc up to 18 MHz). Useful frequency range for an op-amp is similar to the F_T (total frequency) term used with discrete transistors (Sec. 1-3). Generally, the high-frequency limit specified for an op-amp is the frequency at which gain drops to unity.

Power bandwidth is a more useful characteristic since it represents the bandwidth of the op-amp in closed-loop operation connected to a normal load. As shown in Fig. 3-24, power bandwidth is given as the peak-to-peak output-voltage capability of the op-amp (working into a given load) across a band of frequencies. Power bandwidth figures usually imply that the output indicated is free of distortion, or that distortion is within limits (such as total harmonic distortion less than 5 per cent). In the op-amp of Fig. 3-24, the output voltage is about 27 V (peak to peak) up to about 20 kHz, and drops to near zero at 300 kHz.

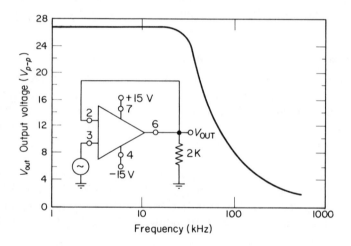

Fig. 3-24. Power bandwidth for MC1556. Courtesy Motorola.

Power output of an op-amp is generally listed in terms of power across a given load (such as 250 mW across 500 ohms). However, power output is usually listed at only one frequency. The same is true of *output current* or *maximum output current* characteristics found on some data sheets. Thus, power bandwidth is the more useful characteristic.

Output voltage swing is defined as the peak or peak-to-peak output-voltage swing (referred to zero) that can be obtained without clipping. A symmetrical voltage swing is dependent upon frequency, load current, output impedance, and slew rate. Generally, an increase in frequency will decrease the possible output voltage swing. Figure 3-25 shows that maximum output

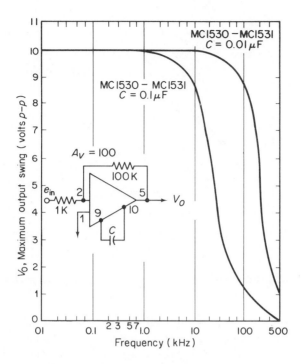

Fig. 3-25. Maximum output voltage swing versus frequency. Courtesy Motorola.

voltage drops as frequency is increased. Also note that phase-compensation capacitance has an effect on maximum output voltage. Figure 3-26 shows than an increase in load resistance will also increase maximum output voltage.

The *slew rate* of an op-amp is the maximum rate of change of the output voltage, with respect to time, that the op-amp is capable of producing while maintaining linear characteristics (symmetrical output without clipping).

Slew rate is expressed in terms of:

$$\frac{\text{difference in output voltage}}{\text{difference in time}} \quad \text{or} \quad \frac{dV_o}{dt}$$

Usually, slew rate is listed in terms of *volts per microsecond*. For example, if the output voltage from an op-amp is capable of changing 7 V in 1 μs, the slew rate is 7. If, after compensation or other change, the op-amp changes a maximum of 3 V in 1 μs, the new slew rate is 3.

Slew rate of an op-amp is the direct function of the phase-shift-compensation capacity. At higher frequencies, the current required to charge and discharge a compensating capacitor can limit available current to succeeding stages or loads, and thus result in lower slew rates. This is one reason why

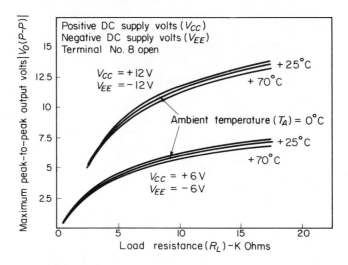

Fig. 3-26. Maximum peak-to-peak output voltage versus load resistance. Courtesy RCA.

op-amp data sheets usually recommend the compensation of early stages in the op-amp where signal levels are still small and little current is required.

Slew rate decreases as compensation capacitance increases. This is shown in Fig. 3-27. Where high frequencies are involved, the lowest values of com-

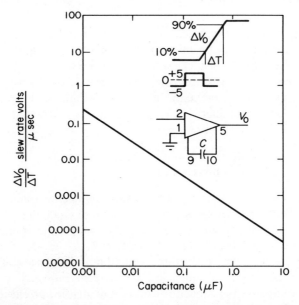

Fig. 3-27. Slew rate of IC versus rolloff compensation capacitance.

pensation capacitor should be used. Figure 3-28 shows the minimum capacitance values that can be used with different closed-loop gain levels for the particular op-amp. The curves of Figs. 3-27 and 3-28 are typical of those found on op-amp data sheets in which slew rate is of particular importance.

The major effect of slew rate in op-amp applications is on output power. All other factors being equal, a lower slew rate results in lower power output. Slew rate and the term *full power response* of an op-amp are directly related. Full power response is the maximum frequency measured in a closed-loop unity-gain configuration for which rated output voltage can be obtained for a sine-wave signal, with a specified load, and without distortion due to slew-rate limiting.

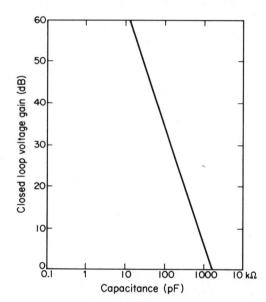

Fig. 3-28. Closed-loop voltage gain of IC versus minimum rolloff capacitance.

The slew rate versus full-power response relationship can be shown as:

$$\text{slew rate (in V/s)} = 6.28 \times F_M \times E_o$$

where F_M is the full-power response frequency in hertz and E_o is the peak output voltage (one-half the peak-to-peak voltage).

For example, using the characteristics shown in Fig. 3-24, the output voltage E_o is about 13 V (one-half the peak-to-peak voltage of 26 V) at a frequency of 30 kHz. Thus, the slew rate is:

$$\text{slew rate} = 6.28 \times 30,000 \times 13 \approx 2,449,200 \text{ V/s} \approx 2.45 \text{ V}/\mu\text{s}$$

The equation can be turned around to find the full-power response frequency. For example, assume that an op-amp is rated as having a slew rate of 2.5 V/μs, and a peak-to-peak output of 20 V ($E_o = 10$ V). Find the full-power response frequency F_M as follows:

$$F_M = \frac{2.5 \text{ V}/\mu s}{6.28 \times 10} = \frac{2,500,000}{62.8} \approx 40,000 \text{ Hz} \approx 40 \text{ kHz}$$

Of course, if curves such as shown in Fig. 3-24 are available, it is not necessary to calculate the maximum frequency for a given output. Simply follow the 20-V line until it crosses the curve at 40 kHz.

The graph of Fig. 3-29 shows the relationship among slew rate, full-power response frequency, and output voltage. For example, if slew rate is 6 V/μs, the maximum output voltage (peak to peak) is about 20 V at a frequency of 100 kHz, and vice versa. If slew rate is 10 V/μs, the peak-to-peak output is increased to 30 V at 100 kHz, and vice versa.

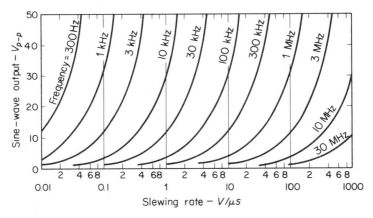

Fig. 3-29. Slew-rate curve. Courtesy RCA.

With a constant output load, the power output of an op-amp is dependent upon output voltage. In turn, all other factors being equal, output voltage is dependent upon the slew rate. Since slew rate depends upon phase-compensation capacitance, op-amp power output is also dependent upon compensation. Some data sheets omit slew rate but provide a graph similar to that shown in Fig. 3-30. This graph shows the direct relationship between full-power output frequency and phase-compensation capacitance. For example, with a phase-compensation capacitance of 0.01 μF and a 500-Ω load, the op-amp shown in Fig. 3-30 will deliver full-rated output power up to a frequency of 80 kHz.

3-3.4. Output Impedance

Output impedance (Z_{out}) is defined as the impedance seen by a load at the output of the IC amplifier (see Fig. 3-31). Excessive output impedance can reduce the gain since, in conjunction with the load and feedback resistors, output impedance forms an attenuator network. In general, output imped-

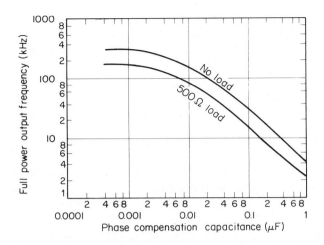

Fig. 3-30. Frequency for full-power output as a function of phase-compensating capacitance.

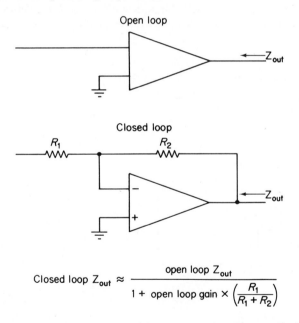

$$\text{Closed loop } Z_{out} \approx \frac{\text{open loop } Z_{out}}{1 + \text{open loop gain} \times \left(\dfrac{R_1}{R_1 + R_2}\right)}$$

Fig. 3-31. Open- and closed-loop output-impedance relationships.

ance of ICs used as operational amplifiers is less than 200 Ω. Generally, input resistances are at least 1000 Ω, with feedback resistances several times higher than 1000 Ω. Therefore, the output impedance of a typical IC will have little effect on gain.

If the IC is serving primarily as a voltage amplifier (which is usually the case), the effect of output impedance will be at a minimum. Output impedance has a more significant effect in design of power devices that must supply large amounts of load current.

Closed-loop output impedance is found by:

$$\text{closed-loop output impedance} = \frac{\text{output impedance (of the IC)}}{1 + \text{open-loop gain} \times [R_1/(R_1 + R_2)]}$$

where R_1 and R_2 are the input impedance and feedback impedance, respectively, as shown in Fig. 3-31.

Therefore, it will be seen that output impedance will increase as frequency increases, since open-loop gain decreases.

3-3.5. Input Impedance

Input impedance (Z_{in}) is defined as the impedance seen by a source looking into one input of the IC amplifier with the other input grounded (see Fig. 3-32). The primary effect of input impedance on design is to reduce amplifier loop gain. Input impedance will change with temperature and frequency. Generally, input impedance is listed on the data sheet at 25°C and 1 kHz.

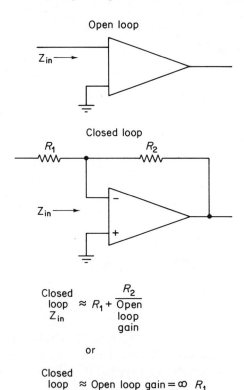

Fig. 3-32. Open- and closed-loop input-impedance relationships.

If input impedance is quite different from the impedance of the device driving the IC, there will be a loss of input signal due to the mismatch. However, in practical terms, it is not possible to alter the IC input impedance. Therefore, if impedance match is critical, either the IC or the driving source must be changed to effect a match.

3-3.6. Input Common-Mode Voltage Swing

Input common-mode voltage swing (V_{iCM}) is defined as the maximum peak input voltage that can be applied to either input terminal of the IC without causing abnormal operation or damage (see Fig. 3-33). Some IC data sheets list a similar term: *common-mode input voltage range* (V_{CMR}). Usually, V_{iCM} is listed in terms of peak voltage, with positive and negative peaks being equal. V_{CMR} is often listed as a positive and negative voltage of different value (such as $+1$ V and -3 V).

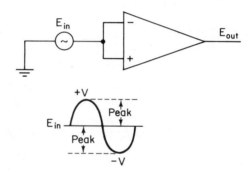

Fig. 3-33. Input common-mode voltage-swing relationships.

In practical design considerations, either of these parameters limit the differential signal amplitude which can be applied to the IC input. As long as the input signal does not exceed the V_{iCM} or V_{CMR} values (in either the positive or negative direction), no design problem should be encountered.

Note that some ICs also list *single-ended* input voltage signal limits, where the differential input is not to be used.

3-3.7. Common-Mode Rejection Ratio

All IC manufacturers do not agree on the exact definition of common-mode rejection. One manufacturer defines common-mode rejection (CMR or CM_{rej}) as the ratio of differential gain (usually large) to common-mode gain (usually a fraction). Another manufacturer defines CMR as the change in output voltage to the change in the input common-mode voltage producing it, divided by the open-loop gain. For example, as shown in Fig. 3-34, assume that the common-mode input (applied to both input terminals simultaneously) was 1 V, the resultant output was 1 mV, and the open-loop gain was 100. The

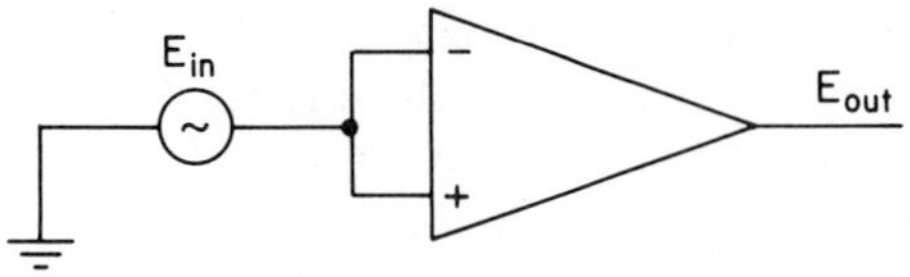

$$\text{Common mode rejection} = \frac{\dfrac{E_{out}}{E_{in}}}{\text{Open loop gain}}$$

or

$$\frac{E_{out}}{\text{Open loop gain}} = \text{Equivalent differential input signal}$$

$$\text{Common mode rejection} = \frac{E_{in}}{\text{Equivalent differential input signal}}$$

Fig. 3-34. Common-mode rejection ratio.

CMR would then be:

$$\frac{0.001/1}{100} = 100{,}000 \quad \text{or} \quad 100 \text{ dB}$$

Another method to calculate the CMR is to divide the output signal by the open-loop gain to find an *equivalent differential input signal*. Then the common-mode input signal is divided by this equivalent differential input signal. Using the same figures as the previous CMR calculation:

$$1 \text{ mV} \div 100 = 0.00001 \text{ equivalent differential input signal}$$

$$1 \text{ V} \div 0.00001 = 100{,}000 \quad \text{or} \quad 100 \text{ dB}$$

No matter what basis is used for calculation, the CMR is an indication of the degree of circuit balance of the differential stages of the amplifier, since a common-mode input signal applied to the input terminals should be amplified identically in both sides of the IC (in theory). A large output for a given *common-mode input* is an indication of large unbalance, or poor common-mode rejection. As a rule of thumb, the CMR should be at least 20 dB greater than the open-loop gain.

Under differential drive conditions, the common-mode rejection has no drastic effects on the performance of an op-amp, unless the rejection ratio is extremely low. However, in a common-mode drive application, such as in a comparator, high common-mode rejection is essential. For example, if an op-amp with a 60-dB differential gain and a 50-dB common-mode rejection is used to compare a 1-V signal against a 1-V reference, the output will be 3.2 V when it should be zero. Such results would be totally unacceptable for a comparator-type application. This is why the common-mode rejection should be at least 20 dB greater than the differential (open-loop) gain.

Note that the CMR decreases as frequency increases in all ICs. The CMR can also be temperature-sensitive, but this is usually not of major importance in practical design.

3-3.8. Input-Bias Current

Input-bias current is defined as the average value of the two input-bias currents of the op-amp differential input stage. This is shown in Fig. 3-35, which illustrates input bias for two typical IC op-amps. Note that input-bias current decreases as temperature increases. Input-bias current is essentially a function of the large-signal current gains of the input stage.

In use, the significance of input-bias current is the resultant voltage drop across input resistors or other source resistances. This voltage drop can restrict the input common-mode voltage range at higher impedance levels. The voltage drop must be overcome by the input signal. Also, a large input-bias current is undesirable in applications where the source cannot accommodate a significant dc current. Examples of such applications are those in which the source resistance is very large (resulting in a large voltage drop), or sources of a magnetic nature that can be severely unbalanced by a flow of dc current (such as transducers that operate on magnetic principles).

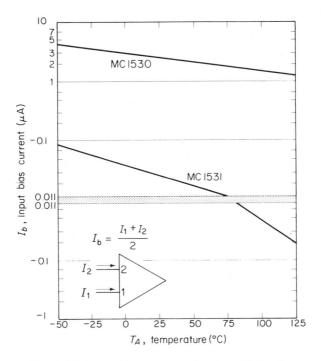

Fig. 3-35. Input-bias current. Courtesy Motorola.

Some op-amps have very low input-bias current and are thus well suited to these applications. Where very low input-bias current is required, the input differential stages of the op-amp often use field-effect transistors or other devices that draw very little current.

3-3.9. Input-Offset Voltage and Current

Input-offset voltage is defined as the voltage that must be applied at the input terminals to obtain zero output voltage (see Fig. 3-36). Input-offset voltage indicates the matching tolerance in the differential-amplifier stages. A perfectly matched amplifier requires zero input voltage to produce zero output voltage. Typically, input-offset voltage is in the order of 1 or 2 mV for an IC op-amp, as shown in Fig. 3-36a. The offset voltage from an op-amp can also be defined as the deviation of the output dc level from the arbitrary

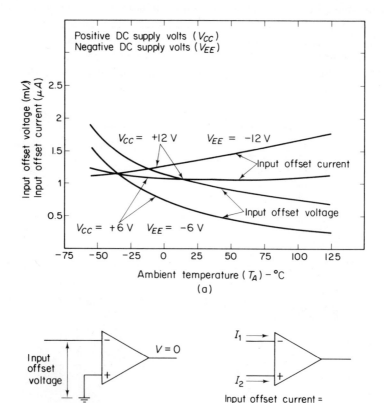

Fig. 3-36. Input-offset voltage and current. Courtesy Motorola.

input–output level, usually taken as a ground reference when both inputs are shorted together.

Input-offset current is defined as the difference in input-bias current into the input terminals of an op-amp (Fig. 3-36c). Input-offset current is an indication of the degree of matching of the input differential stage. Typically, input-offset current is on the order of 1 or 2 μA for an IC op-amp, as shown in Fig. 3-36a. The offset current for an op-amp can also be defined as the deviation when the inputs are driven by two identical dc input-bias-current sources.

Offset voltage and current are usually referred back to the input because their output values are dependent upon feedback. (That is, data sheets rarely list output-offset characteristics.) In normal use, the offset in an op-amp results from a combination of offset voltage and current. For example, if an op-amp has a 1-mA input-offset voltage, and a 1-μA input-offset current, with the inputs returned to ground through 1000-Ω resistors, the total input offset is either zero or 2 mV, depending upon the phase relationship between the two offset characteristics.

The offset of an op-amp is a dc error that should be minimized for numerous reasons, including the following:

1. The use of an op-amp as a true dc amplifier is limited to signal levels much greater than the offset.
2. Comparator applications require that the output voltage be zero (within limits) when the two input signals are equal and in phase.
3. In a dc cascade, the offset of the first stage determines the offset characteristics of the entire system.

Thus, any offset at the input of an op-amp is multipled by the gain at the output. If the op-amp serves to drive additional amplifiers, the increased offset at the op-amp output will be multiplied even further. The gain of the entire system must then be limited to a value that is insufficient to cause limiting in the final output stage.

The *effect of input-offset voltage* on op-amp use is that the input signal must overcome the offset voltage before an output will be produced. For example, if an op-amp has a 1-mV input-offset voltage, and a 1-mV signal is applied, there is no output. If the signal is increased to 2 mV, the op-amp will produce only the peaks. Since input-offset voltage is increased by gain, the effect of input-offset voltage is increased by the ratio of feedback resistance to input resistance, plus unity (or 1), in the closed-loop condition. For example, if the ratio is 100:1 (for a gain of 100), the effect of input-offset voltage is increased by 101.

Input-offset current can be of greater importance than input-offset voltage when high impedances are used in design. If the input-bias current is different for each input, the voltage drops across the input resistors (or input impedance) will not be equal. If the resistance is large, there will be a large unbalance in input voltages. This condition can be minimized by means of a resistance connected between the noninverting input and ground, as shown in Fig. 3-37. The value of this resistor R_3 should equal the parallel equivalent of the input and feedback resistors R_1 and R_2, as shown by the equations. In practical design, the trial value for R_3 is based on the equation of Fig. 3-37. The value of R_3 is then adjusted for minimum voltage difference at both terminals (under normal operating conditions, but with no signal).

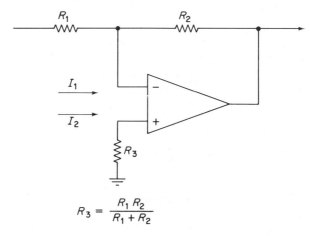

$$R_3 = \frac{R_1 R_2}{R_1 + R_2}$$

Fig. 3-37. Minimizing input-offset current (and input-offset voltage).

Some op-amps (particularly IC op-amps) include provisions to neutralize any offset. Typically, an external voltage is applied through a potentiometer to terminals on the op-amp. The voltage is adjusted until the offset, at the input and output, is zero. Often the terminals are connected to the emitters of the first differential-amplifier stage. Figure 3-38 shows a typical external offset null or neutralization circuit.

For op-amps without offset compensation, the effects of input offset can be minimized by an external circuit. Figure 3-39 shows two such circuits, one for inverting and the other for noninverting op-amps. The equations shown in Fig. 3-39 assume that resistor R_B must be of a value to produce a null range

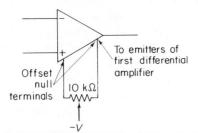

Fig. 3-38. Typical offset null or neutralization circuit.

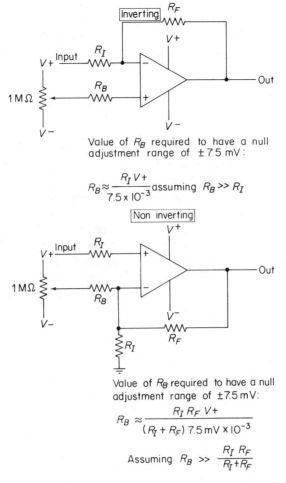

Value of R_B required to have a null adjustment range of ± 7.5 mV:

$$R_B \approx \frac{R_I V+}{7.5 \times 10^{-3}} \text{ assuming } R_B \gg R_I$$

Value of R_B required to have a null adjustment range of ± 7.5 mV:

$$R_B \approx \frac{R_I R_F V+}{(R_I + R_F) 7.5 \text{ mV} \times 10^{-3}}$$

$$\text{Assuming } R_B \gg \frac{R_I R_F}{R_I + R_F}$$

Fig. 3-39. Input-offset minimizing circuits.

of ± 7.5 mV. This is generally sufficient for any op-amp. However, if a different input-offset voltage range is required, simply substitute the desired range for ± 7.5 mV.

One reason for an offset null is that the input and output dc levels of an op-amp should be equal, or nearly equal. This condition is desirable to assure that the resistive feedback network can be connected between the input and output without upsetting the differential or the common-mode dc bias.

The *average temperature coefficient of input-offset voltage*, listed on some data sheets as TCV_{IO}, is dependent upon the temperature coefficients of various components within the op-amp. Temperature changes affect stage gain, match of differential amplifiers, and so forth, and thus change input-offset voltage. From a design standpoint, TCV_{IO} need be considered only if the parameter is large, and the op-amp must be operated under extreme temperatures. For example, if input-offset voltage doubles with an increase to a temperature that is likely to be found during normal operation, the higher input-offset voltage should be considered the "normal" value for design.

3-3.10. Power-Supply Sensitivity

Power-supply sensitivity ($S+$ and $S-$) is defined as the ratio of change in input-offset voltage to the change in supply voltage producing it, with the remaining supply held constant (see Fig. 3-40). Some IC data sheets list a similar parameter: *input-offset voltage sensitivity*. In either case, the parameter is expressed in terms of mV/V or μV/V, representing the change (in mV or μV) of input-offset voltage to a change (in volts) of one power supply. Usually, there is a separate sensitivity parameter for each power supply, with the opposite power supply assumed to be held constant. For example, a typical sensitivity listing is 0.1 mV/V for positive. This implies that with the negative supply held constant, the input-offset voltage will change by 0.1 mV for each 1-V change in positive supply voltage.

The effect of power-supply sensitivity (or input-offset voltage sensitivity) is obvious. If an IC has considerable sensitivity to power-supply variation, the power-supply regulation must be increased to provide correct operation with minimum input signal levels.

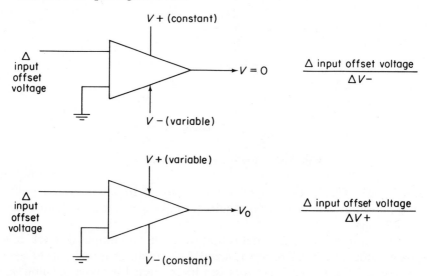

Fig. 3-40. Power-supply sensitivity.

3-3.11. Input-Noise Voltage

There are many systems for measuring noise voltage in an IC amplifier, and equally as many methods used to list the value on data sheets. Some data sheets omit the value entirely. In general, noise is measured with the IC in the open-loop condition, with or without compensation, and with the input shorted, or with a fixed resistance load at the input terminals. The input and/or output voltage is measured with a sensitive voltmeter or oscilloscope. Input noise is of the order of a few microvolts, with output noise usually less than 100 mV (because of the amplifier gain).

Except where the noise value is very high, or the input signal is very low, amplifier noise can be ignored. Obviously, an amplifier with 10-μV noise at the input would mask a 10-μV input signal. If the input signal were raised to 1 mV with the same amplifier, the noise would be unnoticed.

Noise is temperature-dependent, as well as dependent upon the method of compensation used (see Fig. 3-41).

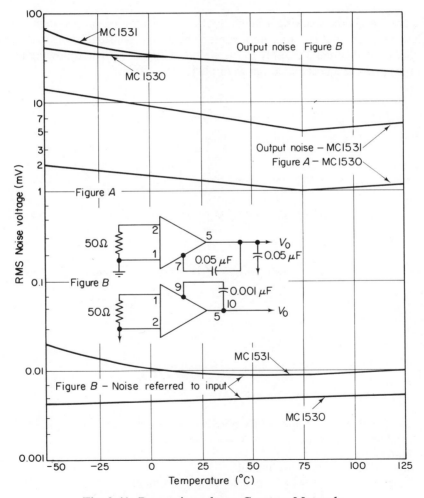

Fig. 3-41. Rms noise voltage. Courtesy Motorola.

3-3.12. Power Dissipation

An IC amplifier data sheet usually lists two power-dissipation ratings. One value is the *total device dissipation*, which includes any load current. The other is *device dissipation* (P_D or P_T), which is defined as the dc power dissipated by the IC itself (with output at zero and no load).

The device dissipation must be subtracted from the total dissipation to calculate the load dissipation. For example, if an IC can dissipate a total of 300 mW (at a given temperature, supply voltage, and with or without a heat sink) and the IC itself dissipates 100 mW, the load cannot exceed 200 mW ($300 - 100 = 200$ mW).

3-4. Integrated-Circuit Operational Amplifier

Figure 3-42 is the working schematic of a closed-loop op-amp system, complete with external circuit components. The design considerations discussed in Secs. 3-1 through 3-3 apply to the circuit of Fig. 3-42. The following paragraphs provide a specific design example for the circuit.

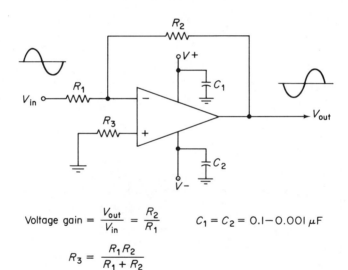

$$\text{Voltage gain} = \frac{V_{out}}{V_{in}} = \frac{R_2}{R_1} \qquad C_1 = C_2 = 0.1 - 0.001 \, \mu F$$

$$R_3 = \frac{R_1 R_2}{R_1 + R_2}$$

Fig. 3-42. Basic IC operational amplifier.

3-4.1. Op-Amp Characteristics

Supply voltage: $+15$ and -15 V nominal, ± 19 V maximum
Total device dissipation: 750 mW, derating 8 mW/°C
Temperature range: 0°C to $+70$°C
Input-offset voltage: 3 mV typical
Input-offset current: 10 nA nominal, 30 nA maximum

Input-bias current: 100 nA nominal, 200 nA maximum
Input-offset-voltage sensitivity: 0.2 mV/V
Device dissipation: 300 mW maximum
Open-loop voltage gain: as shown in Fig. 3-13
Slew rate: 4 V/μs at a gain of 1; 6 V/μs at a gain of 10, 33 V/μs at a gain of 100
Open-loop bandwidth: as shown in Fig. 3-13
Common-mode rejection: 94 dB
Output-voltage swing: 23 V (peak to peak) typical
Input impedance: 1 mΩ
Output impedance: 300 Ω
Input-voltage range: -13 V, $+10$ V
Output power: 250 mW typical

3-4.2. Design Example

Assume that the circuit of Fig. 3-42 is to provide a voltage gain of 100 (40 dB), the input signal is 80 mV rms, the input source impedance is not specified, the output load impedance is 500 Ω, the ambient temperature is 25°C, the frequency range is dc up to 300 kHz, and the power supply is subject to 20 per cent variation.

Frequency/gain relationship. Before attempting to calculate any circuit values, make certain that the op-amp can produce the desired voltage gain at the maximum frequency. This can be done by reference to a graph similar to Fig. 3-13.

Note that the maximum frequency of 300 kHz intersects the phase curve at about 135°. This allows a phase margin of 45°. The op-amp should be stable over the desired frequency range.

Note that the maximum frequency of 300 kHz intersects several capacitance curves above the 40-dB open-loop voltage-gain level. Thus, the op-amp should be able to produce more than 40 dB of open-loop gain.

Supply voltage. The positive and negative supply voltages should both be 15 V, since this is the nominal value listed. Most op-amp data sheets will list certain characteristics as "maximum" (temperature range, total dissipation, maximum supply voltage, maximum input signal, etc.) and then list the remaining characteristics as "typical" with a given "nominal" supply voltage.

In no event can the supply voltage exceed the 19 V maximum. Since the available supply voltage (15 V) is subject to a 20 per cent variation, or 18 V maximum, the supply is within the 19-V limit.

Decoupling or bypass capacitors. The values of C_1 and C_2 should be found on the data sheet. In the absence of a value, use 0.1 μF for any frequency up to 10 MHz. If this value produces a response problem at any frequency (high or low), try a value between 0.001 and 0.1 μF.

Closed-loop resistances. The value of R_2 should be 100 times the value of R_1 to obtain the desired gain of 100. The value of R_1 should be selected so that the voltage drop across R_1 (with the nominal input-bias current) is comparable to the input signal (never larger than the input signal).

A 50-Ω value for R_1 will produce a 10-μV drop with the maximum 200-nA input-bias current. Such a 10-μV drop is less than 10 per cent of the 80-mV input signal. Thus, the fixed drop across R_1 should have no appreciable effect on the input signal.

With a 50-Ω value for R_1, the value of R_2 must be 5 kΩ (50 × 100 gain = 5000).

Offset minimizing resistance. The value of R_3 can be found using the equation of Fig. 3-42 once the values of R_1 and R_2 have been established. Note that the value of R_3 works out to about 49 Ω, using the Fig. 3-42 equation:

$$\frac{R_1 \times R_2}{R_1 + R_2} = \frac{50 \times 5000}{50 + 5000} \approx 49 \; \Omega$$

A simple trial value for R_3 is always *slightly less* than the R_1 value. The final value of R_3 should be such that the no-signal voltages at *each input are equal.*

Comparison of circuit characteristics. Once the values of the external circuit components have been selected, the characteristics of the op-amp and the closed-loop circuit should be checked against the requirements of the design example. The following is a summary of the comparison.

Gain versus phase compensation. The closed-loop gain should always be less than the open-loop gain. As a guideline, the open-loop gain should be at least 20 dB greater than closed-loop gain. Figure 3-13 shows that open-loop gain up to about 66 dB is possible with a proper phase-compensation capacitor. Figure 3-13 also shows that a capacitance of 0.001 μF (1000 pF) will produce an open-loop gain of slightly less than 60 dB at 300 kHz, whereas a 300-pF capacitance will produce a 66-dB open-loop gain. To assure an open-loop gain of 60 dB, use a capacitance value of about 700 pF.

With a 60-dB open-loop gain, and values of 50 and 5000 Ω, respectively, for R_1 and R_2, the closed-loop circuit should have a flat frequency response of 40 dB (gain of 100) from zero up to 300 kHz. A rolloff will start at frequencies above 300. Thus, the closed-loop gain is well within tolerance.

Input voltage. The *peak* input voltage must not exceed the rated maximum input signal. In this case, the rated maximum is +10 V and −13 V, whereas the input signal is 80 mV rms, or approximately 112-mV peak (80 × 1.4). This is well below the +10-V maximum limit.

When the rated maximum input signal is an uneven positive and negative value, always use the *lowest value* for total swing of the input signal. In this case, the input swings from +112 mV to −112 mV, far below +10 V.

An input signal that started from zero could swing as much as $+10$ V and -10 V, without damaging the op-amp. An input signal that started from -2 V could swing as much as ±11 V.

Output voltage. The *peak-to-peak output voltage* must not exceed the rated maximum output voltage swing (with the required input signal and selected amount of gain).

In this case, the rated output voltage swing is 23 V (peak to peak), whereas the actual output is approximately 22.4 V (80 mV rms input × a gain of $100 = 8000$ mV output; 8000 mV × 2.8 = 22.4 V peak to peak). Thus, the anticipated maximum output is within the rated maximum.

However, the actual output voltage is dependent upon slew rate, which, in turn, depends upon compensation capacitance. As shown in the characteristics, the slew rate is given as 33 V/μs when gain is 100. The data sheet does not show a relationship between slew rate and compensation capacitance. However, since slew rate is always maximum with the lowest value of compensation capacitance, it can be assumed that the slew rate will be 33 V/μs with a capacitance of 700 pF (which is near the lowest recommended value of 300 pF, Fig. 3-13).

Using the equations in Sec. 3-3.3, it is possible to calculate the output voltage capability of the op-amp. With a maximum operating frequency of 300 kHz, and an assumed slew rate of 33 V/μs, the peak-output-voltage capability is

$$\frac{33,000,000}{6.28 \times 300,000} \approx 17 \text{ V}$$

Thus, the peak-to-peak output voltage capability is 34 V, well above the anticipated 22.4 V. A quick approximation of the output-voltage capability can also be found by reference to Fig. 3-29.

Output power. The output power of an op-amp is usually computed on the basis of rms output voltage (rather than peak or peak to peak) and output load.

In this example, the output voltage is 8 V rms (80 mV × 100 gain = 8000 mV = 8 V). The load resistance or impedance is 500 Ω, as stated in the design assumptions. Thus, the output power is

$$\frac{(8)^2}{500} = 0.128 \text{ W} = 128 \text{ mW}$$

A 128-mW output is well below the 250-mW typical output power of the op-amp. Also, 128-mW output plus a device dissipation of 300 mW is 428 mW, well below the rated 750-mW total device dissipation. Thus, the op-amp should be capable of delivering full-power output to the load.

Note that power output ratings usually are at some given temperature, 25°C in this case. Assume that the temperature is raised to 50°C. The total device dissipation must be derated by 8 mW/°C, or a total of 200 mW for the 25°C increase in temperature. This derates or reduces the 750-mW total device dissipation to 550 mW. However, the 550 mW is still well above the anticipated 428 mW.

Output impedance. Ideally, the closed-loop output impedance should be as low as possible and always less than the load impedance. The closed-loop output impedance can be found using the equations of Fig. 3-7. In this example, the approximate output impedance will be

$$\frac{300}{1 + 1000 \times [50/(50 + 5000)]} \approx 30 \, \Omega$$

3-5. Integrated-Circuit Summing Amplifier

Figure 3-43 is the working schematic of an IC operational amplifier used as a summing amplifier (also known as an analog adder). Summation of a number of voltages can be accomplished using this circuit. (Voltage summation was one of the original uses for operational amplifiers in computer work.)

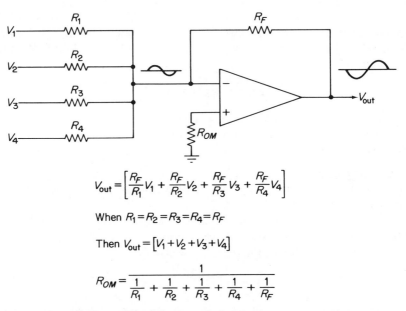

$$V_{out} = \left[\frac{R_F}{R_1} V_1 + \frac{R_F}{R_2} V_2 + \frac{R_F}{R_3} V_3 + \frac{R_F}{R_4} V_4 \right]$$

When $R_1 = R_2 = R_3 = R_4 = R_F$

Then $V_{out} = [V_1 + V_2 + V_3 + V_4]$

$$R_{OM} = \frac{1}{\dfrac{1}{R_1} + \dfrac{1}{R_2} + \dfrac{1}{R_3} + \dfrac{1}{R_4} + \dfrac{1}{R_F}}$$

Fig. 3-43. Summing amplifier.

3-5.1. Design Considerations

All the design considerations for the basic IC operational amplifier described in Secs. 3-1 through 3-4 apply to the summing amplifier, except as follows. When open-loop gain is high, the circuit functions with a minimum of error to sum a number of voltages. One circuit input is provided for each voltage to be summed. The single circuit output is the sum of the various input voltages (a total of four in this case) multiplied by any circuit gain. Generally, gain is set so that the output is at some given voltage value when all inputs are at their maximum value. In other cases, the resistance values are selected for unity gain.

3-5.2. Design Example

Assume that the circuit of Fig. 3-43 is to be used as a summing amplifier to sum four voltage inputs. Each of the voltage inputs varies from 2 to 50 mV rms. The output must be a nominal 1 V rms with full input on all four channels but must not exceed 2 V rms at any time.

To simplify design, make resistors R_1 through R_4 the same value. Note that the input-bias current will then be divided equally, and produce the same voltage drop across each resistor.

The values of R_1 through R_4 should be selected so that the voltage drop (with nominal input-bias current) is comparable with the minimum input signal. Assume a 5-μA input-bias current for the IC. This results in 1.25 μA through each resistor. A 100-Ω value for R_1 through R_4 would produce a 125-μV drop, which is less than 10 per cent of the 2-mV minimum input signal.

The total (or maximum possible) signal voltage at the IC input will be 200 mV (4 × 50 mV). Therefore, the value of R_F should be between 5 and 10 times that of R_1 through R_4, to achieve a nominal 1 V and a maximum 2 V (200 mV × 5 = 1 V, 200 mV × 10 = 2 V).

With a 100-Ω value for R_1 through R_4, the value of R_F could be 500 to 1000 Ω. Assume that the 1000-Ω value is used.

The value of the offset minimizing resistance R_{OM} can be found using the equation of Fig. 3-43, once the values of R_1 through R_4 and R_F have been established.

$$R_{OM} = \frac{1}{1/100 + 1/100 + 1/100 + 1/100 + 1/1000} = 25\ \Omega$$

3-6. Integrated-Circuit Integration Amplifier

Figures 3-44 and 3-45 are the working schematics of IC operational amplifiers used as integration amplifiers (or integrators). Integration of various signals (usually square waves) can be accomplished using this circuit. The output voltage from the amplifier is inversely proportional to the time constant of the feedback network and directly proportional to the integral of the input voltage.

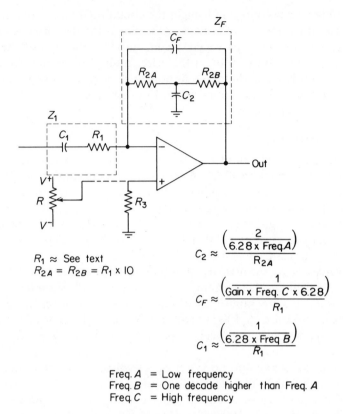

$R_1 \approx$ See text
$R_{2A} = R_{2B} = R_1 \times 10$

$$C_2 \approx \frac{\left(\dfrac{2}{6.28 \times \text{Freq.} A}\right)}{R_{2A}}$$

$$C_F \approx \frac{\left(\dfrac{1}{\text{Gain} \times \text{Freq.} C \times 6.28}\right)}{R_1}$$

$$C_1 \approx \frac{\left(\dfrac{1}{6.28 \times \text{Freq } B}\right)}{R_1}$$

Freq. A = Low frequency
Freq. B = One decade higher than Freq. A
Freq. C = High frequency

Fig. 3-44. Integration amplifier (integrator) for wide frequency-range application. Courtesy Motorola.

The circuit of Fig. 3-44 is best suited to applications in which the integrator must be used over a wide range of frequencies. Keep in mind that the output amplitude will depend upon frequency, once the value are selected. The circuit in Fig. 3-45 is best suited where only one frequency is involved.

3-6.1. Design Considerations

All the design considerations for the basic IC operational amplifier described in Secs. 3-1 through 3-4 apply to the integration amplifier, except as follows. The value of the $R_1 C_F$ time constant should be *approximately equal to the period* of the signal to be integrated. The value of the $R_{\text{shunt}} C_F$ time constant should be *substantially longer than the period* of the signal to be integrated (approximately 10 times longer). Therefore, R_{shunt} should be 10 times R_1. Keep in mind that R_{shunt} and C_F form an impedance which is frequency-sensitive (that will be most noticed at low frequencies).

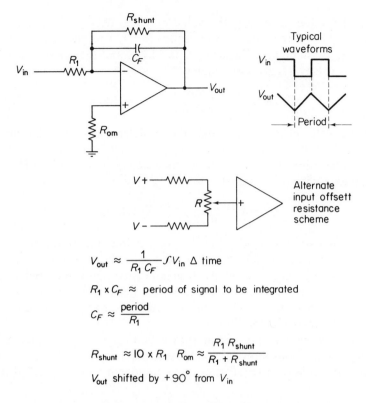

$$V_{out} \approx \frac{1}{R_1 C_F} \int V_{in} \, \Delta \text{ time}$$

$R_1 \times C_F \approx$ period of signal to be integrated

$$C_F \approx \frac{\text{period}}{R_1}$$

$$R_{shunt} \approx 10 \times R_1 \qquad R_{om} \approx \frac{R_1 \, R_{shunt}}{R_1 + R_{shunt}}$$

V_{out} shifted by $+90°$ from V_{in}

Fig. 3-45. Integration amplifier (integrator) for fixed-frequency operation. Courtesy Motorola.

3-6.2. Design Example

In the circuit of Fig. 3-44, the value of R_1 is chosen on the basis of input-bias current and voltage drop, and to produce realistic values for C_F and C_1. For simplicity, assume that R_1 is 10 kΩ.

The time constant of the $R_2 C_F$ combination must be substantially larger than the $R_1 C_F$ combination. For that reason, the values of R_{2A} and R_{2B} are approximately 10 times the value of R_1, or 100 kΩ, as indicated by the equations.

With the values of R_2 set, the value of C_2 is determined by the low-frequency limit of the integrator (frequency A). Typically, frequency A is a fraction of 1 Hz. Assume a frequency A of 0.016 Hz. Using these values, C_2 is approximately

$$C_2 \approx \frac{2/(6.28 \times 0.016)}{100,000} \approx 200 \ \mu F$$

With the value of R_1 set, the value of C_F is determined by the high-frequency limit of the integrator (frequency C), and the gain desired at frequency C. Assume that frequency C is 5 kHz, and the desired gain is 10. Using these values, C_F is approximately

$$C_F \approx \frac{1/(10 \times 5000 \times 6.28)}{10,000} \approx 300 \text{ pF}$$

With the value of R_1 set, the value of C_1 is determined by the intermediate frequency of the integrator (frequency B). Note that frequency B is approximately 1 decade above frequency A. Assume a frequency B of 0.16 Hz. Using these values, C_1 is approximately

$$C_1 \approx \frac{1/(6.28 \times 0.16)}{10,000} \approx 100 \ \mu\text{F}$$

If the input offset voltage of the op-amp is low, the noninverting input can be connected to ground through a fixed resistance. As a first trial value, use the same resistance value as R_1. If greater precision is required, or if the op-amp offset is large, use the potentiometer network shown in Fig. 3-44. With this arrangement, potentiometer R is adjusted so that there is no offset voltage under no-signal conditions.

The voltage gain of the circuit shown in Fig. 3-44 is dependent upon frequency. Figure 3-46 shows a gain versus frequency-response curve for an op-amp (Motorola MC-1531) using the component values specified in this example.

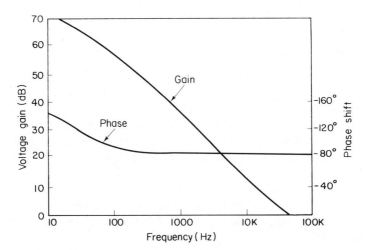

Fig. 3-46. Gain versus frequency response for integration amplifier.

3-6.3. Design Example

In the circuit of Fig. 3-45, the value of R_1 is chosen on the basis of input-bias current and voltage drop. Assume that the input-bias current is 5000 nA and that the desired voltage drop across R_1 is not to exceed 180 mV (say to provide 10 per cent of a 1.8-V input signal). A value of 33 kΩ for R_1 will produce a 165-mV drop.

The value of the $R_1 C_F$ time constant must be *approximately* equal to the period of the signal to be integrated. The value of the $R_{shunt} C_F$ time constant must be *substantially larger* than the period (approximately 10 times longer). Thus, R_{shunt} is 10 times R_1. Note that R_{shunt} and C_F form an impedance that is frequency-sensitive (that is, the impedance is most noticed at low frequencies).

Assume that the circuit of Fig. 3-45 is to be used as an integrator for 1-kHz square waves. This requires a period of approximately 1 ms.

Any combination of C_F and R_1 can be used, provided that the value of C_F times R_1 is approximately 0.001. Using the assumed value of 33 kΩ for R_1, the value of C_F is 0.03 μF.

The value of R_{shunt} must then be at least 330 kΩ. Note that the purpose of R_{shunt} is to provide dc feedback. This feedback is necessary so that an offset voltage cannot continuously charge C_F (which can result in amplifier limiting). If the offset voltage is small or can be minimized by means of R_{om}, it is possible to eliminate R_{shunt}.

Resistor R_{shunt} may have the effect of limiting gain at very low frequencies. However, above about 15 Hz, the effect of R_{shunt} is negligible (because of C_F in parallel).

If greater precision is required, particularly at low frequencies, the input-offset resistance R_{om} can be replaced by the potentiometer network shown in Fig. 3-45. With this arrangement, potentiometer R is adjusted so that there is no offset voltage under no-signal conditions.

3-7. Integrated-Circuit Differentiation Amplifier

Figure 3-47 is the working schematic of an IC operational amplifier used as a differentiation amplifier (or differentiator). Differentiation of various signals (usually square waves, or sawtooth and sloping waves) can be accomplished using this circuit. The output voltage from the amplifier is inversely proportional to the feedback time constant, and is directly proportional to the time rate of change of the input voltage.

3-7.1. Design Considerations

All the design considerations for the basic IC operational amplifier described in Secs. 3-1 through 3-4 apply to the differentiation amplifier, except

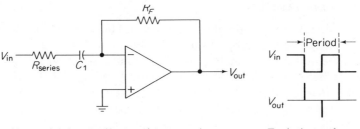

Alternative input offset resistance scheme   Typical waveforms

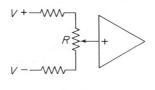

$$V_{out} \approx -R_F C_1 \frac{\Delta V_{in}}{\Delta time}$$

$R_F \times C_1 \approx$ period of signal to be differentiated

$$C_1 \approx \frac{period}{R_F}$$

$R_{series} \approx 50$ ohms

V_{out} shifted by $-90°$ from V_{in}

Fig. 3-47. Differentiation amplifier (differentiator). Courtesy Motorola.

as follows. The value of the $R_F C_1$ time constant should be *approximately equal* to the period of the signal to be differentiated. In practical applications, the time constant will have to be chosen on a trial-and-error basis to obtain a reasonable *output level*.

The main problem in the design of differentiating amplifiers is that the *gain increases with frequency*. Therefore, differentiators are very susceptible to high-frequency noise. The classic remedy for this effect is to connect a small resistor (on the order of 50 Ω) in series with the input capacitor so that the high-frequency gain is decreased. Actually, the addition of the resistor results in a more realistic model of the differentiator because a resistance is always added in series with the input capacitor by the signal-source imped-ance.

Conversely, in some applications a differentiator may be advantageously used to detect the presence of distortion or high-frequency noise in the signal. A differentiator can often detect hidden information that could not be detected in the original signal.

Differentiation permits slight changes in input slope to produce very significant changes in the output. An example of the usefulness of this feature would be in determining the linearity of a sweep sawtooth waveform. Nonlinearity results from changes in slope of the waveform. Therefore, if nonlinearity is present, the differentiated waveform indicates the points of nonlinearity quite clearly. (However, it should be noted that repetitive waveforms with a rise and fall of differing slopes can show erroneous waveforms.)

3-7.2. Design Example

Assume that the circuit of Fig. 3-47 is to be used as a differentiator to differentiate 1 kHz waves. This requires a period of approximately 1 ms. Any combination of C_1 and R_F could be used, provided that the value of C_1 times R_F was approximately 0.001. The values used for the integrator of Sec. 3-6.3 can be used as initial trial values, even though the components are interchanged. Thus, C_1 would be 0.03 μF, with a 33-kΩ value for R_F.

The value of R_{series} (if used) should be an arbitrary 50 Ω. Keep in mind that R_{series} and C_1 form an impedance which is frequency-sensitive (that will be most noticed at high frequencies).

3-8. Integrated-Circuit Narrow-Bandpass Amplifier

Figure 3-48 is the working schematic of an IC operational amplifier used as a narrow-bandpass amplifier (or tuned peaking amplifier).

3-8.1. Design Considerations

All the design considerations for the basic IC operational amplifier described in Secs. 3-1 through 3-4 apply to the narrow-bandpass amplifier, except as follows. Circuit gain is determined by the ratio of R_1 and R_F in the usual manner. However, the frequency at which maximum gain will occur (or the narrow-band peak) is the resonant frequency of the L_1C_1 circuit. Capacitor C_1 and inductance L_1 form a parallel-resonant circuit that rejects the resonant frequency. Therefore, there is minimum feedback (and maximum gain) at the resonant frequency.

3-8.2. Design Example

Assume that the circuit of Fig. 3-48 is to provide 20-dB gain at a peak frequency of 100 kHz. The value of R_1 is chosen on the basis of input-bias current and voltage drop, in the usual manner. Assume an arbitrary value of 3.3 kΩ for R_1. The value of R_{OM} would then be the same, or slightly less. With a value of 3.3 kΩ for R_1, the value of R_F must be 33 kΩ (or R_1 times 10) for a 20-dB gain.

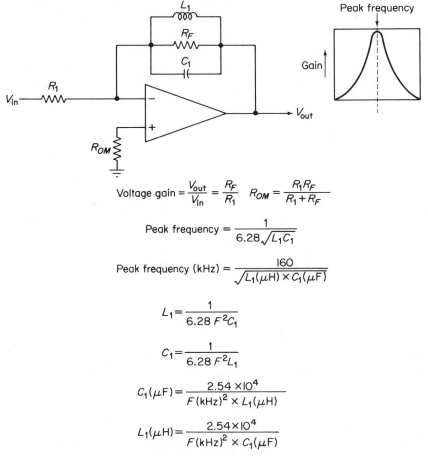

$$\text{Voltage gain} = \frac{V_{out}}{V_{in}} = \frac{R_F}{R_1} \quad R_{OM} = \frac{R_1 R_F}{R_1 + R_F}$$

$$\text{Peak frequency} = \frac{1}{6.28\sqrt{L_1 C_1}}$$

$$\text{Peak frequency (kHz)} = \frac{160}{\sqrt{L_1(\mu H) \times C_1(\mu F)}}$$

$$L_1 = \frac{1}{6.28\, F^2 C_1}$$

$$C_1 = \frac{1}{6.28\, F^2 L_1}$$

$$C_1(\mu F) = \frac{2.54 \times 10^4}{F(kHz)^2 \times L_1(\mu H)}$$

$$L_1(\mu H) = \frac{2.54 \times 10^4}{F(kHz)^2 \times C_1(\mu F)}$$

Fig. 3-48. Narrow-bandpass amplifier (tuned peaking).

Any combination of L_1 and C_1 could be used, provided that the resonant frequency is 100 kHz. For frequencies below 1 MHz, the value of C_1 should be between 0.001 and 0.01 μF. Assume an arbitrary 0.0015 μF for C_1. Using the equations of Fig. 3-48, the value of L_1 would be approximately 1700 μH.

3-9. Integrated-Circuit Wide-Bandpass Amplifier

Figure 3-49 is the working schematic of an IC operational amplifier used as a wide-bandpass amplifier.

3-9.1. Design Considerations

All the design considerations for the basic IC operational amplifier described in Secs. 3-1 through 3-4 apply to the wide-bandpass amplifier, except as follows. Maximum circuit gain is determined by the ratio of R_R

and R_F. That is, the gain of the passband or flat portion of the response curve is set by R_F/R_R. Minimum circuit gain is determined by the ratio of R_1 and R_F, in the usual manner.

The frequencies at which rolloff starts and ends (at both high- and low-frequency limits) are determined by impedances of the various circuit combinations, as shown by the equations of Fig. 3-49.

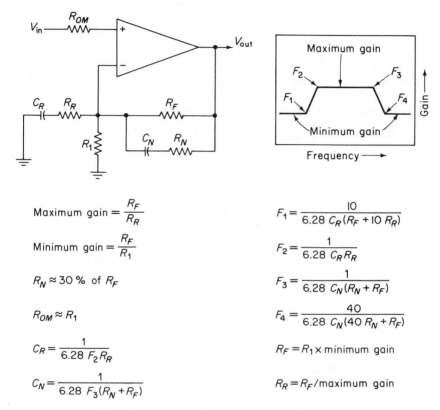

$$\text{Maximum gain} = \frac{R_F}{R_R}$$

$$\text{Minimum gain} = \frac{R_F}{R_1}$$

$$R_N \approx 30\% \text{ of } R_F$$

$$R_{OM} \approx R_1$$

$$C_R = \frac{1}{6.28 \, F_2 R_R}$$

$$C_N = \frac{1}{6.28 \, F_3 (R_N + R_F)}$$

$$F_1 = \frac{10}{6.28 \, C_R (R_F + 10 \, R_R)}$$

$$F_2 = \frac{1}{6.28 \, C_R R_R}$$

$$F_3 = \frac{1}{6.28 \, C_N (R_N + R_F)}$$

$$F_4 = \frac{40}{6.28 \, C_N (40 \, R_N + R_F)}$$

$$R_F = R_1 \times \text{minimum gain}$$

$$R_R = R_F / \text{maximum gain}$$

Fig. 3-49. Wide-bandpass amplifier.

3-9.2. Design Example

Assume that the circuit of Fig. 3-49 is to provide approximately 20-dB minimum gain at all frequencies, and approximately 30-dB gain at the passband. Gain is to start increasing at frequencies above 10 kHz, and rise to an approximate 30-dB passband at 40 kHz. The passband must extend to 200 kHz, and then drop back to 20 dB at 800 kHz and above.

Note that if phase compensation is required for the basic IC, the compensation values should be based on the minimum gain of 20 dB, and not on the passband gain of 30 dB.

The value of R_1 is chosen on the basis of input-bias current and voltage drop, in the usual manner. Assume an arbitrary value of 1 kΩ for R_1. The value of R_{OM} would then be the same or slightly less.

With a value of 1 kΩ for R_1, a value of 10 kΩ should be used for R_F. With a value of 10 kΩ for R_F, the value of R_R should be 330 Ω. These relationships will produce gains of 20 dB and 30 dB, respectively. In practice, it may be necessary to reduce both of these trial values to get the desired gain relationship.

With a value of 330 Ω for R_R, the value of C_R should be 0.012 μF.

With a value of 10 kΩ for R_F, the value of R_N should be 3 kΩ.

With a value of 13 kΩ for $(R_N + R_F)$, the value of C_N should be 61 pF.

With these values established, the remaining equations in Fig. 3-49 can be used to confirm the four frequencies.

3-10. Integrated-Circuit Unity-Gain Amplifier

Figure 3-50 is the working schematic of an IC operational amplier used as a unity-gain amplifier (also known as a voltage follower).

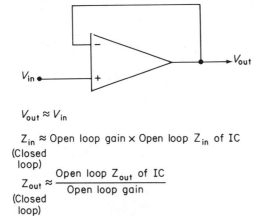

$V_{out} \approx V_{in}$

$Z_{in} \approx$ Open loop gain $\times$ Open loop Z_{in} of IC
(Closed
loop)

$Z_{out} \approx \dfrac{\text{Open loop } Z_{out} \text{ of IC}}{\text{Open loop gain}}$
(Closed
loop)

Fig. 3-50. Unity-gain amplifier.

3-10.1. Design Considerations

All the design considerations for the basic IC operational amplifier described in Secs. 3-1 through 3-4 apply to the unity-gain amplifier, except as follows. Note that feedback and input resistances are completely eliminated from the circuit. With this arrangement, the output voltage equals the input voltage (or may be slightly less than the input voltage). However, the input impedance is very high, while the output impedance is very low (as shown by the equations in Fig. 3-50).

In effect, the input impedance of the op-amp is multiplied by the open-loop gain, whereas the output impedance is divided by the open-loop gain. Keep in mind that open-loop gain varies with frequency. Thus, the closed-loop impedances are frequency-dependent.

Another consideration that is sometimes overlooked in this application is the necessity of supplying the input bias current to the op-amp. In a conventional circuit, bias is supplied through the input resistances. In the circuit of Fig. 3-50, the input bias must be supplied through the signal source. This may alter input impedance.

One more problem with the circuit of Fig. 3-50 is that the total input voltage is a common-mode voltage. That is, a voltage equal to the total input voltage appears across the two inputs. Should the input signal consist of a large dc value, plus a large-signal variation, the common-mode input-voltage range may be exceeded. One solution for the problem is to capacitively couple the input signal to the noninverting input. This eliminates any dc voltage, and only the signal appears at the inputs. This will solve the common-mode problem but will require the use of a resistor from the noninverting input to ground. The resistance provides a path for the input-bias current. However, the resistance also sets the input impedance of the op-amp. If the resistance is large, the input-bias current will produce a large offset voltage (both input and output) across the resistance.

Some manufacturers recommend that resistances (of equal value) be used in the feedback loop (from output to inverting input) and at the noninverting input. This will still result in unity gain, and the circuit will perform as a voltage or source follower. However, the input and output impedances will then be set (primarily) by the resistance values rather than the op-amp characteristics, as is the case with the circuit of Fig. 3-50.

3-10.2. Design Example

Assume that the circuit of Fig. 3-50 is to provide unity gain with high input impedance and low output impedance. Also assume that the IC has the following open-loop characteristics: gain of 1000 (60 dB); output impedance, 200 Ω; input impedance, 15 kΩ. With these characteristics, the closed-loop input impedance would be

$$Z_{\text{in}} \approx 1000 \times 15{,}000 = 15 \text{ M}\Omega$$

The closed-loop output impedance would be

$$Z_{\text{out}} \approx \frac{200}{1000} = 0.2 \ \Omega$$

3-11. Unity Gain with Fast Response

One of the problems of a unity-gain amplifier is that the slew rate is very poor. That is, the response time is very slow and the power bandwidth is decreased. The reason for poor bandwidth with unity gain is that most op-amp data sheets recommend a large-value compensating capacitor for unity gain.

As an example, assume that the op-amp has the characteristics shown in Fig. 3-13 and that the desired operating frequency is 200 kHz (the unity-gain op-amp must have a full-power bandwidth up to 200 kHz). The recommended compensation for 60-dB gain is about 0.001 μF, whereas the unity-gain compensation is 1 μF. Now assume that the op-amp also has the characteristics of Fig. 3-30 and that the load is 100 Ω. With a compensation of 0.001 μF, full power can be delivered at frequencies well above 200 kHz. However, if the compensating capacitor is 1 μF, the maximum frequency at which full power can be delivered is below 4 kHz.

Several methods are used to provide fast response time (high slew rate) and a power bandwidth with unity gain. Two such methods are described next.

3-11.1. Design Examples

Using op-amp data-sheet phase compensation. The circuit of Fig. 3-51 shows a method of connecting an op-amp for unity gain, but with high slew rate (fast response and good power bandwidth). With this circuit, the phase compensation recommended on the data sheet is used, but with modification. Instead of using the unity-gain compensation, use the data-sheet phase compensation recommended for a gain of 100. Then select values of R_1 and R_3 to provide unity gain ($R_1 = R_3$). As shown by the equations, the values of R_1 and R_3 must be approximately 100 times the value of R_2. Thus, R_1 and R_3 must be fairly high values for practical design.

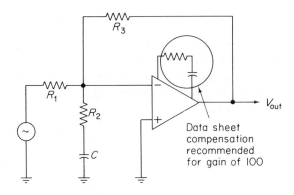

$$R_1 = R_3 = R_2 \times 100 \text{ (gain)}$$
$$R_2 = \frac{R_1}{100} \quad C = \frac{1}{6.28 \times R_2 \times F}$$

Slew rate ≈ Slew rate for gain of 100

Fig. 3-51. Unity-gain op-amp with fast response (good slew rate) using data-sheet phase compensation. Courtesy Motorola.

Assume that the circuit of Fig. 3-51 provides unity gain but with a slew rate approximately equal to that which results when a gain of 100 is used. Also assume that the input–output signal is 8 V and that the op-amp has an input-bias current of 200 nA and compensation characteristics similar to those of Fig. 3-13 but with higher gain.

The compensation capacitance recommended for a gain of 100 is 0.001 μF. With this established, select the values of R_1, R_2, and R_3. The value of R_1 (and, consequently, that of R_3) is selected so that the voltage drop with nominal input bias current is comparable (preferably 10 per cent or less) to the input signal. Ten per cent of the 8-V input is 0.8 V. Using the 0.8-V value and the 200-nA input bias, the resistance of R_1 is 4 MΩ. With R_1 at 4 MΩ, R_3 must also be 4 MΩ, and R_2 must be 40 kΩ (4 MΩ/100 = 40 kΩ).

The value of C_1 is found using the equations of Fig. 3-51 once the value of R_2 and the open-loop rolloff point are established. Assume that the op-amp has characteristics similar to those in Fig. 3-10, where the 6 dB/octave rolloff starts at about 200 kHz. With these figures, the value of C is

$$C \approx \frac{1}{6.28 \times 40\ \text{k}\Omega \times 200\ \text{kHz}} \approx 19\ \text{pF}$$

Using input phase compensation. The circuit of Fig. 3-52 shows a method of connecting an op-amp for unity gain, but with high slew rate, using input compensation. With this circuit, the phase compensation recommended on

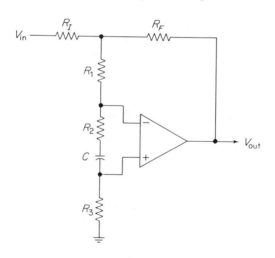

R_1, R_2, R_3, C = See text

$R_I = R_F = 0.25 \times R_1$

Fig. 3-52. Unity-gain op-amp with fast response (good slew rate) using input phase compensation. Courtesy Motorola.

the op-amp data sheet is not used. Instead, the input-phase-compensation system in Sec. 3-2.6 is used.

The first step is to compensate the op-amp by modifying the open-loop input impedance, as described in Sec. 3-2.6 and Fig. 3-11. The recommended values are $R_1 = R_3 = 5\text{ k}\Omega$, $R_2 = 2100\ \Omega$, $C = 0.0004\ \mu\text{F}$.

Next, select values of R_I and R_F to provide unity gain (both R_I and R_F must be the same value). The values of R_I and R_F are not critical, but they must be identical. Using the equations of Fig. 3-52, the values of R_I and R_F should be 0.25 times the value of R_1, or $0.25 \times 5\text{ k}\Omega = 1250\ \Omega$.

3-12. Integrated-Circuit High-Input-Impedance Amplifier

Figure 3-53 is the working schematic of an IC operational amplifier used as a high-input-impedance amplifier.

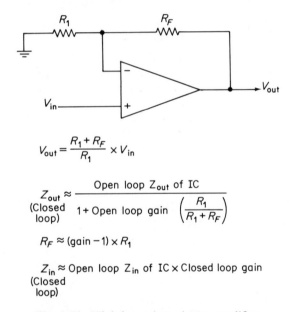

$$V_{out} = \frac{R_1 + R_F}{R_1} \times V_{in}$$

$$Z_{out} \approx \frac{\text{Open loop } Z_{out} \text{ of IC}}{1 + \text{Open loop gain}\left(\dfrac{R_1}{R_1 + R_F}\right)}$$
(Closed loop)

$$R_F \approx (\text{gain} - 1) \times R_1$$

$Z_{in} \approx \text{Open loop } Z_{in} \text{ of IC} \times \text{Closed loop gain}$
(Closed loop)

Fig. 3-53. High-input-impedance amplifier.

3-12.1. Design Considerations

All the design considerations for the basic IC operational amplifier described in Secs. 3-1 through 3-4 apply to the high-input-impedance amplifier, except as follows. The high-input-impedance feature of the unity-gain amplifier (Sec. 3-10) is combined with gain, as shown by the equations of

Fig. 3-53. Note that the circuit of Fig. 3-53 is similar to that of the basic op-amp, except that there is no input offset compensating resistance (in series with the noninverting input) for the high-input-impedance circuit. This results in a tradeoff of higher input impedance, with some increase in output offset voltage.

In the basic op-amp, an offset compensating resistance is used to nullify the input offset voltage of the op-amp. This (theoretically) results in no offset at the output. The output of the basic op-amp is at zero volts in spite of the tremendous gain.

In the unity-gain amplifier (Sec. 3-10), there is no offset compensating resistance, but since there is no gain, the output is at the same offset as the input. In a typical IC op-amp, the input offset is less than 10 mV. This figure should not be critical for the output of a typical unity-gain-amplifier application.

In the circuit of Fig. 3-53, the offset compensation resistance is omitted. The output is offset by an amount equal to the input offset voltage of the op-amp multiplied by the closed-loop gain. However, since the circuit of Fig. 3-53 is to be used for modest gains, modest output offset results.

3-12.2. Design Example

Assume that the circuit of Fig. 3-53 is to provide a gain of 10, with high input impedance and low output impedance. Also assume that the IC has the following open-loop characteristics: output impedance, 200 Ω; input impedance, 15 kΩ; gain of 1000 (60 dB).

The value of R_1 is chosen on the basis of input bias current and voltage drop, in the usual manner. Assume an arbitrary value of 100 Ω for R_1.

With 100 Ω for R_1 and a gain of 10, the value of R_F would be

$$R_F \approx (10 - 1) \times 100 = 900 \ \Omega$$

With a gain of 10, the closed-loop input impedance would be

$$Z_{in} \approx 15 \ k\Omega \times 10 = 150 \ k\Omega$$

The closed-loop output impedance would be

$$Z_{out} \approx \frac{200}{1 + 1000 \times [100/(100 + 900)]} \approx 2 \ \Omega$$

3-13. Integrated-Circuit Difference Amplifier

Figure 3-54 is the working schematic of an IC operational amplifier used as a difference amplifier and/or subtractor. One signal voltage can be subtracted from another through simultaneous application of signals to both inputs.

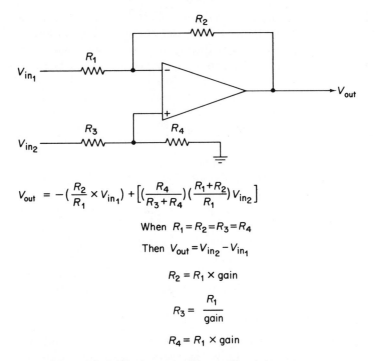

$$V_{out} = -\left(\frac{R_2}{R_1} \times V_{in_1}\right) + \left[\left(\frac{R_4}{R_3+R_4}\right)\left(\frac{R_1+R_2}{R_1}\right)V_{in_2}\right]$$

When $R_1 = R_2 = R_3 = R_4$

Then $V_{out} = V_{in_2} - V_{in_1}$

$R_2 = R_1 \times$ gain

$$R_3 = \frac{R_1}{\text{gain}}$$

$R_4 = R_1 \times$ gain

Fig. 3-54. Difference amplifier and/or subtractor.

3-13.1. Design Considerations

All the design considerations for the basic IC operational amplifier described in Secs. 3-1 through 3-4 apply to the difference and/or subtractor amplifier, except as follows. If the values of all resistors are the same, the output will be equal to the voltage at the positive (noninverting) input, less the voltage at the negative (inverting) input. The output also represents the difference between the two input voltages. Therefore, the circuit can be used as a subtractor or difference amplifier, whichever is required.

If the values of all resistors are not the same, the output will be the algebraic sum of the gains for the two input voltages, as shown by the equations of Fig. 3-54. Generally, it is simpler to use the same values for all resistors.

3-13.2. Design Example

Assume that the circuit of Fig. 3-54 is to be used as a difference amplifier for two input voltages. Each of the voltage inputs varies from 2 to 50 mV rms. The output must vary between 20 and 500 mV rms.

This example requires a gain of 10. Therefore, all resistor values cannot be the same. To provide a gain of 10 for the negative inputs, R_2 must be 10 times R_1.

The value of R_1 should be selected so that the voltage drop (with nominal input bias current) is comparable with the minimum input signal, in the usual manner. Assume an arbitrary value of 100 Ω for R_1. With a 100-Ω value for R_1, the value of R_2 must be 1000 Ω, to provide a negative input gain of 10. With a 100-Ω value for R_1, the values of R_3 and R_4 should be 10 and 1000 Ω, respectively.

3-14. Integrated-Circuit Voltage-to-Current Converter

Figure 3-55 is the working schematic of an IC operational amplifier used as a voltage-to-current converter (transadmittance amplifier). This circuit can be used to supply a current to a load that is proportional to the voltage applied to the input of the amplifier. The current supplied to the load is relatively independent of the load characteristics. This circuit is essentially a *current feedback* amplifier.

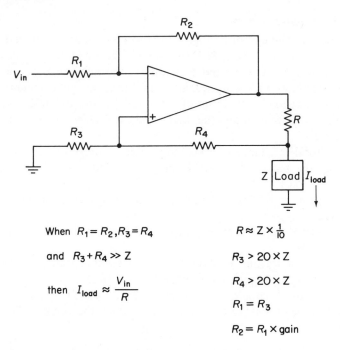

When $R_1 = R_2, R_3 = R_4$

and $R_3 + R_4 \gg Z$

then $I_{load} \approx \dfrac{V_{in}}{R}$

$R \approx Z \times \frac{1}{10}$

$R_3 > 20 \times Z$

$R_4 > 20 \times Z$

$R_1 = R_3$

$R_2 = R_1 \times gain$

Fig. 3-55. Voltage-to-current converter.

3-14.1. Design Considerations

All the design considerations for the basic IC operational amplifier described in Secs. 3-1 through 3-4 apply to the voltage-to-current converter, except as follows. Current sampling resistor R is used to provide the feedback to the positive input. When R_1, R_2, R_3, and R_4 are all the same value, the feedback maintains the voltage across R at the same value as the input voltage. If a constant input voltage is applied to the amplifier, the voltage across R also remains constant, regardless of the load (within very close tolerance). If the voltage across R remains constant, the current through R must also remain constant. With R_3 and R_4 normally much higher than the load impedance, the current through the load must remain nearly constant, regardless of a change in impedance.

The values of R_1 through R_4 should normally be the same. The current sampling resistor R is then selected for the desired load current. The value of R should be selected to limit the output power $[I^2 \times (R + \text{load})]$ to a value within the capability of the IC. For example, if the IC is rated at 600 mW total dissipation, with 100-mW dissipation for the basic IC, the total output power must be limited to 500 mW. As a rule of thumb, the value of R should be approximately $\frac{1}{10}$ of the load (Z).

3-14.2. Design Example

Assume that the circuit of Fig. 3-55 is to be used as a voltage-to-current converter. The output load is 45 Ω (nominal). The maximum power output of the IC is 500 mW. It is desired to maintain the maximum output current, regardless of variation in output load, with a constant input voltage of 5 mV.

With a 45-Ω load, the value of R should be approximately 4.5 Ω. The combined resistance of R and the load is then 49.5 Ω (rounded off to 50 Ω for convenience).

With a total resistance of 50 Ω, and a maximum power output of 0.5 W for the IC, the maximum possible output current is 0.1 A ($I = \sqrt{P/R} = \sqrt{0.5/50} = \sqrt{0.01} = 0.1$ A).

With a value of 4.5 Ω for R, and 0.1 A through R, the drop across R must be 450 mV.

With 450 mV required at the output and 5 mV at the input, the amplifier gain must be 90.

The values of R_3 and R_4 should be at least 950 Ω each (as shown by the equations of Fig. 3-55), with a nominal load of 50 Ω.

The value of R_1 should be the same as R_3, or 950 Ω.

With a value of 950 Ω for R_1, and a gain of 90, the value of R_2 should be 85.5 kΩ.

3-15. Integrated-Circuit Voltage-to-Voltage Amplifier

Figure 3-56 is the working schematic of an IC operational amplifier used as a voltage-to-voltage converter (voltage-gain amplifier). This circuit is similar

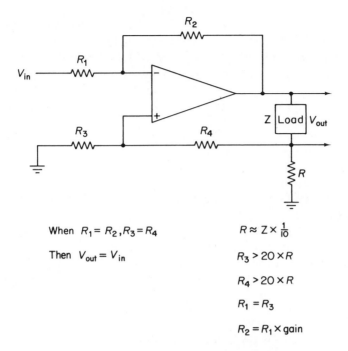

When $R_1 = R_2, R_3 = R_4$

Then $V_{out} = V_{in}$

$R \approx Z \times \frac{1}{10}$

$R_3 > 20 \times R$

$R_4 > 20 \times R$

$R_1 = R_3$

$R_2 = R_1 \times gain$

Fig. 3-56. Voltage-to-voltage converter or amplifier.

to the voltage-to-current converter (Sec. 3-14) except that the load and the current-sensing resistor are transposed. The voltage across the load is relatively independent of the load characteristics.

3-15.1. Design Considerations

All the design considerations for the basic IC operational amplifier described in Secs. 3-1 through 3-4 apply to the voltage-to-voltage converter, except as follows. The values of R_1 through R_4 should normally be the same. The current sampling resistor R is then selected for the desired load current.

The value of R should be selected to limit the output power [$I^2 \times (R +$ load)] to a value within the capability of the IC. For example, if the IC is rated at 600-mW total dissipation, with 100-mW dissipation for the basic IC, the total output power must be limited to 500 mW. As a rule of thumb, the value of R should be approximately $\frac{1}{10}$ of the load (Z).

3-15.2. Design Example

Assume that the circuit of Fig. 3-56 is to be used as a voltage-to-voltage converter. The output load is 100 Ω (nominal). The maximum power output of the IC is 500 mW. It is desired to maintain the maximum output voltage across the load, regardless of variation in load (within the current capabilities of the IC), with a constant input voltage of 50 mV.

With a 100-Ω load, the value of R should be approximately 10 Ω. The combined resistance of R and the nominal load is then 110 Ω.

With a total resistance of 110 Ω, and a maximum power output of 0.5 W for the IC, the maximum possible output current is 0.07 A ($I = \sqrt{P/R} \times \sqrt{0.5/110} \approx \sqrt{0.0045} \approx 0.07$ A).

With a nominal value of 100 Ω for the load, and 0.07 A through the load, the maximum drop across the load is 7 V. This value may be used, provided that the IC is capable of a 7-V output. If not, use a voltage output that is within the IC capabilities.

With 7 V required at the output, and 50 mV at the input, the amplifier gain must be 140.

The values of R_3 and R_4 should be at least 200 Ω each (as shown by the equation of Fig. 3-56), with a value of 10 Ω for R.

The value of R_1 should be the same as R_3, or 200 Ω.

With a value of 200 Ω for R_1, and a gain of 140, the value of R_2 should be 28 kΩ.

3-16. Integrated-Circuit Low-Frequency Sine-Wave Generator

Figure 3-57 is the working schematic of an IC operational amplifier used as a low-frequency sine-wave generator. This circuit is a parallel-T oscillator. Feedback to the negative input becomes positive at the frequency indicated in the equation, while positive feedback is applied at all times. The amount of positive feedback (set by the ratio of R_1 to R_2) is sufficient to cause the IC amplifier to oscillate. In combination with the feedback to the negative input, feedback to the positive input can be used to stabilize the amplitude of oscillations.

3-16.1. Design Considerations

All the design considerations for the basic IC operational amplifier described in Secs. 3-1 through 3-4 apply to the sine-wave generator, except as follows. The value of R_1 should be approximately 10 times the value of R_2. The ratio of R_1 and R_2, as set by the adjustment of R_2, controls the amount of positive feedback. Thus, the setting of R_2 determines the stability of oscillation.

The amplitude of oscillation is determined by the peak-to-peak output capability of the IC and the values of Zener diodes CR_1 and CR_2. As shown by the equations, the Zener voltage should be approximately 1.5 times the desired peak-to-peak output voltage. The nonlinear resistance of the back-

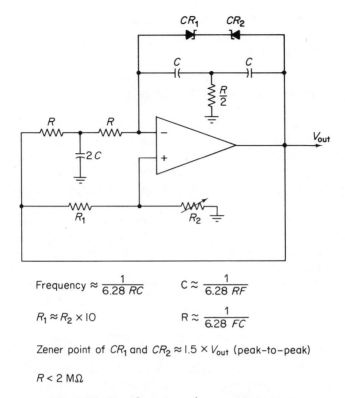

$$\text{Frequency} \approx \frac{1}{6.28\ RC} \qquad C \approx \frac{1}{6.28\ RF}$$

$$R_1 \approx R_2 \times 10 \qquad R \approx \frac{1}{6.28\ FC}$$

Zener point of CR_1 and $CR_2 \approx 1.5 \times V_{out}$ (peak-to-peak)

$R < 2\ M\Omega$

Fig. 3-57. Low-frequency sine-wave generator.

to-back Zener diodes is used to limit the output amplitude and maintain good linearity.

The frequency of oscillation is determined by the values of C and R. The upper-frequency limit is approximately equal to the bandwidth of the basic IC. That is, if the open-loop gain drops 3 dB at 100 kHz, the circuit should provide full-voltage output up to 100 kHz.

3-16.2. Design Example

Assume that the circuit of Fig. 3-57 is to provide 10-V sine-wave signals at 1 Hz. The value of R_2 should be chosen on the basis of bias current, in the usual manner. Assume a value of 10 kΩ for R_2.

With a value of 10 kΩ for R_2, the value of R_1 should be 100 kΩ.

With a required 10-V peak-to-peak output, the values (Zener voltage) of CR_1 and CR should be 15 V. It is assumed that the basic IC is capable of 10-V peak-to-peak output.

The values of R and C are related to the desired frequency of 1 Hz. Any combination of R and C could be used, provided that the combination worked out to a frequency of 1 Hz. However, for practical design, the value of R should not exceed about 2 MΩ (for a typical IC). Using a value of 1.6 MΩ for R, the value of C is 0.1 μF.

3-17. Operational-Amplifier Test Procedures

The following sections describe test procedures for operational amplifiers. The basic procedures are essentially the same as for audio amplifiers (Chapter 2). Therefore, the reader should make reference to Secs. 2-11 and 2-12. The following sections describe the test *differences* required for operational amplifiers, as well as special test procedures unique to IC operational amplifiers.

As a minimum, the following tests should be made on the basic IC, with power sources connected as described in Sec. 3-1.2, operating in an open-loop circuit. This will confirm (or deny) the IC characteristics found on the data sheet. The procedures can also be used to establish a set of characteristics for an IC where the data sheet is missing or inadequate.

3-17.1. Frequency Response

Frequency response for an operational amplifier is tested in essentially the same way as for an audio amplifier (Sec. 2-12) except that the frequency range is extended beyond the audio range. Generally, an oscilloscope is a better instrument for response measurement at higher frequencies. However, an electronic voltmeter can be used. Both open- and closed-loop frequency response should be measured with the same load.

3-17.2. Voltage Gain

Voltage-gain measurement for an operational amplifier is essentially the same as for an audio amplifier (Sec. 2-12.2). Keep in mind that the basic IC has a maximum input and output voltage limit neither of which can be exceeded without possible damage to the IC and/or clipping of the waveform. In general, the maximum rated input should be applied and the actual output measured. Check the output for clipping at the maximum level. If clipping occurs, decrease the input until clipping just stops, and note the input voltage. Record these values as a basis for design.

Note the frequency at which the open-loop voltage gain drops 3 dB from the low-frequency value. This is the open-loop bandwidth. Keep in mind that the open-loop voltage gain and bandwidth are characteristics of the basic IC. Closed-loop gain is (or should be) dependent upon the ratio of feedback and input resistances, while closed-loop bandwidth is essentially dependent upon phase compensation values.

Closed-loop characteristics are generally lower than open-loop characteristics (voltage gain is lower, frequency response is narrower, etc.). However, closed-loop characteristics are modifications of open-loop characteristics.

3-17.3. Power Output, Gain, and Bandwidth

Most IC operational amplifiers are not designed as power amplifiers. However, their power output, gain, and bandwidth can be measured in the same way as audio amplifiers (Secs. 2-12.3 and 2-12.4).

Keep in mind that an IC has a power dissipation of its own which must be subtracted from the *total device dissipation* to find the available power output.

3-17.4. Load Sensitivity

Since an IC operational amplifier is generally not used as a power amplifier, load sensitivity is not critical. However, if it should be necessary to measure the load sensitivity, use the procedure of Sec. 2-12.5.

3-17.5. Input and Output Impedance

Dynamic input and output impedance of an IC operational amplifier (Figs. 3-31 and 3-32) can be found using the procedures of Secs. 2-12.6 and 2-12.7. Keep in mind that closed-loop impedances will differ from open-loop impedances.

3-17.6. Distortion

Distortion measurements for IC operational amplifiers are the same as for audio amplifiers (Secs. 2-12.9 through 2-12.12). However, distortion requirements for operational amplifiers are usually not as critical.

3-17.7. Background Noise

Background-noise measurements for IC operational amplifiers are the same as for audio amplifiers (Sec. 2-12.13). Generally, background noise should be measured under open-loop conditions. Some data sheets specify that both input and output voltages be measured. When input voltage is to be measured, a fixed resistance (usually 50 Ω) is connected between the input terminals.

3-17.8. Feedback Measurement

Since operational amplifier characteristics are based on the use of feedback signals, it is often convenient to measure feedback voltage at a given frequency with given operating conditions.

The basic feedback-measurement connections are shown in Fig. 3-58. While it is possible to measure the feedback voltage as shown in Fig. 3-58a,

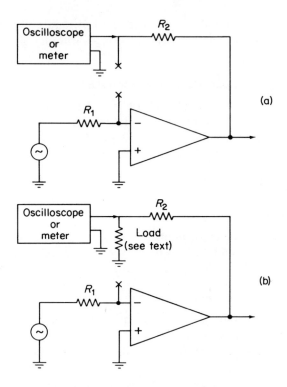

Fig. 3-58. Feedback measurement.

a more accurate measurement will be made when the feedback lead is terminated in the normal operating impedance.

If an input resistance is used in the normal circuit, and this resistance is considerably lower than the IC input impedance, use the resistance value.

If in doubt, measure the input impedance of the circuit (Sec. 3-17.5), then terminate the feedback lead in that value to measure feedback voltage.

3-17.9. Input Bias Current

Input bias current (Fig. 3-35) can be measured using the circuit of Fig. 3-59. Any resistance value for R_1 and R_2 can be used, provided that the value produces a measurable voltage drop. A value of 1 kΩ is realistic for both R_1 and R_2.

Once the voltage drop is found, the input bias current can be calculated. For example, if the voltage drop is 3 mV across 1 kΩ, the input bias current is 3 μA.

In theory, the input bias current should be the same for both inputs. In practice, the bias currents should be *almost equal*. Any great difference in

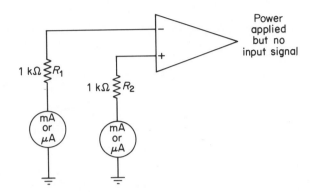

Fig. 3-59. Input-bias current measurement.

input bias is the result of unbalance in the input differential amplifier of the IC and can seriously affect design.

3-17.10. Input-Offset Voltage and Current

Input-offset voltage and current (Fig. 3-36) can be measured using the circuit of Fig. 3-60. As shown, the output is alternately measured with R_3 shorted and with R_3 in the circuit. The two output voltages are recorded as E_1 (S_1 closed, R_3 shorted), and E_2 (S_1 open, R_3 in the circuit).

With the two output voltages recorded, the input-offset voltage and input-offset current can be calculated using the equations of Fig. 3-60. For example, assume that: $R_1 = 51\ \Omega$, $R_2 = 5.1\ \mathrm{k\Omega}$, $R_3 = 100\ \mathrm{k\Omega}$, $E_1 = 83\ \mathrm{mV}$, and $E_2 = 363\ \mathrm{mV}$ (all typical values).

$$\text{input-offset voltage} = \frac{83\ \mathrm{mV}}{100} = 0.83\ \mathrm{mV}$$

$$\text{input-offset current} = \frac{280\ \mathrm{mV}}{100\ \mathrm{k\Omega}(1 + 100)} = 0.0277\ \mu\mathrm{A}$$

3-17.11. Common-Mode Rejection

Common-mode rejection (Fig. 3-34) can be measured using the circuit of Fig. 3-61. First find the open-loop gain under identical conditions of frequency, input, *et cetera*, as described in Secs. 3-17.1 and 3-17.2.

Then connect the IC in the common-mode circuit of Fig. 3-61. Increase the common-mode voltage (V_{in}) until a measurable output V_{out} is obtained. Be careful not to exceed the maximum specified input common-mode voltage swing. If no such value is specified, do not exceed the normal input voltage of the IC.

To simplify calculation, increase the input voltage until the output is 1 mV. With an open-loop gain of 100, this will provide a differential input

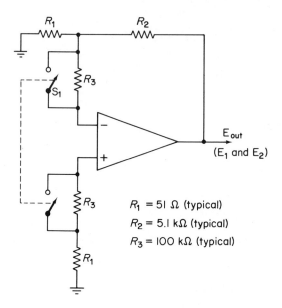

$E_1 = V_{out}$ with S_1 closed (R_3 shorted)

$E_2 = V_{out}$ with S_1 open (R_3 in circuit)

$$\text{Input offset voltage} = \frac{E_1}{\left(\dfrac{R_2}{R_1}\right)}.$$

$$\text{Input offset current} = \frac{(E_2 - E_1)}{R_3\left(1 + \dfrac{R_2}{R_1}\right)}$$

Fig. 3-60. Input-offset voltage and current measurements.

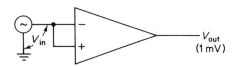

$$\frac{V_{out}\ (1\,\text{mV})}{\text{open loop gain}} = \begin{array}{l}\text{equivalent} \\ \text{differential} \\ \text{input signal}\end{array}$$

$$\begin{array}{l}\text{Common} \\ \text{mode} \\ \text{rejection}\end{array} = \frac{V_{in}}{\begin{array}{c}\text{equivalent differential} \\ \text{input signal}\end{array}}$$

Fig. 3-61. Common-mode rejection measurement.

signal of 0.00001 V. Then measure the input voltage. Move the input voltage decimal point over five places to find the CMR.

3-17.12. Slew Rate

An easy way to observe and measure the slew rate of an operational amplifier is to measure the slope of the output waveform of a square-wave input signal, as shown in Fig. 3-62. The input square wave must have a rise time that exceeds the slew-rate capability of the amplifier. Therefore, the output will not appear as a square wave, but as an integrated wave.

In the example shown, the output voltage rises (and falls) about 40 V in 1 μs.

Note that slew rate is usually measured in the closed-loop condition. Also note that the slew rate increases with higher gain.

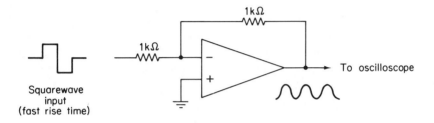

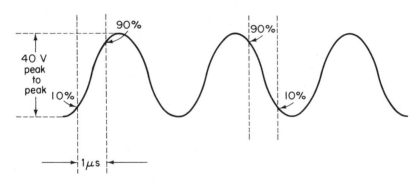

Example shows a slew rate of approximately 40 (40 Volts /μs) at unity gain.

Fig. 3-62. Slew-rate measurement.

3-17.13. Power-Supply Sensitivity

Power-supply sensitivity (Fig. 3-40) can be measured using the circuit of Fig. 3-60 (the same test circuit as for input-offset voltage, Sec. 3-17.10). The procedure is the same as for measurement of input-offset voltage, except that one supply voltage is changed (in 1-V steps) while the other supply voltage is held constant. The amount of change in input-offset voltage for a 1-V change in one power supply is the *power-supply sensitivity* (or *input-offset voltage sensitivity*).

For example, assume that the normal positive and negative supplies are 10 V, and the input offset voltage is 7 mV. With the positive supply held constant, the negative supply is reduced to 9 V. Under these conditions, assume that the input-offset voltage is 5 mV. This means that the negative power-supply sensitivity is 2 mV/V. With the negative power supply held constant at 10 V, the positive supply is reduced to 9 V. Now assume that the input-offset voltage drops to 4 mV. This means that the positive power-supply sensitivity is 3 mV/V.

The test can be repeated over a wide range of power supply voltages (in (1-V steps), if the IC is to be operated under conditions where the power supply may vary by a large amount.

3-17.14. Phase Shift

The phase shift between input and output of an operational amplifier is far more critical than with an audio amplifier. This is because an operational amplifier uses the principle of feeding back output signals to the input. All the phase-compensation schemes are based on this principle. Under ideal open-loop conditions, the output should be 180° out of phase with the negative input, and in phase with the positive input.

The following sections describe two procedures for the measurement of phase shift between input and output of an operational amplifier. The same procedures can be used for any amplifier, provided that the signals are of a frequency that can be measured on an oscilloscope.

The oscilloscope is the ideal tool for phase measurement. The most convenient method requires a dual-trace oscilloscope or an electronic switching unit to produce a dual trace. If neither of these is available, it is still possible to provide accurate phase measurements up to about 100 kHz using the single-trace (or X-Y) method.

3-17.14.1. Dual-Trace Phase Measurement

The dual-trace method of phase measurement provides a high degree of accuracy at all frequencies but is especially useful at frequencies above 100 kHz, where X-Y phase measurements may prove inaccurate owing to the inherent internal phase shift of the oscilloscope.

The dual-trace method also has the advantage of measuring phase differences between signals of different amplitudes and waveshape, as is usually the case with input and output signals of an amplifier. The dual-trace method can be applied directly to those oscilloscopes having a built-in dual-trace feature or to a conventional single-trace oscilloscope using an electronic switch or "chopper" unit. Either way, the procedure is essentially one of displaying both input and output signals on the oscilloscope screen simultaneously, measuring the distance (in screen scale divisions) between related points on the two traces, then converting this distance into phase.

The test connections for dual-trace phase measurement are shown in Fig. 3-63. For the most accurate results, the cables connecting input and output signals should be of the same length and characteristics. At higher frequencies, a difference in cable length or characteristics could introduce a phase shift.

The oscilloscope controls are adjusted until one cycle of the input signal occupies exactly nine divisions (9 cm horizontally) of the screen. Then the phase factor of the input signal is found. For example, if 9 cm represents one complete cycle, or 360°, 1 cm represents 40° (360° ÷ 9 = 40).

With the phase factor established, the horizontal distance between corresponding points on the two waveforms (input and output signals) is measured. The measured distance is then multiplied by the phase factor of 40°/cm to find the exact amount of phase difference. For example, assume a horizontal difference of 0.6 cm with a phase factor of 40° as shown in Fig. 3-63. Multiply the horizontal difference by the phase factor to find the phase difference (0.6 × 40 = 24° phase shift between input and output signals).

If the oscilloscope is provided with a sweep magnification control in which the sweep rate is increased by some fixed amount (5×, 10×, etc.) and only a portion of one cycle can be displayed, more accurate phase measurements can be made. In this case, the phase factor and the approximate phase difference is found as described. Without changing any other controls, the sweep rate is increased (by the sweep magnification control or the sweep rate control), and a new horizontal distance measurement is made, as shown in Fig. 3-63d.

For example, if the sweep rate were increased 10 times, the adjusted phase factor would be 40° ÷ 10 = 4°/cm. Figure 3-63d shows the same signal as used in Fig. 3-63c, but with the sweep rate set to 10×. With a horizontal difference of 6 cm, the phase difference would be 6 × 4° = 24°.

3-17.14.2. Single-Trace (X-Y) Phase Measurement

The single-trace (or X-Y) phase-measurement method can be used to measure the phase difference between input and output of an amplifier, at frequencies up to about 100 kHz. Above this frequency, the inherent phase shift (or difference) between the horizontal and vertical systems of the oscilloscope makes accurate phase measurements difficult.

In the X-Y method, one of the signals (usually the input) provides horizontal deflection (X), and the other signal provides the vertical deflection (Y). The phase angle between the two signals can be determined from the resulting pattern.

The test connections for single-trace phase measurement are shown in Fig. 3-64. Figure 3-64a shows the test connection necessary to find the inherent phase shift (if any) between the horizontal and vertical deflection systems of the oscilloscope. Inherent phase shift (if any) should be checked

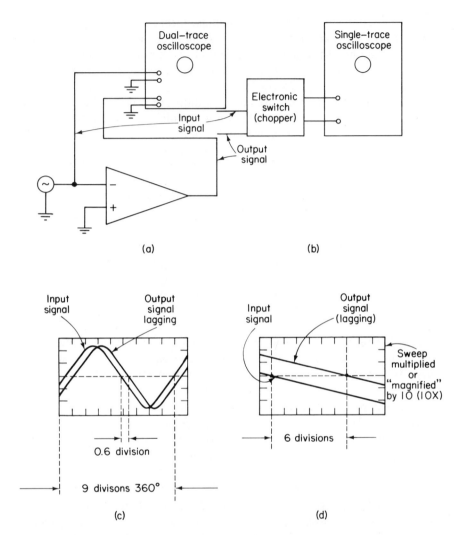

Fig. 3-63. Dual-trace phase measurement.

and recorded. If there is excessive phase shift (in relation to the signals to be measured), the oscilloscope should not be used. A possible exception exists when the signals to be measured are of sufficient amplitude to be applied directly to the oscilloscope deflection plates, bypassing the horizontal and vertical amplifiers.

The oscilloscope controls are adjusted until the pattern is centered on the screen as shown in Fig. 3-64c. With the amplifier output connected to the vertical input, it is usually necessary to reduce vertical channel gain (to compensate for the increased gain through the amplifier). With the display

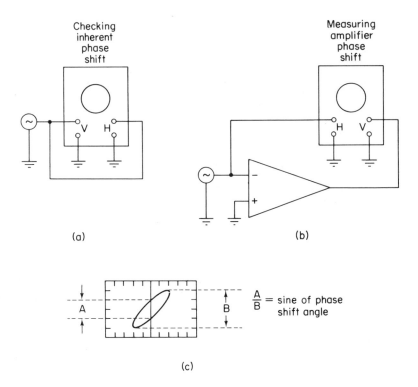

Fig. 3-64. Single-trace (*X-Y*) phase measurement.

centered in relation to the vertical line, distances *A* and *B* are measured, as shown in Fig. 3-64c. Distance *A* is the vertical measurement between two points where the traces cross the vertical center line. Distance *B* is the maximum vertical height of the display. Divide *A* by *B* to find the *sine of the phase angle* between the two signals. This same procedure can be used to find inherent phase shift (Fig. 3-64a) or phase angle (Fig. 3-64b).

If the display appears as a diagonal straight line, the two signals are either in phase [tilted from the upper right to the lower left (positive slope)] or 180° out of phase [tilted from the upper left to lower right (negative slope)]. If the display is a circle, the signals are 90° out of phase. Figure 3-65 shows the displays produced between 0° and 360°. Notice that above a phase shift of 180°, the resultant display will be the same as at some lower frequency. Therefore, it may be difficult to tell whether the signal is leading or lagging. One way to find correct phase polarity (leading or lagging) is to introduce a small, known phase shift into one of the signals. The proper angle may then be found by noting the direction in which the pattern changes.

Once the oscilloscope's inherent phase shift has been established (Fig. 3-64a), and the amplifier phase shift measured (Fig. 3-64b), subtract the

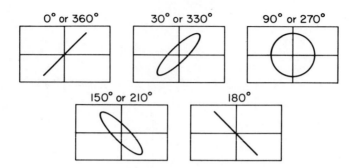

Fig. 3-65. Approximate phase of typical X-Y displays.

inherent phase difference from the phase angle to find the true phase difference. For example, assume an inherent phase difference of 2°, a display as shown in Fig. 3-64c, where A is 2 cm and B is 4 cm. Sine of phase angle = $A \div B$, or $2 \div 4$, or 0.5. From a table of sines, 0.5 = 30° phase angle. To adjust for the phase difference between X and Y oscilloscope channels, subtract the inherent phase factor (30° − 2°) = 28°, true phase difference.

4

RADIO-FREQUENCY CIRCUITS

4-1. Resonant Circuits

RF design is based on the use of resonant circuits (or "tank" circuits), consisting of a capacitor and a coil (inductance) connected in series or parallel as shown Fig. 4-1. At the resonant frequency, the inductive and capacitive reactances are equal, and the circuit acts as a high impedance (if it is a parallel circuit) or a low impedance (if it is a series circuit). In either case, any combination of capacitance and inductance will have some resonant frequency.

Either (or both) the capacitance or inductance can be variable to permit tuning of the resonant circuit over a given frequency range. When the inductance is variable, tuning is usually done by means of a metal slug inside the coil. The metal slug is screwdriver-adjusted to change the inductance (and thus the inductive reactance) as required.

Typical RF circuits used in receivers (AM, FM, communications, etc.) often include two resonant circuits in the form of a transformer (RF or IF transformer, etc.) Either the capacitance or inductance can be variable. Since such transformers are available commercially, their design will not be discussed here. However, measurement of RF (and IF) transformer resonant values, as they affect design, is discussed at the end of the chapter.

In the case of RF transmitters, it is sometimes necessary to design the *coil portion* of the resonant circuit. This is because coils of a given inductance and physical size may not be available from commercial sources. Therefore, design data is included for basic RF coils.

Resonance and impedance

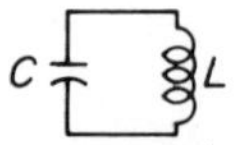

Parallel *
(Infinite impedance)

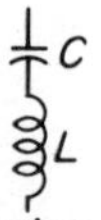

Series
(zero impedance)

$$F = \frac{1}{6.28 \sqrt{LC}}$$

$$F(kHz) = \frac{10^6}{6.28 \sqrt{L(\mu H) \times C(pF)}}$$

$$F(kHz) = \frac{159}{\sqrt{L(\mu H) \times C(\mu F)}}$$

$$L(\mu H) = \frac{2.54 \times 10^4}{F(kHz)^2 \times C(\mu F)}$$

$$F(MHz) = \frac{0.159}{\sqrt{L(\mu H) \times C(\mu F)}}$$

$$C(\mu F) = \frac{2.54 \times 10^4}{F(kHz)^2 \times L(\mu H)}$$

*Approximate; accurate when circuit Q is 10 or higher

Inductive reactance

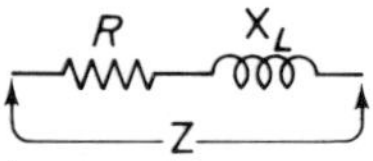

$$Z = \sqrt{R^2 + X_L^2} \qquad Q = \frac{X_L}{R}$$

$$L = \frac{X_L}{6.28 F}$$

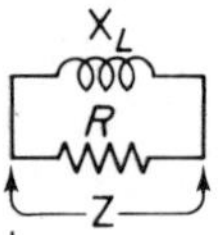

Series

$$Z = \frac{R X_L}{\sqrt{R^2 + X_L^2}} \qquad Q = \frac{R}{X_L}$$

$$F = \frac{X_L}{6.28 L}$$

Parallel

$$X_L = 6.28 \times F(Hz) \times L(H)$$
$$X_L = 6.28 \times F(kHz) \times L(mH)$$
$$X_L = 6.28 \times F(MHz) \times L(\mu H)$$

Capacitive reactance

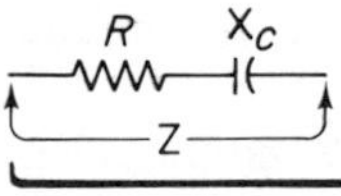

$$Z = \sqrt{R^2 + X_C^2} \qquad Q = \frac{X_C}{R}$$

$$F = \frac{1}{6.28 C X_C}$$

Series

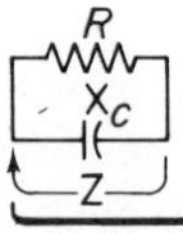

$$Z = \frac{R X_C}{\sqrt{R^2 + X_C^2}} \qquad Q = \frac{R}{X_C}$$

$$C = \frac{1}{6.28 F X_C}$$

Parallel

$$X_C = \frac{1}{6.28 \times F(Hz) \times C(F)}$$
$$X_C = \frac{159}{F(kHz) \times C(\mu F)}$$

Fig. 4-1. Resonant-circuit equations.

4-1.1. Design Considerations for Resonant Circuits

The two basic design considerations for RF resonant circuits are resonant frequency and the Q (or quality) factor.

4-1.1.1. Resonant Frequency

Figure 4-1 contains equations which show the relationships among capacitance, inductance, reactance, and frequency, as they relate to resonant circuits. Note that there are two sets of equations. One set of equations includes reactance (inductive and capacitive). The other set omits reactance. The reason for two sets of equations is that some design approaches require the reactance to be calculated for resonant networks. Solid-state RF transmitter circuits are a classic example of this.

4-1.1.2. Q or Selectivity

A resonant circuit has a Q, or quality, factor. The circuit Q is dependent upon the ratio of reactance to resistance. If a resonant circuit had pure reactance, the Q would be high (actually infinite). However, this is not practical. For example, any coil will have some dc resistance, as will the leads of a capacitor. Also, as frequency increases, the ac resistance presented by the leads will increase as a result of the skin effect. The sum total of these resistances is usually lumped together and is considered as a resistor in series or parallel with the circuit. The total resistance is usually termed the *effective resistance* and is not to be confused with the reactance.

The resonant circuit Q is dependent upon the individual Q factors of inductance and capacitance used in the circuit. For example, if both the inductance and capacitance have a high Q, the circuit will have a high Q, provided that a minimum of resistance is produced when the inductance and capacitance are connected to form a resonant circuit.

From a practical design standpoint, a resonant circuit that has a high Q will produce a sharp resonance curve (narrow bandwidth), whereas a low Q will produce a broad resonance curve (wide bandwidth). For example, a high-Q resonant circuit will provide good harmonic rejection and efficiency, in comparison with a low-Q circuit, all other factors being equal. The *selectivity* of a resonant circuit is therefore related directly to Q. A very high Q (or high selectivity) is not always desired. Sometimes it is necessary to add resistance to a resonant circuit to broaden the response (increase the bandwidth, decrease the selectivity).

Usually, resonant-circuit Q is measured at the points on either side of the resonant frequency where the signal amplitude is down 0.707 of the peak resonant value, as shown in Fig. 4-2. (Resonant-circuit Q measurements are included at the end of this chapter.)

Note that Q must be increased for increases in resonant frequency if the same bandwidth is to be maintained. For example, if the resonant frequency

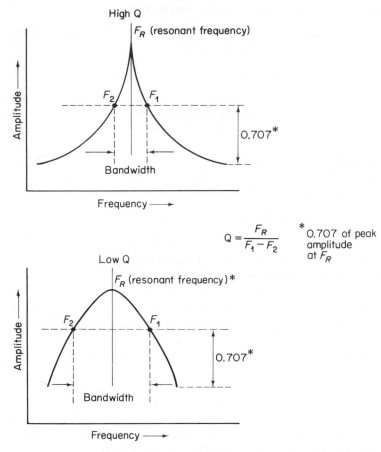

High Q

F_R (resonant frequency)

$$Q = \frac{F_R}{F_1 - F_2}$$

*0.707 of peak
amplitude
at F_R

Fig. 4-2. Relationship of bandpass characteristics to Q of resonant circuit.

is 10 MHz with a bandwidth of 2 MHz, the required circuit Q would be 5. If the resonant frequency was increased to 50 MHz, with the same 2-MHz bandwidth, the required Q would be 25. Also, note that Q must be decreased for increases in bandwidth if the same resonant frequency is to be maintained. For example, if the resonant frequency is 30 kHz, with a bandwidth of 2 kHz, the required circuit Q would be 15. If the bandwidth were increased to 10 kHz, with the same 30-kHz resonant frequency, the required Q would be 3.

4-1.2. Design Examples

Assume that it is desired to find the resonant frequency of a 0.002-μF capacitor and a 0.02-mH inductance. Using the equations of Fig. 4-1, first convert the 0.02 mH to 20 μH. Then

$$F = \frac{0.159}{\sqrt{20\,(\mu H) \times 0.002\,(\mu F)}}$$

$$= \frac{0.159}{\sqrt{0.4}}$$

$$= \frac{0.159}{0.2}$$

$$= 0.795\,\text{MHz} \quad \text{or} \quad 795\,\text{kHz}$$

Assume that it is desired to design a circuit that will be resonant at 400 kHz, with an inductance of 10 μH. What value of capacitor is necessary? Using the equations of Fig. 4-1,

$$C = \frac{2.54 \times 10^4}{400^2 \times 10}$$

$$= 0.0158;\; 0.016\,\mu F \text{ nearest standard value}$$

Assume that it is desired to design a circuit that will resonate at 2.65 MHz, with a capacitance of 360 pF. What value of inductance is necessary? Using the equations of Fig. 4-1,

$$L = \frac{2.54 \times 10^4}{2650^2 \times (360 \times 10^{-6})}$$

$$= 10\,\mu H$$

Assume that an RF power amplifier network must operate at 40 MHz, with a bandwidth of 8 MHz. What circuit Q would be required? Using the equations of Fig. 4-2,

$$Q = \frac{40}{8} = 5$$

4-1.3. Design Considerations for Coils

The equations necessary to calculate the self-inductance of a single-layer, air-core coil are given in Fig. 4-3. Note that maximum inductance is obtained when the ratio of coil radius to coil length is 1.25, that is, when the length is 0.8 of the radius. RF coils wound for this ratio are the most efficient (maximum inductance for minimum physical size).

4-1.4. Design Example

Assume that it is desired to design a coil with 0.5-μH inductance on a 0.25-in radius (air-core, single-layer). Using the equations of Fig. 4-3, for maximum efficiency, the coil length should be 0.8R, or 0.2 in. Then

$$N = \sqrt{\frac{17 \times 0.5}{0.25}}$$

$$= \sqrt{34}$$

$$= 5.8\,\text{turns}$$

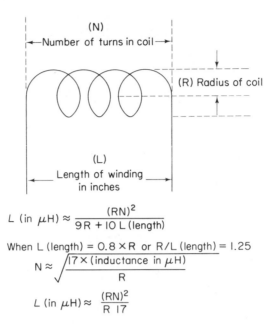

$$L \text{ (in } \mu H) \approx \frac{(RN)^2}{9R + 10 L \text{ (length)}}$$

When L (length) $= 0.8 \times R$ or R/L (length) $= 1.25$

$$N \approx \sqrt{\frac{17 \times \text{(inductance in } \mu H)}{R}}$$

$$L \text{ (in } \mu H) \approx \frac{(RN)^2}{R \; 17}$$

Fig. 4-3. Calculations for self-inductance of single-layer air-core coil.

For practical purposes, use 6 turns and spread the turns slightly. The additional part of a turn will increase inductance, but the spreading will decrease the inductance. After the coil is made, the inductance should be checked with an inductance bridge or as described at the end of this chapter.

4-2. Basic RF-Amplifier Design Approaches

Solid-state RF amplifiers can be designed using *two-port networks*. Basically, the method consists of characterizing the transistor as a linear active two-port network (LAN) with admittances (*y*-parameters), and using the parameters to solve design equations for stability, gain, and input–output admittances. The two-port design approach is best suited for *voltage amplifiers* using *small-signal* characteristics and is recommended for low-power two-junction RF amplifiers. Design based on large-signal characteristics (transistor input–output resistances and capacitance) is recommended for RF power amplifiers (using high-power transistors). Both design approaches are discussed in this chapter.

With either approach, it is difficult at best to provide simple, step-by-step procedures for designing RF amplifiers to meet all possible circuit conditions. In practice, there are several reasons why this procedure often results in considerable trial and error.

First, not all the characteristics are always available in data-sheet form.

For example, input and output admittances may be given at some low frequency, but not at the desired operating frequency.

Often, manufacturers do not agree on terminology. A good example of this is in y-parameters, where one manufacturer uses letter subscripts (y_{fs}) and another uses number subscripts (y_{21}). Of course, this type of variation can be eliminated by conversion.

In some cases, manufacturers will give the required information on data sheets but not in the required form. For example, some manufacturers may give the input capacitance in farads rather than listing the input admittance in mhos. The input admittance is found when the input capacitance is multiplied by $6.28F$ (where F is the frequency of interest). This is based on the assumption that the input admittance is primarily capacitive, and thus dependent on frequency. The assumption is not always true for the frequency of interest; therefore, it may be necessary to use complex admittance measuring equipment to make actual tests of the transistor.

The input and output tuning circuit of an RF amplifier must perform three functions. Obviously, the circuits (capacitors and coils) must tune the amplifier to the desired frequency. In addition, the circuits must match the input and output impedances of the transistor to the impedances of the source and load; otherwise, there will be considerable loss of signal. Finally, as in the case with any amplifier, there is some feedback between output and input. If the admittance factors are just right, the feedback will be of sufficient amplitude and of proper phase to cause oscillation of the amplifier. The amplifier is considered as *unstable* when this occurs.

Amplifier instability in any form is always undesirable and can be corrected by feedback (called *neutralization*) or by changes in the input/output tuning networks. (Generally, the changes involve introducing some slight mismatch to improve stability.) Although the neutralization and tuning circuits are relatively simple, the equations for determining stability (or instability) and impedance matching are long and complex. Generally, such equations are best solved by computer-aided design methods.

In an effort to cut through the maze of information and complex equations, we shall discuss all the steps involved in RF amplifier design. Armed with this information, the reader should be able to interpret data sheets or test information, and use the data to design tuning networks that will provide stable RF amplification at the frequencies of interest. With each step we shall discuss the various alternative procedures and types of information available. Specific design examples are used to summarize the information. On the assumption that not all readers are familiar with two-port networks, we shall start with a summary of the y-parameter system.

4-2.1. y-Parameters

Impedance (Z) is a combination of resistance (R, the real part) and reactance (X, the imaginary part). Admittance (y) is the reciprocal of impedance and is composed of conductance (g, the real part) and susceptance (jb, the imaginary part). Thus, g is the reciprocal of R, and jb is the reciprocal of X. (To find g, divide R into 1; to find R, divide g into 1.) Z is expressed in ohms. Y, being a reciprocal, is expressed in mhos or millimhos (mmhos). For example, an impedance Z of 50 Ω is equal to 20 mmhos ($1 \div 50 = 0.02$; 0.02 mho $= 20$ mmhos).

A y-parameter is an expression for admittance in the form

$$y_i = g_i + jb_i$$

where g_i = real (conductive) part of input admittance

jb_i = imaginary (susceptive) part of input admittance

y_i = input admittance (the reciprocal of Z_i)

The term $y_i = g_i + jb_i$ expresses the y-parameter in *rectangular form*. Some manufacturers describe the y-parameter in *polar form*. For example, they will give the *magnitude* of the input admittance as $|y_i|$ and the *angle* of input admittance as $\underline{/y_i}$. Quite often, manufacturers will mix the two systems of *vector algebra* on their data sheets.

Conversion of vector-algebra forms. It is assumed that the readers are already familiar with the basics of vector algebra. However, the following notes summarize the steps necessary to manipulate vector-algebra terms. With this background, the reader should be able to perform all the calculations involved in the simplified design of RF amplifier networks using y-parameters.

Converting from rectangular to polar form

1. Find the magnitude from the square root of the sum of the squares of the components.

$$\text{polar magnitude} = \sqrt{g^2 + jb^2}$$

2. Find the angle from the ratio of the component values.

$$\text{polar angle} = \arctan\left(\frac{jb}{g}\right)$$

The angle is leading if the jb term is positive and lagging if the jb term is negative.

For example, assume that the y_{fs} is given as $g_{fs} = 30$ and $jb_{fs} = 70$. This is converted to polar form by:

$$|y_{fs}| \text{ polar magnitude} = \sqrt{(30)^2 + (70)^2} = 76$$

$$\underline{/y_{fs}} \text{ polar angle} = \arctan\left(\frac{70}{30}\right) = 67°$$

Converting from polar to rectangular form:

1. Find the real (conductive, or g) part when polar magnitude is multiplied by the cosine of the polar angle.

2. Find the imaginary (susceptance, or jb) part when polar magnitude is multiplied by the sine of the polar angle.

If the angle is positive, the jb component is also positive. When the angle is negative, the jb component is also negative.

For example, assume that the y_{fs} is given as $|y_{fs}| = 20$ and $y_{fs} = -33°$. This is converted to rectangular form by

$$20 \times \cos 33° = g_{fs} = 16.8$$
$$20 \times \sin 33° = jb_{fs} = 11$$

The four basic y-parameters. A y-equivalent circuit is shown in Fig. 4-4. This equivalent circuit is for a *field-effect transistor* (FET). However, a similar circuit can be drawn for a two-junction transistor when analyzing small-signal characteristics.

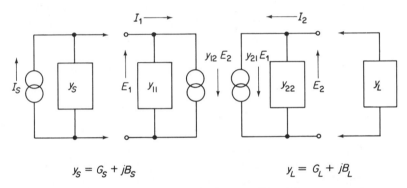

$$y_S = G_S + jB_S \qquad\qquad y_L = G_L + jB_L$$

Fig. 4-4. A y-equivalent circuit (for a FET) with source and load.

Note that y-parameters can be expressed with number subscripts or letter subscripts. The number subscripts are universal because they can apply to two-junction transistors, FETs, and even IC amplifiers. However, the letter subscripts are most popular on FET data sheets.

The following notes can be used to standardize y-parameter nomenclature. Note that the letter s in the letter subscript refers to common-source operation of an FET amplifier and is equivalent to a common-emitter two-junction amplifier.

y_{11} is input admittance and can be expressed as y_{is}.
y_{12} is reverse transadmittance and can be expressed as y_{rs}.
y_{21} is forward transadmittance and can be expressed as y_{fs}.
y_{22} is output admittance and can be expressed as y_{os}.

Input admittance, with $Y_L =$ infinity (a short circuit of the load), is expressed as

$$y_{11} = g_{11} + jb_{11} = \frac{di_1}{de_1} \qquad \text{(with } e_2 = 0\text{)}$$

This means that y_{11} is equal to the difference in current i_1, divided by the difference in voltage e_1, with voltage e_2 at 0. The voltages and currents involved are shown in Fig. 4-4.

Some data sheets do not show y_{11} at any frequency, but give input capacitance instead. If one assumes that the input admittance is entirely (or mostly) capacitive, then the input impedance can be found when input capacitance is multiplied by $6.28F$ ($F =$ frequency in hertz) and the reciprocal is taken. Because admittance is the reciprocal of impedance, admittance is found when input capacitance is multiplied by $6.28F$ (where admittance is capacitive). For example, if the frequency is 100 MHz and the input capacitance is 8 pF, the input admittance is $6.28 \times (100 \times 10^6) \times (8 \times 10^{-12}) \approx 5$ mmhos.

This assumption is accurate only if the real part of y_{11} (or g_{11}) is negligible. Such an assumption is reasonable for an FET but not necessarily for a two-junction transistor. The real part of two-junction transistor input admittance can be quite large in relation to the imaginary jb_{11} part.

Forward transadmittance, with $Y_L =$ infinity (a short circuit of the load), is expressed as

$$y_{21} = g_{21} + jb_{21} = \frac{di_2}{de_1} \qquad \text{(with } e_2 = 0\text{)}$$

This means that y_{21} is equal to the difference in output current i_2, divided by the difference in input voltage e_1, with voltage e_2 at 0. In other words, y_{21} represents the difference in output current for a difference in input voltage.

Two-junction transistor data sheets often do not give any value for y_{21}. Instead, they show forward transadmittance by means of the hybrid system of notation using h_{fe} or h_{21} (which means hybrid forward transmittance with common emitter). No matter what system is used, it is essential that the values of forward transadmittance be considered at the frequency of interest.

Output admittance, with $Y_S =$ infinity (a short circuit of the source), is expressed as

$$y_{22} = g_{22} + jb_{22} = \frac{di_2}{de_2} \qquad \text{(with } e_1 = 0\text{)}$$

Reverse transadmittance, with $Y_S =$ infinity (a short circuit of the source), is expressed as

$$y_{12} = g_{12} + jb_{12} = \frac{di_1}{de_2} \qquad \text{(with } e_1 = 0\text{)}$$

y_{12} is not considered an important two-junction transistor parameter. However, it may appear in equations related to RF design.

y-parameter measurement. It is obvious that *y*-parameter information is
not always available or in a convenient form. In practical design, it may be
necessary to measure the *y*-parameters using laboratory equipment. The
main concern in measuring *y*-parameters is that the measurements are made
under conditions simulating those of the final circuit. For example, if supply
voltages, bias voltages, and operating frequency are not identical (or close)
to the final circuit, the tests may be misleading.

Although the data sheets for transistors to be used as power amplifiers
usually contain input and output admittance information, it may be helpful
to know how this information is obtained.

A typical test-amplifier circuit for two-junction transistors is shown in
Fig. 4-5. During a test, the transistor is placed in the test circuit designed with
variable components to provide wide tuning capabilities. This feature is
necessary to ensure correct matching while characterizing a transistor at

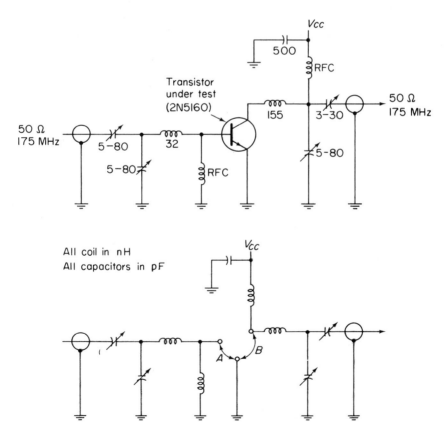

Fig. 4-5. Typical test-amplifier circuit for two-junction
transistors.

several power levels. The circuit is tuned for maximum power gain at each power level for which admittance data are desired.

After the test amplifier has been tuned for maximum power gain, the dc power, signal source, circuit load, and test transistors are disconnected from the circuit. For total circuit impedance to remain the same, the signal source and output load circuit connections are terminated at their characteristic resistances. After these substitutions are performed, complex admittances are measured at the base- and collector-circuit connections of the test transistor (points A and B, respectively, in Fig. 4-5).

The two-junction transistor input and output admittances are the *conjugates* of the base-circuit connection and the collector-circuit connection admittances, respectively. For example, if the base-circuit connection admittance is $8 + j3$, the input admittance of the transistor is $8 - j3$.

In some systems of two-junction transistor RF amplifier design, the networks are calculated on the basis of input–output resistance and capacitance, instead of admittance. In such cases, the admittances measured in the circuit of Fig. 4-5 must be converted to resistance and capacitance.

Admittances are expressed in mhos (or millimhos, mmhos). Resistance can be found by dividing the real part of the admittance into 1. Capacitance can be found by dividing the imaginary part of the admittance into 1 (to find reactance); then the reactance is used in the equation $C = 1/(6.28FX_C)$ to find the actual capacitance.

4-2.2. Stability Factors

There are two factors used to determine the potential stability (or instability) of transistors in RF amplifiers. One factor is known as the Linvill C factor; the other is the Stern k factor. Both factors are calculated from equations requiring y-parameter information (to be taken from data sheets or by actual measurement at the frequency of interest).

The main difference between the two factors is that the Linvill C factor assumes that the transistor is not connected to a load. The Stern k factor includes the effect of a specific load.

The Linvill C factor is calculated from

$$C = \frac{y_{12}y_{21}}{2g_{11}g_{22} - R_e(y_{12}y_{21})}$$

where $R_e(y_{12}y_{21})$ is the real part of $y_{12}y_{21}$.

If C is less than 1, the transistor is unconditionally stable. That is, using a conventional (unmodified) circuit, no combination of load and source admittance can be found which will cause oscillation. If C is greater than 1, the transistor is potentially unstable. That is, certain combinations of load and source admittance will cause oscillation.

The Stern k factor is calculated from

$$k = \frac{2(g_{11} + G_S)(g_{22} + G_L)}{y_{12}y_{21} + R_e(y_{12}y_{21})}$$

where G_S and G_L are source and load conductance, respectively. ($G_S = 1$/source resistance; $G_L = 1$/load resistance.)

If k is greater than 1, the amplifier circuit is stable (the opposite of Linvill). If k is less than 1, the amplifier is unstable. In practical design, it is recommended that a k factor of 3 or 4 be used, rather than 1, to provide a margin of safety. This will accommodate parameter and component variations (particularly with regard to bandpass response).

Note that both equations are fairly complex and require considerable time for their solution. In practical work, computer-aided design techniques are used for stability equations.

Some manufacturers provide alternative solutions to the stability and load-matching problems, usually in the form of a data-sheet graph, as shown in Fig. 4-6. Note that the device is unconditionally stable at frequencies above 250 MHz. At frequencies below about 50 MHz, the device becomes highly unstable.

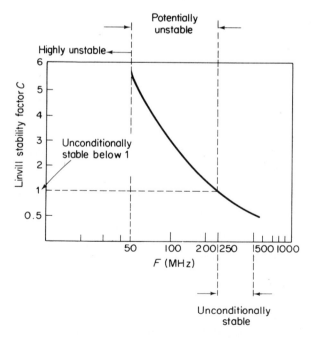

Fig. 4-6. Linvill stability factor C, for typical FET.

4-2.3. Solutions to Stability Problems

There are two basic design solutions to the problem of unstable RF amplifiers. First, the amplifier can be neutralized; that is, part of the output can be fed back (after it is shifted in phase) to the input so as to cancel oscillation. When an amplifier has been neutralized, it can then be matched perfectly to the source and load. This type of match is known as a *conjugate match*. In perfect conjugate match the transistor input and source, as well as the transistor output and load, are matched resistively, and all reactance is tuned out. Neutralization requires extra components and creates a problem when frequency is changed.

The second solution is to introduce some mismatch into either the source or load tuning networks. This solution, sometimes known as the *Stern solution*, requires no extra components but does produce a reduction in gain.

A comparison of these two methods is shown in Fig. 4-7. The higher gain curve represents the unilateralized (or neutralized) operation. The lower gain curve represents the circuit power gain when the Stern k factor is 3.

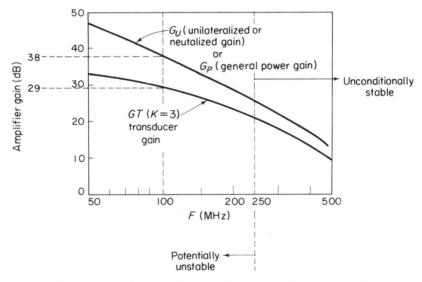

Fig. 4-7. Comparison of neutralized gain (G_U) versus gain with mismatch (G_T) when the Stern k factor is 3.

Assume that the frequency of interest is 100 MHz. If the amplifier is matched directly to the load (perfect conjugate match) without regard to stability or using neutralization to produce stability, the top curve applies, and the power gain is about 38 dB. If the amplifier is matched to a load and source where the Stern k factor is 3 (resulting in a mismatch with the actual load and source), the lower curve applies, and the power gain is about 29 dB.

The upper curve of Fig. 4-7 is found by the *general power-gain equation,*

$$G_P = \frac{\text{power delivered to load}}{\text{power delivered to input}}$$

$$= \frac{(y_{21})^2 G_L}{(Y_L + y_{22})^2 R_e[y_{11} - y_{12}y_{21}/(y_{22} + Y_L)]}$$

The general power-gain equation applies to circuits with no external feedback and to circuits that have external feedback (neutralization), provided the composite *y*-parameters of both transistor and feedback networks are substituted for the transistor *y*-parameters in the equation.

The lower curve in Fig. 4-7 is found by the *transducer gain equation,*

$$G_T = \frac{\text{power delivered to load}}{\text{maximum power available from source}}$$

$$= \frac{4G_S G_L (y_{21})^2}{[(y_{11} + Y_S)(y_{22} + Y_L) - y_{12}y_{21}]^2}$$

The transducer gain expression includes input mismatch. The lower curve in Fig. 4-7 assumes that the input mismatch produces a Stern *k* factor of 3. That is, the circuit tuning networks are adjusted for admittances that produce a Stern *k* factor of 3. The transducer gain expression considers the input and output networks as part of the source and load.

With either gain expression, the input and output admittances of the transducer are modified by the load and source admittances.

The *input admittance* of the transistor is given by

$$Y_{IN} = y_{11} - \frac{y_{12}y_{21}}{y_{22} + Y_L}$$

The *output admittance* of the transistor is given by

$$Y_{OUT} = y_{22} - \frac{y_{12}y_{21}}{y_{11} + Y_S}$$

At low frequencies, the second term in the input and output admittance equations is not particularly significant. At VHF, the second term makes a very significant contribution to the input and output admittances.

The imaginary parts of Y_S and Y_L (B_S and B_L, respectively) must be known before values can be calculated for power gain, transducer gain, input admittance, and output admittance. Exact solutions for B_S and B_L almost always consist of time-consuming complex algebraic manipulations.

To find fairly good simplifying approximations for the equations, let $B_S \approx -b_{11}$ and $B_L \approx -b_{22}$, so that

General power-gain expression:

$$G_P \approx \frac{(y_{21})^2 G_L}{(G_L + g_{22})^2 R_e[y_{11} - y_{12}y_{21}/(g_{22} + G_L)]}$$

Transducer gain expression:

$$G_T \approx \frac{4G_s G_L (y_{21})^2}{(g_{11} + G_s g_{22} + G_L - y_{12} y_{21})^2}$$

Input admittance:

$$Y_{IN} \approx y_{11} - \frac{y_{12} y_{21}}{g_{22} + G_L}$$

Output admittance:

$$Y_{OUT} \approx y_{22} - \frac{y_{12} y_{21}}{g_{11} + G_L}$$

The other gain expressions sometimes found on the data sheets of transistors used in RF applications are maximum available gain (MAG) and maximum usable gain (MUG).

MAG is usually applied as the gain in a conjugately matched, neutralized circuit and is expressed as

$$\text{MAG} = \frac{(y_{21})^2 R_{IN} R_{OUT}}{4}$$

where R_{IN} and R_{OUT} are the input and output resistances, respectively, of the transistor.

An alternative MAG expression is

$$\text{MAG} = \frac{(y_{21})^2}{4R_e(y_{11})R_e(y_{22})}$$

where $R_e(y_{11})$ is the real part (g_{11}) of the input admittance, and $R_e(g_{22})$ is the real part (g_{22}) of the output admittance.

MUG is usually applied as the *stable gain* which may be realized in a *practical* (neutralized or unneutralized) RF amplifier. In a typical unneutralized circuit, MUG is expressed as

$$\text{MUG} \approx \frac{0.4 y_{21}}{6.28F \times \text{reverse transfer capacitance}}$$

MAG and MUG are often omitted on data sheets for two-junction transistors; instead, gain is listed as h_{fe} at a given frequency. This is supplemented with graphs that show available power output at given frequencies with a given input.

4-2.4. Neutralized Solution

There are several methods for neutralization of RF amplifiers. The most common method is the *capacitance-bridge* technique, as shown in Fig. 4-8a. Capacitance-bridge neutralization becomes more apparent when the circuit is redrawn, as shown in Fig. 4-8b. The condition for neutralization is that $I_F = I_N$. That is, the neutralization current I_N must be equal to the feedback current I_F in amplitude, but of opposite phase.

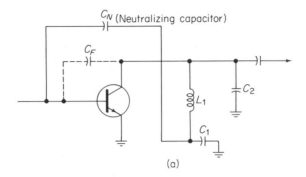

(a)

C_F = Reverse capacitance of transistor

$$C_N \approx C_F \times \left(\frac{C_1}{C_2}\right)$$

$$C_1 \gg C_2$$

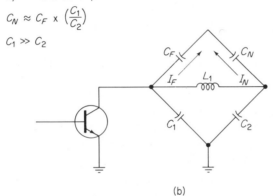

(b)

Fig. 4-8. Capacitance-bridge neutralization circuit.

The equations normally used to find the value of the feedback neutralization capacitor are long and complex. However, for practical work, if the value of C_1 is made quite large in relation to C_2 (at least 4 times), the value of C_N can be found by

$$C_N \approx C_F \frac{C_1}{C_2}$$

where C_F is the reverse capacitance of the transistor.

In simple terms, the value of C_N is approximately equal to the value of reverse capacitance times the ratio of C_1/C_2. For example, if reverse capacitance (sometimes listed as collector-to-base capacitance) is 7 pF, C_1 is 30 pF and C_2 is 3 pF, the C_1/C_2 ratio is 10, and $C_N \approx 10 \times 7$ pF = 70 pF.

4-2.5. The Stern Solution

A stable design with a potentially unstable transistor is possible without external feedback (neutralization) by proper choice of source and load admittances. This can be seen by inspection of the Stern k factor equation; G_S and G_L can be made large enough to yield a stable circuit, regardless of the degree of potential instability. Using this approach, a circuit stability factor k (typically $k = 3$) is selected, and the Stern k factor equation is used to arrive at values of G_S and G_L which will provide the desired k.

Of course, the actual G of the source and load cannot be changed; instead, the input and output tuning circuits are designed as if the actual G values were changed. This results in a mismatch and a reduction in power gain but does produce the desired degree of stability.

To get a particular circuit stability factor, the designer may choose any of the following combinations of matching and mismatching of G_S and G_L to the transistor input and output conductances, respectively:

G_S matched and G_L mismatched
G_L matched and G_S mismatched
Both G_S and G_L mismatched

Other performance requirements or practical considerations often dictate the decision on which combination to use. For example, it may not be practical to mismatch to some extreme value of G_S or G_L.

Once G_S and G_L have been chosen, the remainder of the design may be completed by using the relationships that apply to the amplifier without feedback. Power gain and input–output admittances may be computed using the appropriate equations (refer to Sec. 4-2.3).

Simplified Stern approach. Although the procedure above may be adequate in many cases, a more systematic method of source and load admittance determination is desirable for designs that demand maximum power gain per degree of circuit stability. Stern has analyzed this problem and developed equations for computing the best G_S, G_L, B_S, and B_L for a particular circuit stability factor (Stern k factor). Unfortunately, these equations are very complex and become quite tedious when they are used frequently. The complete Stern solution is best applied by computer.

Programs have been written to provide essential information for transistors used as RF amplifiers, including the effects of various specific sources and loads. These programs permit the designer to experiment with theoretical breadboard circuits in a matter of seconds.

When a Stern solution must be obtained *without the aid of a computer*, it is best to use one of the many shortcuts that have been developed over the years. The following shortcut is by far the simplest and most widely accepted, yet provides an accuracy close to that of the computer solutions.

1. Let $B_S \approx -b_{11}$ and $B_L \approx -b_{22}$, as in the case of the Sec. 4-2.3 equations. This method permits the designer to closely approximate the exact Stern solution for Y_s and Y_L, while avoiding that portion of the computations

which is the most complex and time consuming. Further, the circuit can be designed with tuning adjustments for varying B_S and B_L, thereby creating the possibility of achieving the true B_S and B_L (by experiment) for maximum gain as accurately as if all the Stern equations had been solved.

2. Mismatch G_S to g_{11} and G_L to g_{22} by an *equal ratio*. That is, find a ratio that produces the desired Stern k factor, then mismatch G_S to g_{11} (and G_L to g_{22}). For example, if the ratio is 4 : 1, make G_S 4 times the value of g_{11} (and G_L 4 times the value of g_{22}).

If the mismatch ratio, R, is defined as

$$R = \frac{G_L}{g_{22}} = \frac{G_S}{g_{11}}$$

the R may be computed for any particular circuit stability k factor using the equation

$$R = \sqrt{k \frac{y_{21}y_{12} + R_e(y_{12}y_{21})}{2g_{11}g_{22}}} - 1$$

As an example, assume that it is desired to mismatch input and output circuit so there is a Stern k factor of 4, using a transistor that has the following characteristics: $y_{12}y_{21} = 0.5$, $g_{11} = 5.0$, $g_{22} = 0.05$, and $R_e(y_{12}y_{22}) = 0.2$ (all values in mmhos).

$$R = \sqrt{4 \cdot \frac{0.5 + 0.2}{(2)(5)(0.05)}} - 1 \approx 1.37$$

Using the value of 1.37 for R, and the equation

$$1.37 = \frac{G_S}{g_{11}} = \frac{G_L}{g_{22}}$$

then

$$G_S = (1.37)(5)(10^{-3}) = 6.85 \text{ mmhos}$$

and

$$R_S = \frac{1}{G_S} \approx 146 \, \Omega$$

$$G_L = (1.37)(0.05)(10^{-3}) = 0.0685 \text{ mmho}$$

and

$$R_L = \frac{1}{G_L} \approx 14{,}600 \, \Omega$$

The shortcut Stern method may be advantageous if the source and load admittances and power gains for several different values of k are desired. Once the R for a particular k has been determined, the R for any other k may be quickly found from the equation

$$R = \frac{(1 + R_1)^2}{(1 + R_2)^2} = \frac{k_1}{k_2}$$

where R_1 and R_2 are values of R corresponding to k_1 and k_2, respectively.

The Stern solution with data-sheet graphs. It is obvious that the Stern solution, even with the shortcut method, is somewhat complex. For this reason, some manufacturers have produced data-sheet graphs that show the best source and load admittances for a particular transistor over a wide range of frequencies. Examples of these graphs are shown in Figs. 4-9 and 4-10.

Figure 4-9 shows both the real (G_S) and imaginary (B_S) values that will produce maximum gain, but with a stability (Stern k) factor of 3 at frequencies from 50 to 500 MHz. Figure 4-10 shows corresponding information for G_L and B_L.

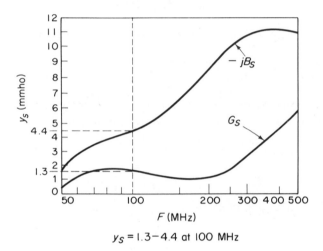

$y_S = 1.3 - 4.4$ at 100 MHz

Fig. 4-9. Best source admittance, $Y_S = G_S + jB_S$.

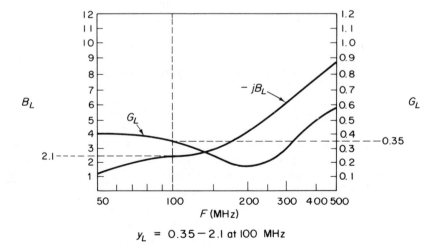

$y_L = 0.35 - 2.1$ at 100 MHz

Fig. 4-10. Best load admittance, $Y_L = G_L + jB_L$.

To use these figures, simply select the desired frequency, and note where the corresponding G and B curves cross the frequency line. For example, assuming a frequency of 100 MHz, $Y_L = 0.35 - j2.1$ mmhos, and $Y_S = 1.3 - j4.4$ mmhos.

If the tuning circuits are designed to match these admittances rather than the actual admittances of the source and load, the circuit will be stable. Of course, the gain will be reduced. Use the *transducer gain expression* (G_T) of Sec. 4-2.3 to find the resultant power gain.

4-3. Power Amplifiers and Multipliers

Figure 4-11 shows the working schematics of typical RF power amplifiers. The same basic circuits can be used as frequency multipliers. However, in a multiplier circuit, the output must be tuned to a multiple of the input. A multiplier may or may not provide amplification. Usually, most of the ampli-

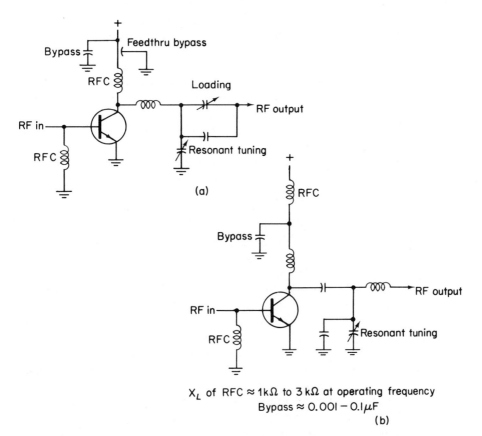

X_L of RFC $\approx 1\,\mathrm{k}\Omega$ to $3\,\mathrm{k}\Omega$ at operating frequency
Bypass $\approx 0.001 - 0.1\,\mu\mathrm{F}$
(b)

Fig. 4-11. Typical RF power-amplifier and multiplier circuits.

fication is supplied by the final amplifier stage, which is not operated as a multiplier. That is, the input and output of the final stage are at the same frequency. A typical RF transmitter will have three stages: an oscillator to provide the basic signal frequency, an intermediate stage that provides amplification and/or frequency multiplication, and a final stage for power amplification.

Oscillators (RF, AF, etc.) are discussed in Chapter 5.

4-3.1. Design Considerations for Power Amplifiers and Multipliers

All the design considerations in Chapter 1 apply to RF power amplifiers and multipliers. Of particular importance are interpreting data sheets (Sec. 1-2), determining parameters at different frequencies (Sec. 1-3), and temperature-related design problems (Sec. 1-4). In addition to these basic design considerations, the following problems must be considered in RF power amplifier design.

Note that circuit (a) of Fig. 4-11 has two tuning controls (variable capacitors) in the output network, while the network circuit (b) of Fig. 4-11 has only one adjustment control. The circuit of Fig. 4-11a is typical for power amplifiers, where the output must be tuned to the resonant frequency by one control, and adjusted for proper impedance match by the other control (often known as the loading control). In practice, both controls affect tuning and loading (impedance matching). The circuit of Fig. 4-11b is typical for multipliers or intermediate amplifiers where the main concern is tuning to the resonant frequency.

Also note that variable capacitors are connected in parallel with fixed capacitors in both networks. This parallel arrangement serves two purposes. First, it provides a minimum fixed capacitance in case the variable capacitor is adjusted to its minimum value. In some cases, if a minimum capacitance were not included in the network, a severe mismatch could occur when the variable capacitor is at its minimum, resulting in damage to the transistor. The second purpose for a parallel capacitor is to reduce required capacitance rating (and thus the physical size) of the variable capacitor.

When designing networks such as shown in Fig. 4-11, use a capacitor with a midrange capacitance equal to the desired capacitance. For example, if the desired capacitance is 25 pF (to produce resonance at the normal operating frequency), use a variable capacitor with a range of 1 to 50 pF. If such a capacitor were not readily available, use a fixed capacitor of 15 pF in parallel with a 15-pF variable capacitor. This would provide a capacitance range of 16 to 30 pF, with a midrange of about 23 pF. Of course, the maximum capacitance range is dependent upon the required tuning range of the circuit. (A wide frequency range requires a wide capacitance range.)

The transistors remain cut off until a signal is applied. Therefore, the transistors are never conducting for more than 180° (half a cycle) of the 360° input signal cycle. In practice, the transistors conduct for about 140° of the input cycle, either on the positive half or negative half, depending on the transistor type (*NPN* or *PNP*). No bias, as such, is required for this class of operation.

The emitter is connected directly to ground. In those transistors where the emitter is connected to the case (typical in many RF power transistors), the case can be mounted on a chassis that is connected to the ground side of the supply voltage. A direct connection between emitter and ground is of particular importance in high-frequency applications. If the emitter were connected to ground through a resistance (or a long lead), an inductive or capacitive reactance could develop at high frequencies, resulting in undesired changes in the network.

The transistor base is connected to ground through an RF choke (RFC). This provides a dc return for the base, as well as RF signal isolation between base and emitter or ground. The transistor collector is connected to the supply voltage through an RFC and (in some cases) through the coil portion of the resonant network. The RFC provides dc return, but RF signal isolation, between collector and power supply. When the collector is connected to the power supply through the resonant network, the coil must be capable of handling the full collector current. For this reason, final amplifier networks should be chosen so that collector current does not pass through the coil (such as Fig. 4-11a). The circuit of Fig. 4-11b should be used for power applications where current is low.

The ratings for RF chokes are sometimes confusing. Some manufacturers list a full set of characteristics: inductance, dc resistance, ac resistance, Q, current capability, and nominal frequency range. Other manufacturers give only one or two of these characteristics. Alternating Current (ac) resistance and Q are usually frequency-dependent. A nominal frequency-range characteristic is a helpful, but usually not critical, design parameter. All other factors being equal, the dc resistance should be at a minimum for any circuit carrying a large amount of current. For example, a large dc resistance in the collector of a final power amplifier can result in a large voltage drop between power supply and collector. Usually, the selection of a trial value for an RFC is based on a tradeoff between inductance and current capability. The minimum current capacity should be greater (by at least 10 per cent) than the maximum anticipated direct current. The inductance is dependent upon operating frequency. As a trial value, use an inductance that will produce a reactance between 1000 and 3000 Ω at the operating frequency.

The power-supply circuits of power amplifiers and multipliers must be bypassed, as shown in Fig. 4-11. The feed-through bypass capacitors shown in Fig. 4-11 are used at higher frequencies where the RF circuits are physically shielded from the power-supply and other circuits. The feed-through capacitor permits direct current to be applied through the shield, but prevents RF from passing outside the shield (RF is bypassed to the ground return). As a trial value, use a total bypass capacitance range of 0.001 to 0.1 μF.

From a practical standpoint, the best test for adequate bypass capacitance and RF choke inductance is the presence of RF signals on the power-supply side of the dc voltage line. If RF signals are present on the power-supply side of the line, the bypass capacitance and/or the RFC inductance is not adequate. (A possible exception to this is where the RF signals are being picked up due to inadequate shielding.) If the shielding is good, and RF signals present in the power supply, increase the bypass capacitance value. As a second step, increase the RFC inductance. Of course, circuit performance must be checked with each increase in capacitance or inductance value. For example, too much bypass capacitance can cause undesired feedback and oscillation; too much RFC inductance can reduce amplifier output and efficiency. The procedures for the measurement of RF signals are described at the end of this chapter.

A class C RF amplifier will have a typical efficiency of about 65 to 70 per cent. That is, the RF power output will be 65 to 70 per cent of the dc input power. To find the required dc input power, divide the desired RF power output by 0.65 or 0.7. For example, if the desired RF output is 50 W, the dc input power would be 50 ÷ 0.7, or approximately 70 W. Since the collector of an RF amplifier is at a dc potential approximately equal to the power supply (slightly less due to a drop across the RFC and/or coil), divide the input power by the power-supply voltage to find the collector current. For example, with a dc input of 70 W and a 28-V power supply, the collector current would be approximately 2.7 A.

It is obvious that the transistors must be capable of handling the full power-supply voltage at their collectors, and that its current and/or power rating is greater than the maximum calculated values. Likewise, the transistor must be capable of producing the necessary power output, at the operating frequency. These problems are discussed in Secs. 1-2, 1-3, and 1-4.

It is also obvious that the transistors must provide the necessary power gain at the operating frequency. Likewise, the input power to an amplifier must match the desired output and gain. For example, assume that a 50-W, 50-MHz transmitter is to be designed, and that transistors with a power gain of 10 are available. The final amplifier must produce 50-W output, with 5-W input. Generally, a transistor oscillator produces less than 1-W output. Therefore, an intermediate amplifier would be required to deliver an output of 5 W to the final amplifier. The intermediate amplifier would require about 7-W dc input (50 ÷ 7 $\cong$ 7). Assuming a gain of 10 for the intermediate amplifier, an input of 0.5 W would be required from the oscillator.

When an intermediate amplifier is also used as a frequency multiplier, the efficiency drops from the 65 to 70 per cent value. As rules of thumb, the efficiency of a second harmonic amplifier (output at twice the input frequency) is 42 per cent, third harmonic 28 per cent, fourth harmonic 21 per cent, and fifth harmonic 18 per cent. Therefore, if an intermediate amplifier is to be operated at the second harmonic and produce 5-W RF output, the dc input power would have to be approximately 12 W (5 ÷ 0.42 = 12).

Another problem to be considered in frequency multiplication is that power gain (as listed on the data sheet) may not remain the same as when amplifier input and output are at the same frequency. Some data sheets specify power gain at the basic frequency and then derate the power gain for second harmonic operation. As a rule of thumb, always use the minimum power-gain factor when calculating power input and output values.

4-3.1.1. Resonant Network Design

Now we come to the most critical design consideration for RF power amplifiers: the *resonant network*. This network must be resonant at the desired frequency. (Inductive and capacitive reactance must be equal at the selected frequency.) Second, the network must match the transistor output impedance to the load. Generally, an antenna load impedance is on the order of 50 Ω, while output impedance of a typical transistor at radio frequencies is a few ohms. In the case where one amplifier feeds into another amplifier, the network must match the output impedance of one transistor to the input impedance of another transistor. Any mismatch can result in a loss of power between stages, or to the final load.

Transistor impedance (both input and output) has both resistive and reactive components, and therefore varies with frequency. To design a resonant network for the output of a transistor, it is necessary to know the output reactance (usually capacitive), the output resistance at the operating frequency, and the output power. Likewise, it is necessary to know the input resistance and reactance of a transistor at a given frequency and power, when designing the resonant network of the stage feeding into the transistor.

Generally, the input resistance, the input capacitance, and the output capacitance of RF power transistors are shown by means of graphs similar to Fig. 4-12. The reactance can then be found using the corresponding frequency and capacitance. For example, the output capacitance shown on the graph of Fig. 4-12 is about 15 pF at 80 MHz. This produces a capacitive reactance of about 130 Ω, at 80 MHz. The reactance and resistance can then be combined to find impedance as shown in Fig. 4-12.

Input and output transistor impedances are generally listed on data sheets in parallel form. That is, the data sheets assume that the resistance is in parallel with the capacitance. However, some networks require that the impedance be calculated in series form. It is therefore necessary to convert between series and parallel impedance forms. The necessary equations are listed in Fig. 4-12. The output resistance of RF power transistors is usually

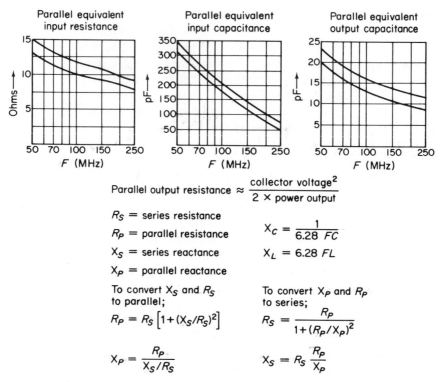

$$\text{Parallel output resistance} \approx \frac{\text{collector voltage}^2}{2 \times \text{power output}}$$

R_S = series resistance

R_P = parallel resistance $X_C = \dfrac{1}{6.28 \ FC}$

X_S = series reactance $X_L = 6.28 \ FL$

X_P = parallel reactance

To convert X_S and R_S To convert X_P and R_P
to parallel;   to series;

$$R_P = R_S \left[1 + (X_S/R_S)^2 \right] \qquad R_S = \frac{R_P}{1 + (R_P/X_P)^2}$$

$$X_P = \frac{R_P}{X_S/R_S} \qquad\qquad X_S = R_S \frac{R_P}{X_P}$$

Fig. 4-12. Typical RF power-amplifier transistor characteristics.

not shown on data sheets, but may be calculated using the equation of Fig. 4-12.

Figures 4-13 through 4-17 show five typical resonant networks, together with the equations necessary to find component values. The networks can be used as amplifiers and/or multipliers. Note that the network of Fig. 4-13 is similar to that of Fig. 4-11a, while Fig. 4-15 is similar to Fig. 4-11b.

The resistor and capacitor shown in the box labeled "transistor to be matched" represent the *complex output impedance* of a transistor. When the network is to be used with a final amplifier, the resistor labeled R_L is the antenna impedance or other load. When the network is used with an intermediate amplifier, R_L represents the input impedance of the following transistor. It is therefore necessary to calculate the input impedance of the transistors being fed by the network, using the data and equations of Fig. 4-12.

The complex impedances have been represented in series form in some cases and parallel form in others, depending on which form is the most convenient for network calculation. The resultant impedance of the network, when terminated with a given load, must be equal to the conjugate of the impedance in the box. For example, assume that the transistor has a series output impedance of $7.33 - j3.87$. That is, the resistance (real part of impedance) is $7.33 \, \Omega$, while the capacitance reactance (imaginary part of impedance) is 3.87. For maximum power transfer from the transistor to the load, the load impedance must be the conjugate of the output impedance, or $7.33 + j3.87$. If the transmitter were designed to operate into a 50-Ω load, the network must transfer the $(50 + j0)$-Ω transmitter load to the $7.33 + j3.87$ transistor load. In addition to performing this transformation, the network provides harmonic rejection (unless a harmonic is needed in a multiplier stage), low loss, and provisions for adjustment of both loading and tuning.

Each of the networks has its advantages and limitations. The following is a summary of the five networks.

The network of Fig. 4-13 is applicable to most RF power amplifiers, and is especially useful where the series real part of the transmitter output imped-

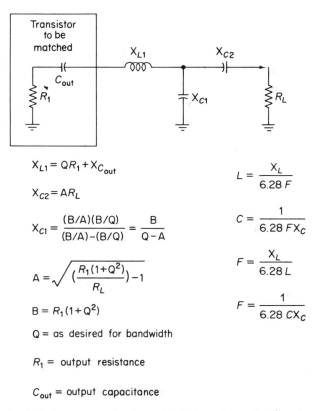

$$X_{L1} = QR_1 + X_{C_{out}}$$

$$X_{C2} = AR_L$$

$$X_{C1} = \frac{(B/A)(B/Q)}{(B/A)-(B/Q)} = \frac{B}{Q-A}$$

$$A = \sqrt{\left(\frac{R_1(1+Q^2)}{R_L}\right) - 1}$$

$$B = R_1(1+Q^2)$$

$Q =$ as desired for bandwidth

$R_1 =$ output resistance

$C_{out} =$ output capacitance

$$L = \frac{X_L}{6.28\,F}$$

$$C = \frac{1}{6.28\,F X_C}$$

$$F = \frac{X_L}{6.28\,L}$$

$$F = \frac{1}{6.28\,C X_C}$$

Fig. 4-13. RF network where R_1 is less than R_L. Courtesy Motorola.

ance, or R_1, is less than 50 Ω. With a typical 50-Ω load, the required reactance for C_1 rises to an impractical value when R_1 is close to 50 Ω.

The network of Fig. 4-14 [often called a *pi network*] is best suited where the parallel resistor R_1 is high (near the value of R_L, typically 50 Ω). If the network of Fig. 4-14 is used with a low value of R_1 resistance, the inductance of L_1 must be very small, while C_1 and C_2 become very large (beyond practical limits).

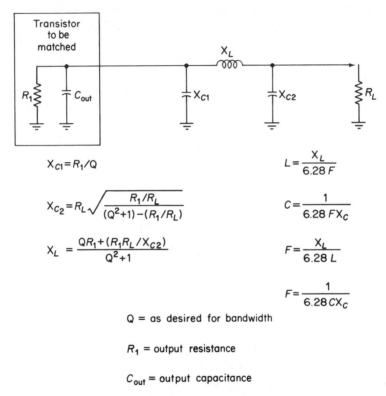

$$X_{C1} = R_1/Q$$

$$X_{C2} = R_L\sqrt{\frac{R_1/R_L}{(Q^2+1)-(R_1/R_L)}}$$

$$X_L = \frac{QR_1+(R_1R_L/X_{C2})}{Q^2+1}$$

$$L = \frac{X_L}{6.28\,F}$$

$$C = \frac{1}{6.28\,FX_C}$$

$$F = \frac{X_L}{6.28\,L}$$

$$F = \frac{1}{6.28\,CX_C}$$

Q = as desired for bandwidth

R_1 = output resistance

C_{out} = output capacitance

Fig. 4-14. RF network where R_1 is approximately equal to R_L. Courtesy Motorola.

The networks of Figs. 4-15 and 4-16 will produce practical values for C and L, especially where R_1 is very low. The main limitation for the networks of Figs. 4-15 and 4-16 is that R_1 must be substantially lower than R_L. These networks, or variations thereof, are often used with intermediate stages where a low output impedance of one transistor must be matched to the low input impedance of another transistor.

The network of Fig. 4-17 (often called a *tee network*) is best suited where R_1 is much less or much greater than R_L.

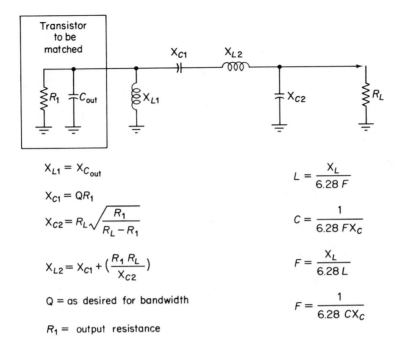

$$X_{L1} = X_{C_{out}}$$

$$X_{C1} = QR_1$$

$$X_{C2} = R_L\sqrt{\frac{R_1}{R_L - R_1}}$$

$$X_{L2} = X_{C1} + \left(\frac{R_1 R_L}{X_{C2}}\right)$$

Q = as desired for bandwidth

R_1 = output resistance

C_{out} = output capacitance

$$L = \frac{X_L}{6.28\,F}$$

$$C = \frac{1}{6.28\,FX_C}$$

$$F = \frac{X_L}{6.28\,L}$$

$$F = \frac{1}{6.28\,CX_C}$$

Fig. 4-15. RF network where R_1 is very small in relation to R_L. Courtesy Motorola.

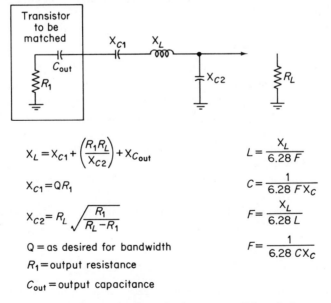

$$X_L = X_{C1} + \left(\frac{R_1 R_L}{X_{C2}}\right) + X_{C_{out}}$$

$$X_{C1} = QR_1$$

$$X_{C2} = R_L\sqrt{\frac{R_1}{R_L - R_1}}$$

Q = as desired for bandwidth

R_1 = output resistance

C_{out} = output capacitance

$$L = \frac{X_L}{6.28\,F}$$

$$C = \frac{1}{6.28\,FX_C}$$

$$F = \frac{X_L}{6.28\,L}$$

$$F = \frac{1}{6.28\,CX_C}$$

Fig. 4-16. RF network where R_1 is very small in relation to $R_{(alternate)}$. Courtesy Motorola.

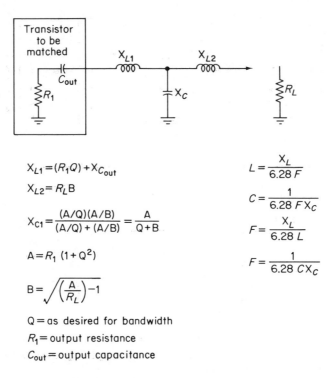

$$X_{L1} = (R_1 Q) + X_{Cout}$$

$$X_{L2} = R_L B$$

$$X_{C1} = \frac{(A/Q)(A/B)}{(A/Q) + (A/B)} = \frac{A}{Q+B}$$

$$A = R_1 (1 + Q^2)$$

$$B = \sqrt{\left(\frac{A}{R_L}\right) - 1}$$

$$L = \frac{X_L}{6.28\,F}$$

$$C = \frac{1}{6.28\,F X_C}$$

$$F = \frac{X_L}{6.28\,L}$$

$$F = \frac{1}{6.28\,C X_C}$$

$Q =$ as desired for bandwidth

$R_1 =$ output resistance

$C_{out} =$ output capacitance

Fig. 4-17. RF network where R_1 is very small or very large in relation to R_L. Courtesy Motorola.

4-3.2. Design Example

Assume that it is desired to design a network similar to that of Fig. 4-13. The network must match the output of a transistor to a 50-Ω antenna. The transistor shows an output capacitance of 200 pF at the operating frequency (obtained from the data sheet). The required output is 50 W at 50 MHz with a 28-V power supply. It is assumed that the transistor will produce the required output at the operating frequency, with the available signal, as discussed in Sec. 4-3.1.

With a value of 28 V and an output of 50 W, the equations of Fig. 4-12 show the value of R_1 as $28^2/(2 \times 50)$, or 7.84 Ω.

With an output capacitance of 200 pF, and an operating frequency of 50 MHz, the equations of Fig. 4-12 show the reactance of C_{out} as

$$\frac{1}{6.28 \times (50 \times 10^6) \times (200 \times 10^{-12})} = 16\,\Omega$$

The combination of these two values result in a *parallel output impedance* of $7.84 - j16$.

Usually, the data sheet gives the output capacitance in parallel form with R_1. For the network of Fig. 4-13, the values of R_1 and C_{out} must be converted to series form.

Using the equations of Fig. 4-12, the equivalent series output impedance is:

$$R_{series} = \frac{7.84}{1 + (7.84/16)^2} = 6.32 \, \Omega \qquad (R_1)$$

$$X_{series} = 6.32 \times \left(\frac{7.84}{16}\right) = 3.1 \, \Omega \qquad (X_{C_{out}})$$

The combination of these two values results in a *series output impedance* of $6.32 - j3.1$.

Using the equations of Fig. 4-13, and assuming a Q of 10 for simplicity, the *reactance values* for the network are:

$$X_L = 10 \times (6.32) + 3.1 = 66.6 \, \Omega$$

$$A = \sqrt{\frac{6.32(1 + 10^2)}{50} - 1} = 3.3$$

$$X_{C2} = 3.3 \times 50 = 165 \, \Omega$$

$$B = 6.32(1 + 10^2) = 638.32$$

$$X_{C1} = \frac{(638.32/3.3)(638.32/10)}{(638.32/3.3) - (638.32/10)} \approx \frac{638.32}{10 - 3.3} = 95 \, \Omega$$

Using the equations of Fig. 4-13, the corresponding *inductance and capacitance values* are:

$$L_1 = \frac{66.6}{6.28 \times (50 \times 10^6)} = 0.21 \, \mu\text{H}$$

$$C_1 = \frac{1}{6.28 \times (50 \times 10^6) \times (95)} = 33 \, \text{pF}$$

$$C_2 = \frac{1}{6.28 \times (50 \times 10^6) \times (165)} = 19 \, \text{pF}$$

If C_1 and C_2 are variable, the values obtained should be the midrange values.

4-4. Using Data-Sheet Graphs to Design RF Amplifier Networks

High-frequency characteristics are especially important in the design of RF networks. Unfortunately, the high-frequency information provided on many data sheets in tabular form is not adequate for simplified design. To properly match impedances, both the resistive and reactive components must be considered. The reactive component (either inductive or capacitive) changes with frequency. In practical amplifier work it is necessary to know the reactance values over a wide range of frequencies, not at some specific frequency (unless you happen to be designing for that frequency only).

The best way to show how resistance and reactance vary in relation to frequency for a particular transistor is by means of graphs or curves. Fortunately, manufacturers who are trying to sell their transistors for high-frequency power-amplifier use generally provide a set of curves showing the characteristics over the anticipated frequency range.

The following paragraphs describe the basic design steps for an RF power amplifier, using typical data-sheet curves and the equations of Figs. 4-12 through 4-17.

The amplifier shown in Fig. 4-18 delivers 80-W output at 175 MHz, with a gain of about 26 dB and 50 per cent overall efficiency. The circuit gain varies less than 0.3 dB when the output and driver-stage transistor case temperatures are varied from 25°C to 100°C.

4-4.1. Circuit Analysis

A summary of the amplifier's performance is given in the table of Fig. 4-19. A plot of amplifier power output versus power input appears in Fig. 4-20. The circuit operates from a 12.5-V supply and draws 12.9 A for an 80-W output. All the amplifier stages are of the common-emitter configuration, and except for the input stage, all are operated class C. The input stage is forward-biased at approximately 40 mA of collector current (with no signal applied) to improve performance at extremely low input signal levels.

The high-power output capability is obtained by operating two 2N6084 devices in parallel for the output stage. Ruggedness and reliability are obtained by using an output-stage design that is capable of providing power levels in excess of 120 W (see Fig. 4-21). As a result, the amplifier is capable of withstanding open- and short-circuit load conditions for all load phase angles without damage to any transistor.

4-4.2. Device Description

The transistors used in this design are part of the Motorola VHF land-mobile series optimized for 12.5-V FM operation. The transistors are of balanced-emitter construction. The driver and final-stage devices are manufactured using the Isothermal (Motorola trademark) technology process, which provides an additional measure of ruggedness and reliability. The effectiveness of this fabrication technique is mainly due to the use of nichrome resistors in series with each emitter. The result is a transistor with an optimized current distribution over the emitter-resistor area. This minimizes temperature variations across the chip to assure ruggedness and resist transistor damage over a wide range of thermal and load VSWR swings.

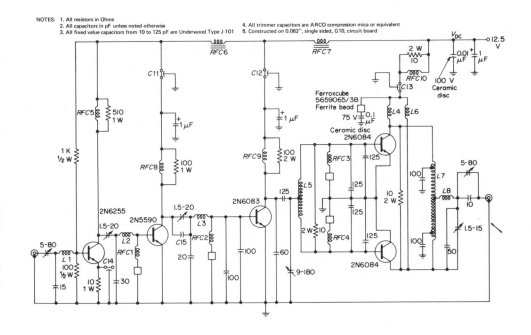

NOTES: 1. All resistors in Ohms
2. All capacitors in pF unless noted otherwise
3. All fixed value capacitors from 10 to 125 pF are Underwood Type J-101
4. All trimmer capacitors are ARCO compression mica or equivalent
5. Constructed on 0.062″, single sided, G10, circuit board

RFC1, 2, 3, 4 — 0.15 μH molded choke with Ferroxcube
5659065/3B ferrite bead on ground lead
RFC5 — 0.15 μH molded choke
RFC6, 7 — Ferroxcube VK-200 19/4B
RFC8 — 4T #16 awg wire, wound on 100 Ω 1 W resistor (75 nH)
RFC9 — 2T #15 awg wire, wound on 100 Ω 2 W resistor (45 nH)
RFC10 — 10T #14 awg wire, wound on 10 Ω 2 W resistor
L1, 2, 3 — 1 T #18 awg wire, ¼″ dia, ¾″ L (25 nH)
L4, 6 — 2T #15 awg wire, ¼″ dia, ½ L (30 nH)
L5, 7 — See outline diagram.
L8 — #12 awg wire approximately 1″ long (9 nH)
C11, 12, 13 — 680 pF, Allen Bradley type FA5C
C14 — 470 pF, Allen Bradley type SS5D
C15 — 5 pF, dipped silvered mica

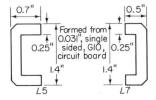

Outline diagrams for coils L5 and L7

Fig. 4-18. RF power-amplifier schematic diagram. Courtesy
Motorola.

259

Parameter	
RF power output	80 W
RF power input	180 mW
Supply voltage	12.5 V
Total current	12.9 A
Output stage collector efficiency	76%
Overall efficiency	49.5%
2nd harmonic	38 dB down
All other harmonics	>50 dB down
Output stage current (both devices)	8.4 A
Driver stage current	3.4 A
Pre–driver stage current	0.94 A
Input stage current	0.16 A
Stability — All spurious outputs are greater than 40 dB below rated power for input drive levels from 0 to 400 mW and V_{CC} values between 6.0 and 15.0 volts Burnout — No damage to any transistor with load open or shorted with $0 \pm 180°$ phase angle	

Fig. 4-19. Amplifier-circuit performance data. Courtesy Motorola.

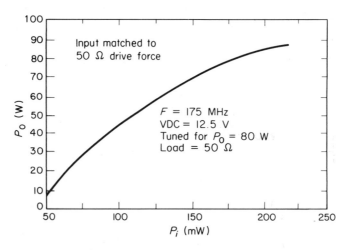

Fig. 4-20. Amplifier power output versus power input. Courtesy Motorola.

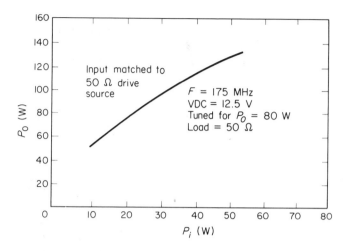

Fig. 4-21. Output-stage-power output versus power input. Courtesy Motorola.

The typical gain and RF power output capabilities of the transistors used in the design are shown in Figs. 4-22 through 4-25. The three higher-power devices are supplied in a radial-lead strip-line stud package for improved RF performance. The 2N6255 device is packaged in a TO-39 case.

The power levels for each stage (when the amplifier is used as a transmit-

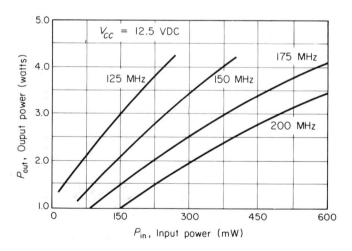

Fig. 4-22. Output power versus input power, 2N6255. Courtesy Motorola.

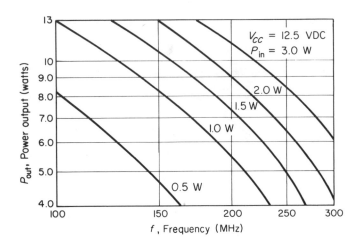

Fig. 4-23. Power output versus frequency, 2N5590. Courtesy Motorola.

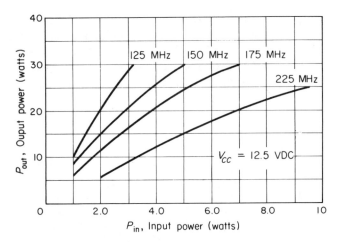

Fig. 4-24. Output power versus input power, 2N6083. Courtesy Motorola.

ter to provide 80 W of output at 175 MHz with a 12.5-V supply) are given on the block diagram of Fig. 4-26. These power levels have been determined by using minimum power gains as specified on the respective device data sheets. Input and interstage power levels lower than those shown can be expected because, typically, the transistor gains will be higher than the minimum specified value.

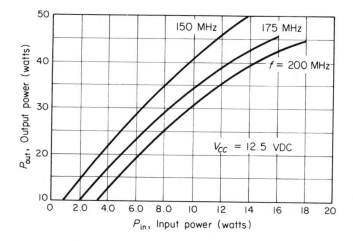

Fig. 4-25. Output power versus input power, 2N6084. Courtesy Motorola.

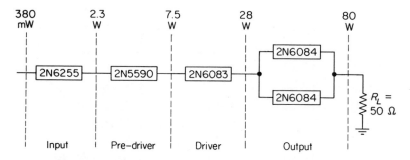

Fig. 4-26. Amplifier block diagram showing power levels obtained with minimum specified device gains. Courtesy Motorola.

4-4.3. Overall Amplifier Design

Large-signal transistor impedances have been used as a basis for synthesis of the amplifier matching networks. These impedances for the 2N6083 and 2N6084 transistors appear in Figs. 4-27 through 4-32. Large-signal data for the 2N6255 and 2N5590 transistors can be found on the respective device data sheets. This impedance information was obtained at the power-supply outputs and supply voltages indicated, with the transistors operating in the common-emitter configuration (both base and emitter dc grounded).

The resistive (R_{in}) and reactive (C_{in}, C_{out}) parallel impedance components are given by the curves. The approximate resistive portion of the load (R_L)

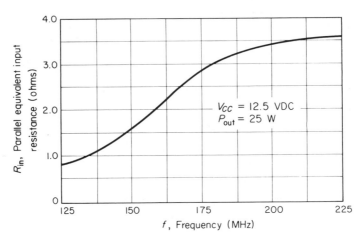

Fig. 4-27. Parallel equivalent input resistance versus frequency, 2N6083. Courtesy Motorola.

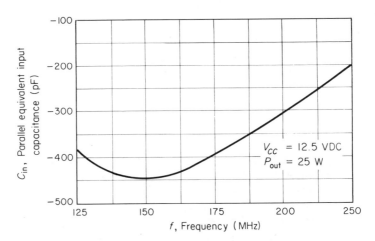

Fig. 4-28. Parallel equivalent input capacitance versus frequency, 2N6083. Courtesy Motorola.

that must be presented to the collector by the associated matching network can be calculated for an output power level, P, and a collector supply voltage, V_{CC}, by the equation

$$R_L = \frac{(V_{CC})^2}{2P}$$

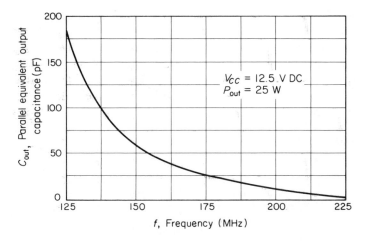

Fig. 4-29. Parallel equivalent output capacitance versus frequency, 2N6083. Courtesy Motorola.

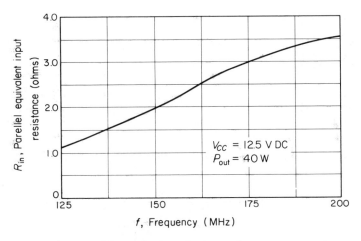

Fig. 4-30. Parallel equivalent input resistance versus frequency, 2N6084. Courtesy Motorola.

as shown in Fig. 4-12. Since $V_{CE(sat)}$ is much less than V_{CC} for the devices used, a peak-to-peak collector voltage swing of twice V_{CC} has been assumed. The actual optimum collector load resistance will vary somewhat from the value computed. However, designing the transmitter matching networks for a conjugate match to the resistance R_L and C_{out} values has yielded satisfactory results for the transistor types being used.

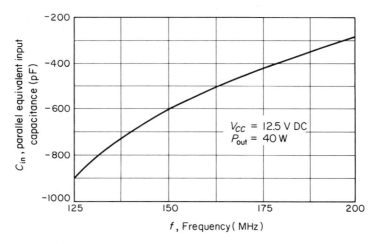

Fig. 4-31. Parallel equivalent input capacitance versus frequency, 2N6084. Courtesy Motorola.

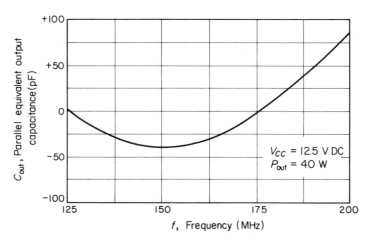

Fig. 4-32. Parallel equivalent output capacitance versus frequency, 2N6084. Courtesy Motorola.

It should be noted that the parallel input or output reactance component (C_{in} or C_{out}) is a capacitance only when the value as given by the curves is positive. When the value is negative, the reactance is inductive. *The value of the inductance is that which will resonate with a capacitance equal to the magnitude of the negative value shown on the curve.* The resonant frequency to use in calculating the inductance is that frequency for which the input or output reactive components are being determined. Likewise, the capacitive reactance equation must be used to find the parallel reactance.

4-4.4. Output Stage

When a single device is not capable of providing the required output power, or when it is desirable to use multiple devices to permit better heat distribution and higher reliability, the circuit designer is faced with combining transistors in a manner that is both efficient and economical. The choices that are available include the use of transformers, one of the many types of hybrid couplers, and the method used here (with conventional LC components). The difficulties usually produced by unequal load sharing and matching to extremely low impedance levels when power transistors are connected directly in parallel have been minimized through the use of *signal splitting* in both the input and output matching networks.

A simplified schematic for the output stage is shown in Fig. 4-33. Coil L_5 is used to step up the input impedance of Q_1 and Q_2 to give a higher impedance at point A and thus facilitate impedance matching to the driver stage. The collector coil L_7 divides the load between Q_1 and Q_2 and permits the power output of each transistor to be combined at a higher impedance level at point B. External capacitance has been added near each base and collector to provide a coarse impedance match at the operating frequency and a low impedance path to ground at the second harmonic frequency for improved efficiency.

The base capacitors C_3 and C_4 are located adjacent to the transistor cases. The collector capacitors C_5 and C_6 are located approximately at the midpoint of each half of inductance L_7. Always use capacitors with a low lead inductance. By using two capacitors in parallel to form C_3 and C_4, the inductance can be minimized, and a symmetrical configuration is obtained.

Resistors R_1 and R_2 help compensate for differences that may occur in transistor power gains and input impedances. Thus, the resistors help equalize load sharing between the two transistors. This produces a significant improvement in amplifier stability when V_{CC} and the input drive levels are varied. For symmetrical conditions, signals equal in both phase and amplitude will appear on each terminal of R_1, and each terminal of R_2, and no current will flow through the resistors. In a practical case, a small current will flow, but the effect on the matching network design will be insignificant.

All fixed-value capacitors from 10 to 125 pF used are Underwood mica dielectric units. The effective capacitance of these components at 175 MHz will deviate only slightly from their low-frequency value for nominal capacitance values up to approximately 60 pF. Larger-valued capacitors of this type have been characterized at 175 MHz and the results tabulated in Fig. 4-34.

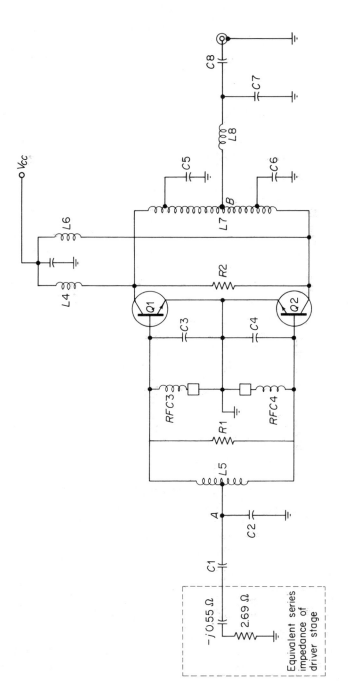

Fig. 4-33. Simplified output-stage schematic. Courtesy Motorola.

Nominal Low frequency Capacitance (pf)	Capacitance 175 MHz (pf)	Parallel reactance Component at 175 MHz Xp (Ohms)
80	88	-10.3
100	115	-7.9
125	151	-6.0
250	372	-2.4
Capacitor type: Type J-101. Underwood electric & Mfg. Co., Inc., Maywood, Ill. 60153		

Fig. 4-34. Capacitor values at 175 MHz. Courtesy Motorola.

4-4.5. Output Network Design

Using Fig. 4-32, the parallel equivalent output capacitance is found to be 2 pF for each 2N6084 when providing 40 W at 175 MHz. This produces a reactance of 455 Ω as shown by the equations of Fig. 4-35. (However, the negative sign on Fig. 4-32 indicates that the actual parallel impedance com-

$$X_C = \frac{1}{6.28\,FC} \qquad X_C \text{ of } Q_1 \text{ Output} = \frac{1}{6.28 \times 175\text{ MHz} \times 2\text{ pF}} \approx 455\ \Omega$$

$$R_L = \frac{(V_{CC})^2}{2p} \qquad R_L \text{ of } Q_1 = \frac{(12.5)^2}{2 \times 40} \approx 1.95\ \Omega$$

$$X_L \text{ of } L_4 = 6.28 \times 175\text{ MHz} \times 30\text{ mH} \approx 33\ \Omega$$

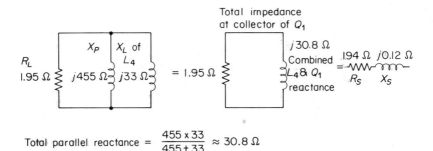

$$\text{Total parallel reactance} = \frac{455 \times 33}{455 + 33} \approx 30.8\ \Omega$$

$$R_S \approx \frac{1.95}{1 + \left(\frac{1.95}{30.8}\right)^2} \approx 1.94\ \Omega \qquad X_S \approx \frac{30.8}{1 + \left(\frac{30.8}{1.95}\right)^2} \approx j0.12\ \Omega$$

Fig. 4-35. Total impedance at the collector of Q_1.

ponent is an inductance.) The R_L for each 2N6084 is found to be 1.95 Ω when providing 40 W with a 12.5-V power supply.

The remaining parallel impedance component to be accounted for at the collector of Q_1 is the dc feed coil L_4. This coil should present a low impedance for frequencies below approximately 30 MHz. This is necessary to assure stable operation, since the device gain is very high and the normal load is essentially removed by capacitor C_8 at these lower frequencies. Coil L_4 must also have low dc resistance to permit efficient operation at the dc current levels involved. (Any appreciable resistance will cause a voltage drop to the collector of Q_1.) An arbitrary inductance value of 30 nH meets these conditions and also permits adequate decoupling from the dc line at 175 MHz.

As shown in Fig. 4-35, the inductive reactance of L_4 is 33 Ω and, when combined with the reactance of Q_1 (455 Ω), produces a total parallel reactance of 30.8 Ω. For the next step in output network design, the parallel impedance components of Q_1 and L_4 must be converted to series components. The necessary equations and values are given in Fig. 4-35.

The value of inductor L_7 and the location and value of capacitor C_5 is a tradeoff among: (1) circuit Q, (2) providing an impedance at point B that can be efficiently matched to the 50-Ω load, and (3) providing a low impedance from collector to ground for the second harmonic of 175 MHz.

Good results have been achieved by using an inductance of 6 nH for one-half of L_7, with C_5 having a nominal (low-frequency) value of 100 pF. As shown in Fig. 4-34, the 100-pF capacitance is raised to about 115 pF when the capacitor is used at 175 MHz. Capacitor C_5 is located approximately 3 nH away from the collector of Q_1. As shown in Fig. 4-36, the L_7-C_5 network is combined with the Q_1-L_4 network to find the impedance at point B of Fig. 4-33. The necessary equations and values are given in Fig. 4-36.

The final parallel impedance shown in Fig. 4-36 is the resulting impedance at point B of Fig. 4-33 contributed by transistor Q_1. A step-up in parallel resistance from 1.95 to 15.2 ohms is obtained by the network. Following the same procedure for transistor Q_2 produces the same result. The combined impedance (Q_1, Q_2, L_4, L_6, L_7, C_5, and C_6) at point B is then as shown in Fig. 4-37.

The output network from point B must be designed to transform the 50-Ω external load to the conjugate of 2.5-Ω resistance in series with 3.6-Ω positive reactance. The required load impedance at B is thus $2.5 - j3.6\ \Omega$, as shown in Fig. 4-38. Note that the network of Fig. 4-38 is quite similar to that of Fig. 4-13, and that the equations of Fig. 4-13 can be used to solve network design. However, the following paragraphs describe a slightly different approach to the problem.

$$X_C \text{ of } C_5 = \frac{1}{6.28 \times 175 \text{ MHz} \times 115 \text{ pF}} \approx -7.9 \ \Omega$$

$$X_L \text{ of } L_7 = 6.28 \times 175 \text{ MHz} \times 6 \text{ MH} \approx 6.6 \ \Omega$$

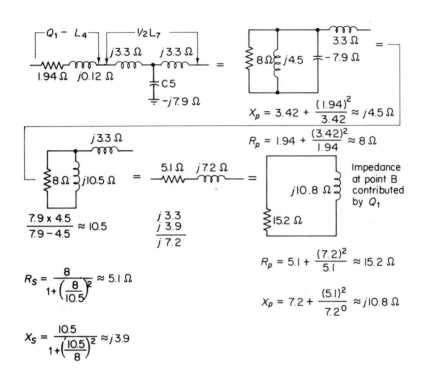

Fig. 4-36. Parallel impedance at point B contributed by Q_1.

For the purposes of this network design example, the loaded Q (or Q_L) of the network in Fig. 4-38 is defined as $(X_{L8} + X_B)/R_B$. When Q_L is so defined, values of 5 to 8 will provide a good tradeoff among harmonic attenuation, low resistance loss, and the voltage levels that can occur during high-output VSWR conditions. For this design, a value of 9 nH was chosen for L_8, yielding a Q_L value of

$$Q_L = \frac{9.9 + 3.6}{2.5} \approx 5.4$$

where

$$X_{LB} = 6.28 \times 175 \text{ MHz} \times 9 \text{ nH} \approx 9.9$$

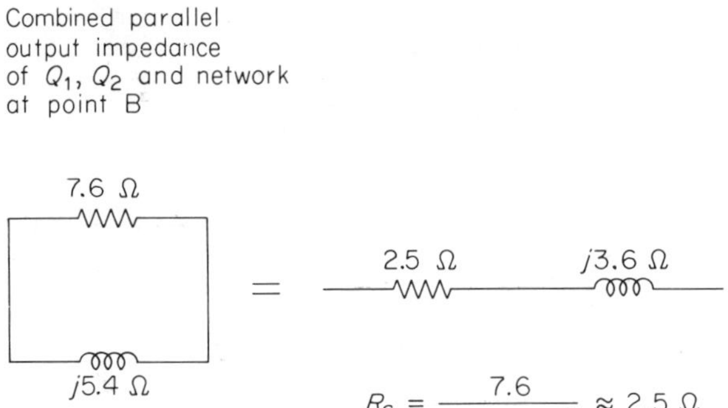

Combined parallel
output impedance
of Q_1, Q_2 and network
at point B

$$R_S = \frac{7.6}{1 + \left(\frac{7.6}{5.4}\right)^2} \approx 2.5 \;\Omega$$

$$X_S = \frac{5.4}{1 + \left(\frac{5.4}{7.6}\right)^2} \approx j3.6 \;\Omega$$

Fig. 4-37. Combined series output impedance of Q_1, Q_2, and the network at point B.

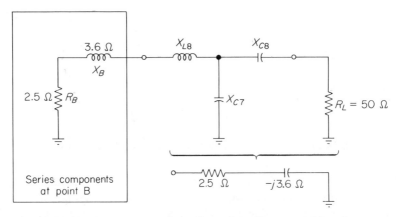

Fig. 4-38. Output network configuration. Courtesy Motorola.

With Q_L established as 5.4, the values of C_7 and C_8 can be found using the Fig. 4-13 equations, as follows:

$$A = \sqrt{\frac{2.5(1 + 5.4^2)}{50} - 1} \approx 0.71$$

$$B = 2.5(1 + 5.4^2) \approx 75.4$$

$$X_{C7} = \frac{75.4}{5.4 - 0.71} \approx 16.1 \;\Omega$$

$$X_{C8} = 0.71 \times 50 \approx 35.5\,\Omega$$

$$C_7 = \frac{1}{6.28 \times 175\text{ MHz} \times 16.1} \approx 56.5\text{ pF}$$

$$C_8 = \frac{1}{6.28 \times 175\text{ MHz} \times 35.5} \approx 25.6\text{ pF}$$

As shown in Fig. 4-18, capacitance C_7 and C_8 consist of low-loss fixed-value capacitors in parallel with trimmer capacitors to minimize power loss and improve tuning resolution.

4-4.6. Driver/Final Network Design

The parallel input impedance components for each 2N6084 transistor, as obtained from Figs. 4-30 and 4-31, are: $R_{in} = 3\,\Omega$, $C_{in} = -420$ pF. A C_{in} of -420 pF produces a reactance of 2.2 Ω at 175 MHz as shown by the equations of Fig. 4-39. (However, the negative sign in Fig. 4-31 indicates that the actual parallel impedance component is an inductance.)

$$X_C \text{ of } Q_1 \text{ input} = \frac{1}{6.28 \times 175\text{ MHz} \times -420\text{ pF}} \approx -2.2\,\Omega$$

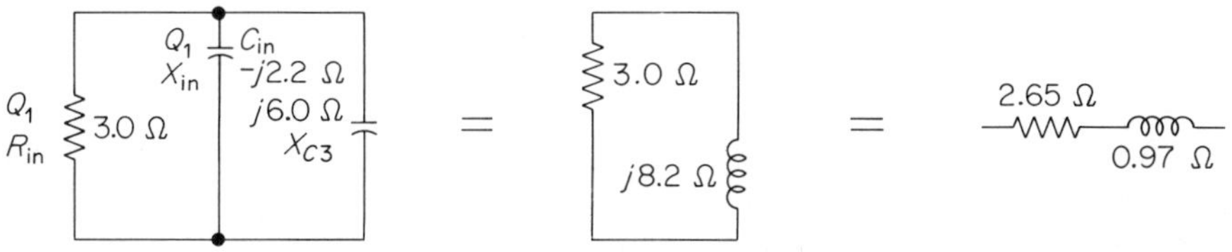

$$\begin{array}{r} -j2.2 \\ -j6.0 \\ \hline j8.2\ \Omega \end{array}$$

$$R_S = \frac{3.0}{1 + \left(\frac{3.0}{8.2}\right)^2} \approx 2.65\,\Omega$$

$$X_S = \frac{8.2}{1 + \left(\frac{8.2}{3.0}\right)^2} \approx 0.97\,\Omega$$

Fig. 4-39. Combination base-to-ground series impedance of one 2N6084 device.

Neglecting the effects of RFC_3 and RFC_4 (Fig. 4-33), but including measured values from Fig. 4-34 for the two external 125-pF capacitors from base to ground, yields a combination base/ground impedance for one 2N6084 transistor of 2.65 Ω, j0.97 Ω, as shown in Fig. 4-39.

A value of 8 nH for one-half of the signal splitting inductor L_5 provides low network insertion loss, adequate harmonic rejection, and a convenient means of matching the driver stage. Converting one-half of L_5 to its reactance value at 175 MHz yields a reactance of 8.8 Ω and gives an impedance at point A (Fig. 4-33) of 38.67 Ω in parallel with $j10.49$ Ω, as shown in Fig. 4-40.

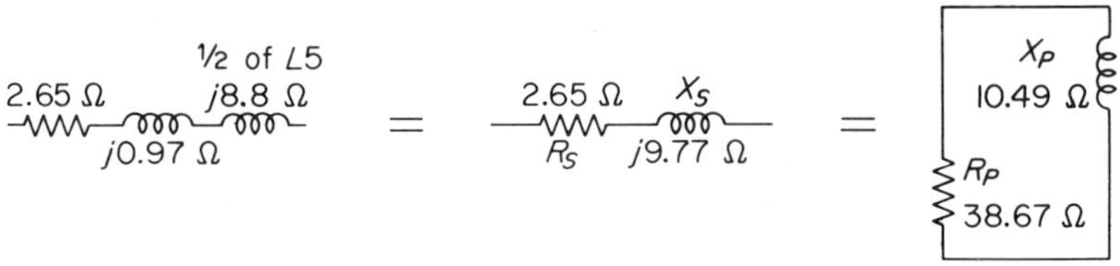

$$R_P = 2.65 + \frac{(9.77)^2}{2.65} \approx 38.67$$

$$X_P = 9.77 + \frac{(2.65)^2}{9.77} \approx 10.49$$

$$\tfrac{1}{2}L_5 = 8 \text{ nH}$$

$$X_L \text{ of } \tfrac{1}{2}L_5 = 6.28 \times 175 \text{ MHz} \times 8 \text{ nH} \approx 8.8 \text{ } \Omega$$

Fig. 4-40. One-half of impedance at point A.

A like impedance will result for the other half of the output stage, giving the total impedance from point A (Fig. 4-33) to ground of 1.33 Ω in series with $j4.89$ Ω, as shown in Fig. 4-41.

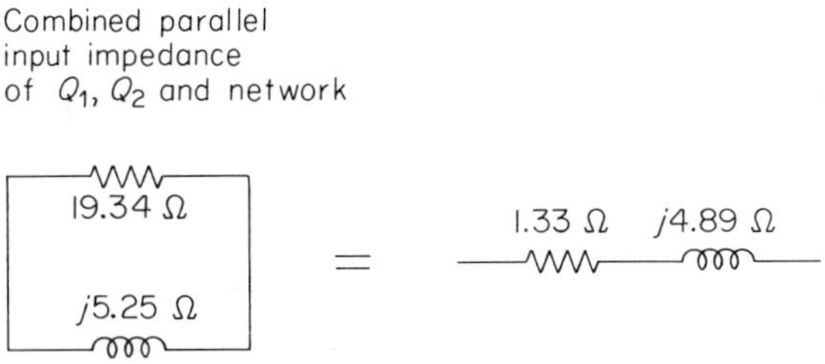

$$R_S = \frac{19.34}{1 + \left(\frac{19.34}{5.25}\right)^2} \approx 1.33 \text{ } \Omega$$

$$X_S = \frac{5.25}{1 + \left(\frac{5.25}{19.34}\right)^2} \approx j4.89 \text{ } \Omega$$

Fig. 4-41. Total series impedance from point A to ground.

Using Fig. 4-29, the parallel equivalent output capacitance of the 2N6083 transistor is found to be 25 pF when providing 28 W at 175 MHz. As shown in Fig. 4-18, a 60-pF capacitor is connected in parallel with the output of the 2N6083 transistor. This results in a combined capacitance of 85 pF (60 + 25), and produces a reactance of 10.7 Ω, as shown by the equations of Fig. 4-42. Also as shown in Fig. 4-18, a 45-nH coil (RFC_9) appears at the collector of the 2N6083 and produces a reactance of 48.5 Ω (Fig. 4-42). The resistive parallel impedance component at the collector of the 2N6083 is 2.8 Ω, as shown by the Fig. 4-42 equations.

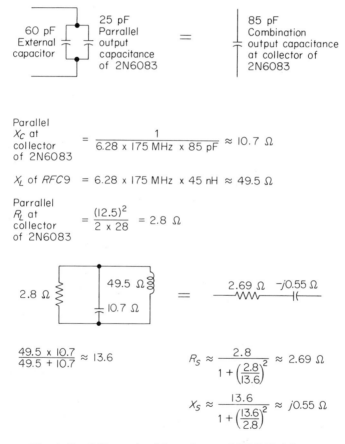

$$\text{Parallel } X_C \text{ at collector of 2N6083} = \frac{1}{6.28 \times 175 \text{ MHz} \times 85 \text{ pF}} \approx 10.7 \ \Omega$$

$$X_L \text{ of } RFC9 = 6.28 \times 175 \text{ MHz} \times 45 \text{ nH} \approx 49.5 \ \Omega$$

$$\text{Parallel } R_L \text{ at collector of 2N6083} = \frac{(12.5)^2}{2 \times 28} = 2.8 \ \Omega$$

$$\frac{49.5 \times 10.7}{49.5 + 10.7} \approx 13.6$$

$$R_S \approx \frac{2.8}{1 + \left(\frac{2.8}{13.6}\right)^2} \approx 2.69 \ \Omega$$

$$X_S \approx \frac{13.6}{1 + \left(\frac{13.6}{2.8}\right)^2} \approx j0.55 \ \Omega$$

Fig. 4-42. Collector load impedance of 2N6083 driver.

The parallel components combine to provide the series components of $2.69 - j0.55 \ \Omega$, also as shown in Fig. 4-42. Thus, the driver/final interstage network must match the impedance at point A of Fig. 4-33 ($1.33 - j4.89 \ \Omega$)

$$X_{C2} = \frac{D}{Q_L - C} = \frac{19.2}{3.7 - 2.5} \approx 16 \ \Omega$$

$$X'_{C1} = CR_S = (2.5)(2.69) \approx 6.72 \ \Omega$$

$$X_{C1} = (X'_{C1}) - (X_S) = 6.72 - 0.55 = 6.17 \ \Omega$$

$$C = \sqrt{\left(\frac{R_L(1 + Q_L^2)}{R_S}\right)} - 1 = \sqrt{\left(\frac{1.33(1 + 3.7^2)}{2.69}\right)} - 1 \approx 2.5$$

$$D = R_L(1 + Q_L^2) = (1.33)(1 + 3.7^2) \approx 19.2$$

$$Q_L = \frac{X_L}{R_L} = \frac{4.89}{1.33} \approx 3.7$$

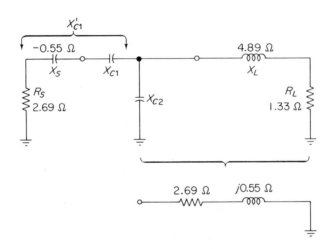

Fig. 4-43. Driver/final matching network.

to the required collector load impedance of the 2N6083 (that is, to the conjugate of $2.69 - j0.55 \ \Omega$) as shown in Fig. 4-43. The proper reactance of capacitors C_1 and C_2 to achieve the impedance can be found by solving the equations with the values of Fig. 4-43.

Converting X_{C_1} and X_{C_2} to capacitance at 175 MHz produces a C_1 of 147 pF and a C_2 of 57 pF. A capacitor having a nominal low-frequency value of 125 pF is used for C_1 since, as noted in Fig. 4-34, such a capacitor will show an effective capacitance of approximately the calculated value (147 pF) at 175 MHz. Capacitor C_2 is made variable (9 to 180 pF) to permit adjustment and tuning.

The output-stage combining and matching networks have been designed for optimum operation near 175 MHz. If it is desirable to operate significantly lower in frequency or to have the capability of tuning across the entire 144 to 175 MHz band, the same basic circuit can be used with some modifications. Coil L_5 should be broken down into two parts, each having an inductance value on the order of 12 to 15 nH. The output combining network should also be modified to use a single 250-pF Underwood type J-101 capacitor from point B (Fig. 4-33) to ground rather than the two 100-pF capacitors C_5 and C_6. This will serve to minimize the flow of circulating harmonic current power that otherwise tends to become excessive for frequencies below 160 MHz.

Output-stage designs of this type have been evaluated for power outputs of 80 W and show excellent performance when tuned from 144 to 175 MHz. Collector efficiencies were greater than 72 per cent, and second harmonic output power typically was more than 20 dB down.

This completes the discussion of the matching networks associated with the output stage. A similar design technique can be followed for the input network and other interstage networks shown in Fig. 4-18. Section 4-3 summarizes design procedures for the networks involved.

4-4.7. Thermal Design

The importance of good thermal design and construction techniques cannot be overemphasized when high RF power levels are involved. Amplifier construction should include the use of:

1. A smooth heat-sink surface to maximize heat sink/transistor case contact area.
2. A proper amount of thermal joint compound at the heat sink/transistor case interface.
3. Torque as specified for the transistor when fastening the transistor to the heat sink.
4. A heat-sink configuration that will permit locating the higher-power-level stages near the maximum heat-transfer position on the heat sink.

Using the measured current values given in Fig. 4-19, and nominal RF power levels for each as shown in Figs. 4-22 through 4-25, the approximate power dissipated in each amplifier stage, P_D, may be calculated as follows:

$$P_D = P_{in(RF)} + P_{in(dc)} - P_{out(RF)}$$

In practical terms, allowing for worst-case conditions, the actual power dissipated by each stage is: input 1.2 W, predriver 6.5 W, driver 22 W, and output 52 W. (Keep in mind that the input stage is forward-biased for about 40-mA collector current, even under no-signal conditions, thus resulting in a minimum 0.5-W dissipation at all times.)

Thermal data for transistors amplifier using devices appear in Fig. 4-44. For any single-transistor stage, the heat sink/ambient thermal resistance

Device	$R\theta_{JC}$	$R\theta_{CS}$	T_J (max)
	°C/W	°C/W	°C
2N6255	35	- - -	200
2N5590	5.58	0.3	200
2N6083	1.92	0.3	200
2N6084	1.56	0.3	200

$R\theta_{JC}$ = junction-to-case thermal resistance

$R\theta_{CS}$ = case-to-heat-sink thermal resistance. Values given apply when using thermal compound and with nut torqued to 6.5 in. lb

$T_{J(max)}$ = maximum junction temperature

Fig. 4-44. Thermal data for amplifier devices. Courtesy Motorola.

$(R\theta_{SA})$ requirement may be computed using the following equations:

$$R\theta_{JS} = R\theta_{JC} + R\theta_{CS}$$

$$R\theta_{SA} = \frac{T_{J(max)} - T_A}{P_D} - R\theta_{JS}$$

where $R\theta_{JS}$ = junction/heat-sink thermal resistance

T_A = ambient temperature

When two transistors are mounted close together on a common heat sink (as in the case for the two 2N6084 output transistors), the equations can be modified to treat the two transistor thermal resistance paths as a parallel circuit. The combined $R\theta_{JS}$ value (using the individual Fig. 4-44 values) for the output stage then calculates to be:

$$R\theta_{JS(output)} = \frac{(1.56 + 0.3)(1.56 + 0.3)}{(1.56 + 0.3) + (1.56 + 0.3)} = 0.93°C/W$$

Heat-sink requirements for the output stage can then be obtained by using the $R\theta_{JS(output)}$. If the maximum ambient-temperature requirement is assumed to be 60°C, this gives:

$$R\theta_{SA(output)} = \frac{200 - 60}{52} - 0.93 = 1.77°C/W$$

The output-stage heat sink must therefore provide a sink to ambient thermal resistance of 1.77°C/W maximum for a maximum T_A of 60°C. If a single higher-power transistor is used for the output stage in place of two lower-power devices operating in parallel, the resulting heat-sink requirements for the same total P_D will be more severe, unless $R\theta_{JS}$ for the single device is equal to (or less than) $(R\theta_{JS})/2$ of each parallel device.

The respective stage P_D values and transistor thermal data of Fig. 4-44 can be used in the equations to calculate heat-sink requirements for the driver and predriver stages. For $T_A = 60°C$, the requirements are:

$$R\theta_{SA(driver)} = \frac{200 - 60}{22} - 2.22 = 4.14°C/W$$

$$R\theta_{SA(predriver)} = \frac{200 - 60}{6.5} - 6.15 = 15.45°C/W$$

No heat sink is normally required for the input-stage (2N6255) transistor if the transistor is mounted on a metal surface. The 2N6255 transistor is provided with a TO-39 type of case.

The following points should be noted when computing heat-sink requirements:

1. The computed $R\theta_{SA}$ values will be conservative since the calculations are based on worst-case junction-to-case thermal resistance values.
2. If operation into mismatched loads is anticipated, the $R\theta_{SA}$ values must be modified to take into account the increase in dissipation that can occur for these operating conditions. This is discussed further in Sec. 4-4.8.
3. For the $R\theta_{JS(output)}$ equation to fully apply, the transistors must be located at the *same point on the heat sink*. Of course, this cannot actually be true, but can be considered a good approximation for the output stage.
4. The $R\theta_{SA}$ values are for continuous amplifier operation. For duty-cycle operation, such as 1 minute on/3 minutes off, the heat-sink requirements will be significantly reduced. Refer to Sec. 1-4.3.6.

4-4.8. Other Design Considerations

Low-impedance emitter ground paths are extremely important for high power gain. A good ground plane is always important in RF amplifiers, and this is especially true in the present case, where high peak currents flow in the output stage.

The low-dc impedance connected from base to ground to establish class C bias must:

1. Be low in Q to minimize resonant conditions.
2. Present a high impedance at the operating frequency to prevent power loss.

These conditions can be obtained by using a high-Q, 0.15-μH choke with ferrite bead inserted over the lead going to ground.

The method and types of components used in bringing dc power to the various stages is very important in preventing oscillations. Multiple bypass capacitors have been used to provide adequate decoupling over a broad band of frequencies. Carbon resistors (swamping resistors) have been used in parallel with collector dc feed coils to lower their Q and lessen the possibility of unwanted resonant conditions.

Even though the amplifier has been subjected to burn-out testing with open and shorted loads at all possible load phase angles, and with no damage to any transistor, the heat-sink requirements for prolonged operation under these conditions can be quite severe. Figure 4-45 indicates the total amplifier current and power dissipation when the output is terminated in an open- or short-circuit air line of variable length.

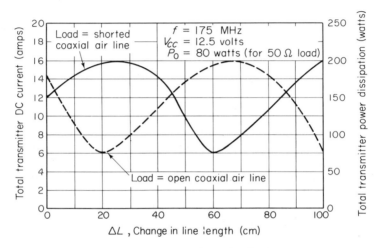

Fig. 4-45. Open- and short-circuit load test. Courtesy Motorola.

Worst-case power dissipation in only the output stage for open and short conditions can increase from the normal 52 W to approximately 160 W. For an ambient temperature of only 30°C, the output-stage heat sink/ambient thermal resistance requirement will now be approximately 0.14°C/W rather than the more realistic 2.35°C/W required with the proper output termination. Designing the heat sink to be adequate for this infrequent mode of operation will result in a sink that is neither economical nor of a practical size. The need for a protection circuit with high power transmitters which will reduce either the RF drive power or the dc supply power to the output stage during high VSWR conditions must therefore still be given serious consideration, even though transistors capable of withstanding infinite VSWR conditions are used.

4-4.9. Amplifier Construction

Care should be exercised during amplifier layout and construction to obtain a symmetrical arrangement for the output stage. That is, the output transistors should be in line with each other and equally spaced from a center line passing through the remaining transistors. Generally, the input, pre-driver, and driver transistor should be in a straight line.

The following precautions should be observed to prevent physical damage to the transistor stud package:

1. The nut should be installed on the stud, and turned to the specified torque, *before* soldering the transistor leads to the circuit. This sequence is recommended to prevent an upward force being applied to the leads near the case body.
2. The maximum torque ratings for the mounting nut must not be exceeded. The ratings are 6.5 in/lb for the 2N5590, 2N6083, and 2N6084 transistors.
3. Capacitors can be expected to deviate somewhat from the calculated values because of finite lead, case, and interconnection inductance. Lead length and interconnection inductance can be minimized by circuit layout and construction procedures. A reduction in case inductance can be realized by replacing the fixed-value capacitors with ceramic-chip capacitors.

4-5. Using Sampled y-Parameter Techniques to Design RF Networks

This section describes the design of a UHF broadband amplifier using sampled *y*-parameter techniques as developed by Motorola. The section also illustrates the use of the Smith chart to calculate *y*-parameters (admittances). The sampled *y*-parameter technique is an extension of the conventional small-signal *y*-parameter design described in previous sections of this chapter, and is particularly useful at ultrahigh frequencies.

The amplifier shown in Fig. 4-46 has a midband gain of 31 dB $\pm$ 1 dB. The lower 3-dB cutoff point occurs at 3.1 MHz and the upper end at 405 MHz. The frequency response is given in Fig. 4-47. The usable sensitivity defined as input signal required to produce a S $+$ N/N $=$ 10 dB is 100 μV. The dynamic range is 40 dB, which corresponds to a root-mean-square output voltage of 1 to 100 mV.

4-5.1. Circuit Analysis

As shown in Fig. 4-46, the 400-MHz broadband amplifier consists of three identical stages. Both input and output are terminated into 50-Ω impedances. Since the stages are identical, only one stage is discussed.

A shunt feedback network composed of L_1, L_3, R_1, R_3, and C_2 is used to stabilize gain over the 400-MHz bandwidth. Capacitor C_2 is used for blocking direct current and could be replaced by an appropriate Zener diode. Inductor L_2 is the interstage coupling used to match the output impedance of one stage to the input impedance of the following stage.

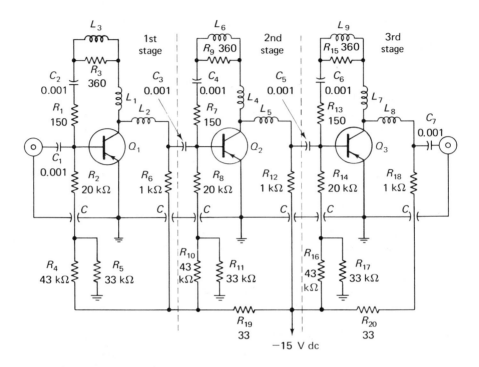

L_1, L_4 and L_7 = 3.5 turns #20 soft-drawn, ID = $\frac{1}{4}$ inch, Length = 1 inch

L_2, L_5 and L_8 = 6 turns #20 soft-drawn, ID = $\frac{1}{4}$ inch, Length = $\frac{3}{4}$ inch

L_3, L_6 and L_9 = 6 turns #24 tinned wire on CF 103, Q_3 toroid with Teflon sleeve

C = 500 pF Erie button capacitor with 0.02 μF ceramic disk. All other capacitors are ceramic disk.

Q_1, Q_2, Q_3, = 2N4957

Bandwidth: 3.1 - 405 MHz at 3-dB point

Gain: 31 dB ± 1 dB

Sensitivity: 100 μV ($\frac{S+N}{N}$ = 10 dB)

I_C = 5 mA, V_{CE} = 10 V dc for all devices

Fig. 4-46. 400-MHz wideband amplifier. Courtesy Motorola.

The choice of the 2N4957 for this application stems from its high gain and very low noise figure at ultrahigh frequencies. At a bias current of 5 mA, the 2N4957 has a current gain of 50. The input impedance is *approximately* equal to the current gain divided by the transconductance, or 250 Ω. A relatively small feedback loop gain (small feedback conductance) is required to reduce the input impedance to 50 Ω, which is the prescribed source value.

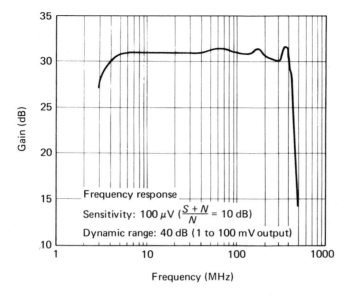

Fig. 4-47. Frequency response of amplifier. Courtesy Motorola.

Thus, the open-loop gain of the amplifier is not excessively sacrificed, allowing a relatively high closed-loop gain.

If a minimum noise figure is a major consideration, the 2N4957 should be biased at 2 mA, which has been found to be an optimum bias point for a noise figure. However, if 2 mA is used and the input impedance is held to 50 Ω, a larger feedback loop gain is needed, and a considerable amount of open-loop gain is sacrificed.

It is not necessary to bias each stage identically. The first stage may be biased for minimum noise, and the following stages for high gain, to achieve overall optimum performance. However, the design work is more time consuming for this case.

4-5.2. Design Analysis

The design of a multistage broadband amplifier, without an overall feedback loop, may be broken down to that of the individual stages. To minimize the interaction between stages, an interstage coupling network is needed to match the output impedance of one stage to the input impedance of the next stage. For example, if the individual stages are designed to have input and output terminations of 50 Ω, cascading of individual stages can be achieved without interaction between stages.

Usually, negative feedback is required to stabilize the gain-frequency characteristics. There are four possible feedback schemes, as shown in Fig. 4-48. Each scheme has its own merit, and can be readily characterized by different two-port parameters: y-parameters for shunt–shunt, g-parameters for shunt–series, h-parameters for series–shunt, and z-parameters for series–series.

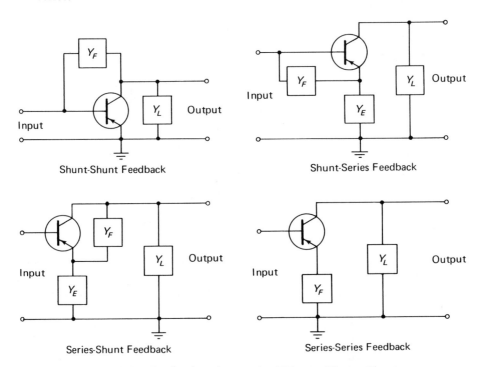

Fig. 4-48. Feedback schemes in RF amplifiers. Courtesy Motorola.

For the transistor having an input impedance higher than that of the input termination, a shunt–shunt feedback scheme is quite suitable for matching the required source to the input impedance of the device. Moreover, y-parameters can be used readily for design calculations. Only the shunt–shunt feedback configuration will be discussed.

The design procedure can be divided into the following steps:

1. Determine the required design specifications of gain, bandwidth, and input and output terminations.
2. Determine the feedback network.
   a. Use a low-frequency transistor model to calculate the value of feedback conductance to meet the gain and input impedance requirements.
   b. Use a constant G_{oo} (which is the power gain when $Y_L = y_{22}$, or the power gain when the load admittance equals the transistor output admittance)

plot to eliminate device high-frequency gain capability under the bias conditions.

  c. Use the constant-gain expression to calculate the feedback inductance to meet gain and bandwidth requirements.
3. Realize a feedback network.
4. Determine the interstage coupling network.

4-5.3. Design Example

The following is an illustration of how the four basic steps can be implemented:

  1. *Design specifications:*

Gain: 30 dB $\pm$ 1 dB
Bandwidth: 400 MHz
Input termination: 50 Ω
Output termination: 50 Ω

  2a. *Construct a low-frequency model* as shown in Fig. 4-49. Note the gain expression and input admittance expression shown in Fig. 4-49.

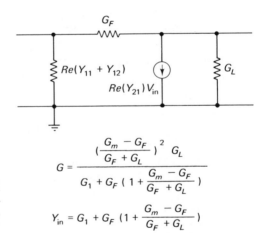

Fig. 4-49. Low-frequency model with corresponding gain expression and input impedance expression. Courtesy Motorola.

$$G = \frac{(\frac{G_m - G_F}{G_F + G_L})^2 \, G_L}{G_1 + G_F \, (1 + \frac{G_m - G_F}{G_F + G_L})}$$

$$Y_{in} = G_1 + G_F \, (1 + \frac{G_m - G_F}{G_F + G_L})$$

Once the load, G_L, is chosen, there is one-to-one correspondence between gain and the value of G_F, and between Y_{IN} and G_F. The choice of G_F for a desired value of gain automatically fixes the input impedance.

The gain and its corresponding input admittance are plotted versus G_F in Fig. 4-50. If the constant-gain function is mapped on the Z_F plane of an impedance chart (known as the Smith chart), it represents a circle of constant R_F, which is equal to the reciprocal of G_F. Figure 4-51 shows two circles of constant gain of 10 and 12 dB plotted on a Smith chart.

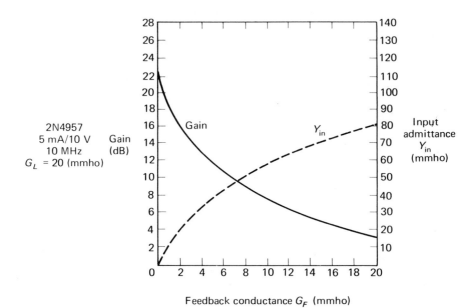

Fig. 4-50. Gain and corresponding input admittance plotted against G_F for the 2N4957. Courtesy Motorola.

2b. *Estimate high-frequency gain capability.* This is done by mapping G_{oo} into the $Y_F = 1/Z_F$ plane of the Smith chart. The equations for calculating G_{oo} are given in Fig. 4-51, along with the values necessary to make the plot. Figure 4-51 shows two 400-MHz constant G_{oo} curves of 10 and 12 dB superimposed on the low-frequency constant-gain curves. If the constant G_{oo} curve intersects the corresponding low-frequency gain circle, the device is capable of yielding the gain and bandwidth at the specified input impedance. Maximum gain-bandwidth capability occurs when the two circles are tangent to each other.

For this particular example, a 2N4957 biased at 5 mA of collector current and 10 V collector voltage has a capability of 12-dB gain at 400 MHz with an input impedance of 30 Ω, and 10 dB at 400 MHz with a 23-Ω input. Figure 4-52 lists the *y*-parameters of the device at the sampled frequencies.

2c. *Approximate determination of feedback network.* The *intersection* of the constant G_{oo} curve, and its corresponding low-frequency gain, represents a feedback network that at least will yield the gain specified at the two extremes. In this example, for 10-dB gain the possible values of Y_F are 5.5 — $j3.0$ and $1.0 + j2.5$. The latter, being a capacitive network, is discarded. The $5.5 — j3.0$ Y_F corresponds to a series feedback network composed of a 150-Ω resistance and an approximate 30 nH inductance. A typical frequency

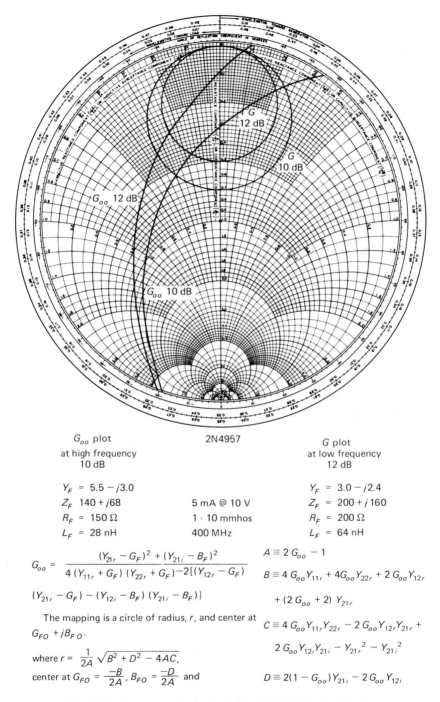

G_{oo} plot	2N4957	G plot
at high frequency		at low frequency
10 dB		12 dB

$Y_F = 5.5 - j3.0$		$Y_F = 3.0 - j2.4$
$Z_F \quad 140 + j68$	5 mA @ 10 V	$Z_F = 200 + j160$
$R_F = 150\ \Omega$	1 - 10 mmhos	$R_F = 200\ \Omega$
$L_F = 28\ \text{nH}$	400 MHz	$L_F = 64\ \text{nH}$

$$G_{oo} = \frac{(Y_{21r} - G_F)^2 + (Y_{21i} - B_F)^2}{4(Y_{11r} + G_F)(Y_{22r} + G_F) - 2[(Y_{12r} - G_F)}$$

$$(Y_{21r} - G_F) - (Y_{12i} - B_F)(Y_{21i} - B_F)]$$

The mapping is a circle of radius, r, and center at $G_{FO} + jB_{FO}$.

where $r = \dfrac{1}{2A}\sqrt{B^2 + D^2 - 4AC}$,

center at $G_{FO} = \dfrac{-B}{2A}$, $B_{FO} = \dfrac{-D}{2A}$ and

$A \equiv 2\,G_{oo} - 1$

$B \equiv 4\,G_{oo}Y_{11r} + 4G_{oo}Y_{22r} + 2\,G_{oo}Y_{12r}$

$\qquad + (2\,G_{oo} + 2)\,Y_{21r}$

$C \equiv 4\,G_{oo}Y_{11r}Y_{22r} - 2\,G_{oo}Y_{12r}Y_{21r} +$

$\qquad 2\,G_{oo}Y_{12i}Y_{21i} - Y_{21r}{}^2 - Y_{21i}{}^2$

$D \equiv 2(1 - G_{oo})Y_{21i} - 2\,G_{oo}Y_{12i}$

Fig. 4-51. Constant G_{oo} and G for 2N4957 in Smith chart form. Courtesy Motorola.

response of the amplifier using this feedback network and $Y_L = y_{22} + Y_F$ at the high-frequency extreme is shown in Fig. 4-53.

2d. *Final determination of feedback network.* In order to flatten the frequency response at the intermediate frequencies (that is, in order to eliminate the dip at about 300 MHz shown in Fig. 4-53), alteration of either Y_F, Y_L, or both should be made. One way of determining the feedback network is by repeating the same steps of 2c at intermediate sampled frequencies. (It is obvious that all these calculations are best accomplished on a computer.)

Figure 4-54 shows the intersections of the 10-dB G_{oo} with the 10-dB low-frequency gain curve at 100, 200, and 300 MHz, with a respective Y_L of 4.2 − $j3.5$, $4.7 − j3.4$, and $5.0 − j3.5$ mmhos.

The corresponding series inductance, L_F, of the feedback network is plotted versus frequency as the dashed line of Fig. 4-55. If Y_F is kept constant, in this case $20 + j0$, L_F varies with frequency as shown by the solid line of Fig. 4-55. The solid line is obtained from calculation of the 10-dB constant-gain expression.

3. *Realization of the feedback network.* The frequency characteristics of a VHF toroidal inductance are similar to that of the desired L_F, so it is par-

f MHz	Y_{11r}	Y_{11i}	Y_{12r}	Y_{12i}	Y_{21r}	Y_{21i}	Y_{22r}	Y_{22i}
10	3.77	0.283	0	−0.033	134	−11.7	0.024	0.076
100	4.50	2.80	0	−0.23	53	−35	0.2	0.72
200	5.90	5.9	0	−0.44	50.5	−46.5	0.2	1.5
300	7.80	8.4	0	−0.66	42.5	−54.25	0.3	2.2
400	9.60	10.5	0	−0.90	34.0	−61.0	0.3	3.0

Fig. 4-52. The *y*-parameters of the transistor at sampled frequencies. Courtesy Motorola.

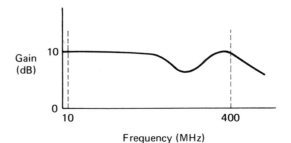

Fig. 4-53. Amplifier frequency response with a series R_L feedback network. Courtesy Motorola.

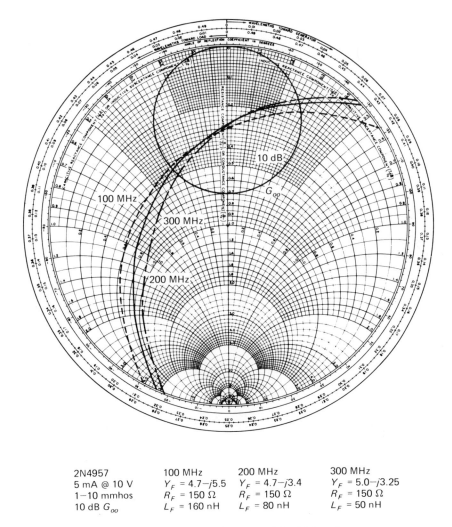

2N4957	100 MHz	200 MHz	300 MHz
5 mA @ 10 V	$Y_F = 4.7-j5.5$	$Y_F = 4.7-j3.4$	$Y_F = 5.0-j3.25$
1–10 mmhos	$R_F = 150\ \Omega$	$R_F = 150\ \Omega$	$R_F = 150\ \Omega$
10 dB G_{oo}	$L_F = 160$ nH	$L_F = 80$ nH	$L_F = 50$ nH

Fig. 4-54. Intersections of 10-dB G_{oo} and 10-dB low-frequency gain curve at 100, 200, and 300 MHz. Courtesy Motorola.

ticularly well suited for this application. Since frequency response is highly dependent upon the feedback network, it is best determined with the physical circuit. Figure 4-56 shows the typical responses obtained when the corresponding feedback networks are used. The measured L_F in the final circuit is given in Fig. 4-56.

4. *Determination of interstage coupling network.* As long as Y_L is not frequency-selective, Y_L does not affect the frequency response significantly. The interstage coupling network is designed to provide some impedance

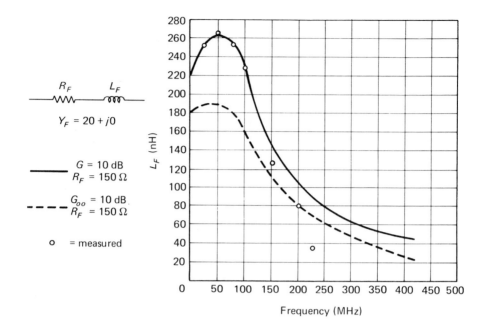

$$G = 10 \text{ dB} \qquad \frac{(Y_{21} - Y_F)^2 \ \text{Re}(Y_L)}{(Y_L + Y_{22} + Y_F)^2 \ \text{Re}\left[Y_{11} + Y_F - \dfrac{(Y_{12} - Y_F)(Y_{21} - Y_F)}{Y_{22} + Y_F + Y_L}\right]}$$

Fig. 4-55. Frequency versus L_F for G and G_{oo}. Courtesy Motorola.

transformation to improve gain at the high-frequency end. At low frequencies, $Y_L = 20 + j0$ mmhos. A series inductance of 180 nH is used for the interstage coupling in this case.

4-6. Voltage Amplifiers

RF voltage amplifiers are used primarily in receivers and receiver-type circuits. An IF (intermediate frequency) amplifier or IF limiter amplifier are examples of voltage amplifiers. The input or first stage of a receiver may include a separate RF voltage amplifier (such as with some communications receivers). However, most solid-state receivers combine the RF voltage-amplifier function with that of the local oscillator. Such circuits are discussed in Sec. 4-7.

Figure 4-57 shows the working schematic of a typical RF voltage amplifier. Such a circuit could be used as an IF amplifier, IF limiter, or separate

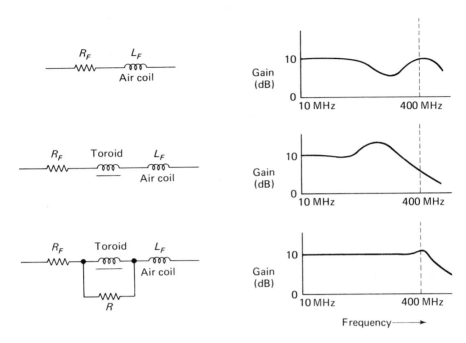

Fig. 4-56. Feedback network response. Courtesy Motorola.

RF amplifier, with a few modifications. Both the input and output are tuned to the desired operating frequency by means of the resonant circuits. In this case, the resonant circuits are composed of transformers with a capacitor across the primary. The capacitors could be variable but are usually fixed. The resonant circuit is tuned by an adjusting slug between the windings.

4-6.1. Design Considerations

The design considerations for transformer-coupled RF amplifiers are similar to those of audio amplifiers, as described in Sec. 2-7. However, the rules of thumb for trial values differ as described in the following paragraphs.

Stage gain. Voltage gain of a fully bypassed RF amplifier is approximately equal to transistor gain (beta) and the turns ratio of the output transformer, as shown in Fig. 4-57. When the turns ratio is considered as primary/secondary, the stage gain equals transistor gain times the inverse of the turns ratio. For example, if the transistor gain was 10, and the transformer had a turns ratio of 10 (primary) to 1 (secondary), the net voltage gain would be 1, or unity.

The required stage gain will depend upon the circuit application. As rules of thumb, a communications receiver RF amplifier requires a voltage gain between 10 and 20, an AM broadcast IF stage requires a gain of between 30

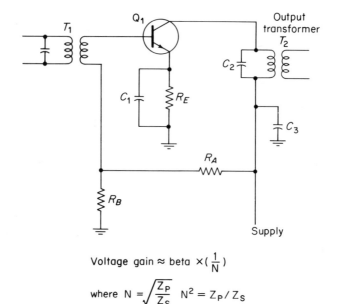

Voltage gain $\approx$ beta $\times(\frac{1}{N})$

where $N = \sqrt{\dfrac{Z_P}{Z_S}}$ $N^2 = Z_P/Z_S$

Voltage drop across	At operating frequency
$R_E \approx$ emitter-base voltage 0.5 for silicon 0.2 for germanium	$X_{C1} \lesseqgtr Z_{in}$ of Q_1 $X_{C3} \lesssim 100\ \Omega$ $X_{C2} \approx X_{primary}$ of T_2

$R_B \approx 10\ R_E$

Voltage drop across R_A = supply – drop across R_B

Supply $\approx$ 3 to 4 times desired output voltage

Fig. 4-57. Basic RF voltage amplifier.

and 40, an IF amplifier for FM requires a gain of between 40 and 50, while a televison IF amplifier (broadband) requires a gain of between 15 and 20.

Transformer characteristics. It is often necessary to design an RF ampli-fier around the characteristics of an existing commercial transformer (inter-stage IF transformer, IF detector transformer, RF/IF transformer, etc.). Such transformers are usually rated as to primary and secondary impedance (rather than turns ratio), and possibly current capacity. However, the typical low currents involved present no design problems. Some commercial trans-formers are provided with a built-in fixed capacitor across the primary (and/or secondary in some cases). When a capacitor is used, the transformers are rated as to the resonant frequency range or midpoint (455 kHz for an AM broadcast IF, 500 to 1600 kHz for a ferrite RF input transformer or "loopstick," 10.7 MHz for an FM broadcast IF, etc.). In other transformers, a fixed or variable capacitor must be connected across the transformer windings. For example, a "loopstick" requires a variable capacitor of the given range to provide full tuning across the broadcast band. When it is necessary to calculate the capacitance value for a given inductance or reactance, use the equations of Fig. 4-1.

Supply voltage. The value of supply voltage for an RF amplifier is not critical. Of course, the supply voltage cannot exceed the transistor collector voltage limits. The supply voltage should be between 3 and 4 times the desired output voltage for the stage. In most RF circuits, the desired output is between 1 and 2 V, so the supply could be between 3 and 9 V. A higher supply voltage can be used, provided that the transistor characteristics are not exceeded.

Transistor characteristics. The design considerations for selection of transistors described in Chapter 1 apply to RF amplifiers. Of particular importance are: interpreting data sheets (Sec. 1-2) and determining parameters at different frequencies (Sec. 1-3). The temperature-related design problems of Sec. 1-4 generally do not apply, since RF voltage amplifiers usually operate at very low power levels. The main concern is that the transistor will provide the required gain at the frequency of interest. In general, the transistor should provide 1.5 times the required gain, at the operating frequency. This will compensate for mismatch, variation in gain due to differences in transistors, etc.

Emitter resistance. When the emitter resistance R_E is bypassed, the resistance value should be chosen on the basis of direct current rather than signal. The value of R_E should provide a voltage drop equal to the emitter–base voltage differential, when the normal (dc operating point) collector current is flowing. A typical silicon emitter–base differential is 0.5 V (0.2 V for germanium transistors). The drop across R_E will serve to stabilize the gain of Q_1, as discussed in Chapters 1 and 2.

To get a perfect impedance match between transistor and output transformer (a practical impossibility) the total impedance presented by R_E and the transistor should match the transformer primary impedance. As a rule of thumb, assume that the impedance represented by the full collector voltage and current (V_C/I_C) is the total transistor and R_E impedance. This will establish the desired collector current I_C and a corresponding voltage drop across R_E. For example, assume that the supply voltage is 10 V and the primary impedance is 10 kΩ. Ignoring the small dc drop across the primary, the collector will be at 10 V. The desired collector current to match impedance would then be

$$I = \frac{E}{R} = \frac{10}{10,000} = 0.001 \text{ A}$$

or 1 mA. With 1 mA flowing through R_E (at the operating point) and a desired 0.5 V drop, the value of R_E is

$$R_E = \frac{E}{I} = \frac{0.5}{0.001} = 500 \text{ }\Omega$$

Bypass capacitors. The value of the emitter-bypass capacitor C_1 should be such that the reactance, at the operating frequency, is less than the input impedance of the transistor. This will effectively remove the emitter resistor from the circuit, insofar as signal is concerned. The input capacitance of a typical transistor for RF applications is on the order of a few ohms and is given on the data sheet. If the input impedance is not known, use a capacitor value that will produce a reactance of less than 10 Ω at the operating frequency.

The value of supply-line bypass capacitor C_3 should be between 0.001 and 0.01 μF. As a first trial value, use that value for C_3 that will produce a reactance of less than 100 Ω at the operating frequency.

Bias resistance network. The values of the bias resistance network should be chosen to place transistor Q_1 at the desired operating point. For example, if the desired collector current is 1 mA, and Q_1 has a nominal gain of 10, the base current must be 0.1 mA. Likewise, if the emitter–base voltage differential is assumed to be 0.5 V, with another 0.5-V drop across R_E, the base should be at 1 V, under no-signal conditions. Any combination of R_B and R_A that produces these relationships would be satisfactory. As a first trial value, make R_B 10 times the value of R_E. Then calculate a corresponding value for R_A, using the equations of Fig. 4-57.

When the circuit of Fig. 4-57 is to incorporate an AVC–AGC (automatic volume control–automatic gain control) function, the bias network is also used as the AVC–AGC line. Therefore, the bias network values must be calculated on that basis. A discussion of AVC–AGC circuits is covered in Sec. 4-10.

4-6.2. Design Example

Assume that the circuit of Fig. 4-57 is to be used as an RF amplifier, without AVC. The input signal voltage (at the secondary of T_1) is 0.1 V. The desired stage output (at the secondary of T_2) is 2 V. This requires a stage gain of 20 (2 ÷ 0.1 = 20). Transformer T_2 is available from commercial sources, complete with the fixed capacitor across the primary. The transformer is slug-tuned to a midrange of 500 kHz, which is the operating frequency of the circuit. The primary impedance of T_2 is 10 kΩ, with 4 kΩ at the secondary. The available source voltage is 9 V from a battery supply.

With 10 kΩ at the primary and 4 kΩ at the secondary, the impedance ratio is 2.5 (10 kΩ ÷ 4 kΩ = 2.5). With an impedance ratio of 2.5, the turns ratio is approximately 1.5 ($\sqrt{2.5}$), and the inverse of the turns ratio is approximately 0.7. With a required 2-V output at the secondary, and an inverse turns ratio of 0.7, the primary signal voltage (output of Q_1) must be approximately 3 V (2 ÷ 0.7).

With a signal input of 0.1 V and a required output of 3 V, the gain of Q_1 must equal 30 (3 ÷ 0.1 = 30). To allow for the variation in gain, Q_1 should have a gain of approximately 45 at the operating frequency (500 kHz).

With a 10-kΩ primary impedance and a dc collector voltage of approximately 9 V, the collector current should be approximately 0.9 mA (9 ÷ 10 kΩ = 0.0009 A).

With 0.9 mA through R_E, and a desired drop of approximately 0.5 V, the value of R_E would be approximately 550 Ω (0.5 ÷ 0.0009 = 550 Ω).

The reactance of C_1 should be less than the input impedance of Q_1 at 500 kHz. Assume that the Q_1 input impedance is 10 Ω. Calculate a trial value for C_1 that will produce a reactance of 5 Ω at 500 kHz:

$$C_1 = \frac{1}{6.28 \times (500 \times 10^3) \times 5} = 0.06 \ \mu\text{F}$$

The value of C_3 should be between 0.001 and 0.1 μF. A value of 0.01 μF would produce approximately a 30-Ω reactance at 500 kHz.

With a desired collector current of 0.9 mA and an approximate gain of 45 of Q_1, the base current for Q_1 should be 20 μA (0.0009 ÷ 45 = 20).

With R_E at 550 Ω, a trial value for R_B is 5500 Ω. Since the base of Q_1 should be at 1 V under no-signal conditions, 1 V is dropped across R_B, and the current flow through R_B is approximately 0.19 mA. This combines with the 20-μA base current, resulting in a current flow of about 0.2 mA through R_A. Since the source is 9 V, R_A must drop 8 V (9 V source − 1 V base voltage). This requires a value for R_A of 40 kΩ (8 ÷ 0.0002 = 40).

4-7. Frequency Mixers and Converters

Figure 4-58 shows the working schematic of a typical frequency mixer and converter. Such a circuit is actually a combination of an RF voltage amplifier and an RF oscillator. The individual outputs of the two sections are combined to produce an intermediate-frequency (IF) output. Usually, the RF oscillator operates at a frequency above the RF amplifier, with the difference in frequency being the intermediate frequency. The resonant circuit of T_1 is tuned to the incoming RF signal, T_2 is tuned to the oscillator frequency (RF + IF), and T_3 is tuned to the intermediate frequency (IF). The resonant circuits of T_1 and T_2 are usually tuned by means of variable capacitors, ganged together so that both the oscillator and RF amplifier will remain at the

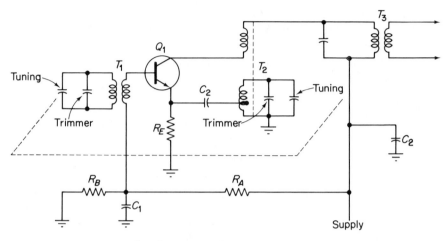

$R_E \approx$ impedance of tap on T_2

Drop across $R_B \approx 0.2 \times$ supply voltage

Drop across $R_A \approx 4 \times R_B$

$R_A + R_B \approx 15 - 20 \times R_E$
at operating frequency X_{C1} and $X_{C2} \approx 50\ \Omega$ (or less)

Power output of $T_2 \approx 0.5 \times \dfrac{\text{collector voltage}^2}{Z \text{ of } T_2 \text{ collector winding}}$

Power output of $T_3 \approx 0.125 \times \dfrac{\text{collector voltage}^2}{Z \text{ of } T_2 \text{ collector winding}}$

Fig. 4-58. Basic RF mixer and converter (RF amplifier and local oscillator).

same frequency relationship over the entire tuning range. For example, if T_1 tuned from 550 to 1600 kHz, and T_3 were at a fixed intermediate frequency of 455 kHz, T_2 would tune from 1005 kHz to 2055 kHz. Usually, trimmer capacitors are placed in parallel with the variable capacitors, to permit adjustment over the tuning range.

4-7.1. Design Considerations

The RF portion of the circuit can be designed on the same basis as the RF voltage amplifier described in Sec. 4-6. However, design of the oscillator section usually sets the operating characteristics for the remainder of the circuit. For example, the oscillator portion may require a power transistor, typically on the order of 0.5-W maximum output. Assuming that the oscillator has 70 per cent efficiency, an output of 0.5 W requires an input of 0.7. Many

transistors are capable of this power dissipation, without heat sinks. However, the transistor power-dissipation characteristics should be checked, as
described in Sec. 1-4.

Transformer characteristics. As in the case of RF amplifiers (Sec. 4-6.1)
it is often necessary (or convenient) to design a converter around the characteristics of existing commerical transformers. The input impedance of T_1
should match the antenna (or other source) impedance. The output impedance of T_1 should match the input impedance of the stage. As a rule of thumb,
the stage input impedance can be considered as the emitter resistance R_E value
(since R_E is unbypassed). However, a mismatch in T_1 is usually not critical.
The output impedance of T_3 should match the input impedance of the
following stage (an IF amplifier), while the input impedance of T_3 should
match the collector winding impedance of T_2. The impedance at the emitter
tap of T_2 should match the oscillator input impedance (approximately equal
to the value of R_E).

Although both signal currents (IF and oscillator) are present in the
collector circuit, the IF signal power is approximately 25 per cent of the
oscillator signal power Also, T_2 and T_3 are resonant at different frequencies.
Therefore, the impedance presented by T_3 has little effect on the oscillator
signal in the collector circuit.

Some commercial transformers specify the fixed capacitance necessary to
provide the desired resonant frequency (or variable capacitance limits for a
given tuning range). If it is necessary to calculate the capacitance values for a
given inductance or reactance, use the equations of Fig. 4-1.

Emitter resistance. Ideally, the emitter resistance R_E should be chosen
to match the output impedance of T_1 and the emitter tap inpedance of T_2.
Where the design must be adapted to existing transformers, and there is a
mismatch between transformers, use a value of R_E that matches T_2 (the
oscillator resonant circuit).

Bias relationships. One problem in design of any oscillator circuit is that
the circuit should be operated as a class A amplifier for starting, and then
switch to class C operation as the oscillations build up. That is, the emitter–
base should be forward-biased initially, and then be reverse-biased except on
peaks of the oscillation. (Ideally, the transistor conducts during approximately 140° of each 360° cycle.) Thus, a variable bias is required. This is
accomplished when the capacitors are charged and discharged through the
emitter resistance and bias resistance. However, if the capacitors are too small
in value, the oscillator may not start, or the output waveform will be distorted. If the capacitors are too large, the change in charge (to produce a
variable bias) will be slow, causing the oscillator to operate intermittently.

A fixed forward bias should be placed on the base by means of R_A and R_B.
As a first trial value, the drop across R_B should be 0.2 the supply voltage at

the collector. Therefore, R_A should be approximately 4 times the value of R_B. Any combination of R_B and R_A that would produce the required bias could be used. However, the total series resistance of $R_A + R_B$ should be 15 to 20 times that of R_E. This will minimize excess current drain by the bias network.

Bypass and coupling capacitors. The value of the bypass and coupling capacitors are typically between 0.001 and 0.1 μF. For a first trial, choose a capacitance value that will produce a reactance of 50 Ω at the operating frequency. As discussed, the final value of the capacitors can be critical (to produce continuous oscillations with good waveforms). Therefore, the only final test of correct capacitor values is the display of the output waveform on an oscilloscope.

Power output. Power output of the oscillator is approximately equal to

$$P_O = 0.5 \times \frac{\text{collector voltage}^2}{\text{impedance of } T_2 \text{ collector winding}}$$

Power output of T_3 to the following IF stage is:

$$P_{T_3} = 0.125 \times \frac{\text{collector voltage}^2}{\text{impedance of } T_2 \text{ collector winding}}$$

Transistor characteristics. The design considerations for selection of transistors described in Chapter 1 apply to converters and mixers. Of particular importance are Secs. 1-2, 1-3, and 1-4. The main concern is that the transistor will provide the required power at the frequency of interest.

4-7.2. Design Example

Assume that the circuit of Fig. 4-58 is to be used as a combination RF amplifier and converter for a broadcast radio receiver. The circuit must be designed around existing transformers and an available supply voltage of 7 V. (If, in actual design practice, the transformers can be ordered to specification, the following values can be used as a guide.) The transformer impedances (in ohms) are: T_1 input 50, T_1 output 1000, T_2 emitter winding tap 3000, T_2 collector winding 1000, T_3 input 1000, and T_3 output 5000. Transformers T_1 and T_2 are supplied with data that show the correct values of variable capacitance to tune across the broadcast range. Transformer T_3 is supplied with a fixed capacitor and is slug-tuned to the intermediate frequency.

Once the trial values have been calculated, find the total power to be dissipated. In actual design practice, the next step would be to consult transistor data sheets and make sure that the selected transistor could provide the required power at the operating frequency.

Since the transformer impedances are stated as part of the design example, the first step is to match the emitter resistance R_E to the emitter winding tap of T_2, a value of 3 kΩ. This will result in a mismatch with T_1, but the effects should not be serious.

The fixed bias at the base of Q_1 should be 0.2 times the supply voltage, or $0.2 \times 7 = 1.4$ V. This voltage is developed by the drop across R_B. The total resistance of R_A and R_B should be between 15 and 20 times R_E, or 45 to 60 kΩ. Using the higher value (60 kΩ) for minimum current drain by the bias network, the value of R_B should be 0.2×60, or 12 kΩ, with a value for R_A of $60 - 12$, or 48 kΩ.

The bypass and coupling capacitors should both produce a reactance of 50 Ω (or less) at operating frequency. Using the lowest frequency involved (about 500 kHz), the values of C_1 and C_2 should be

$$\frac{1}{6.28 \times (500 \times 10^3) \times 50} = 0.006 \ \mu F$$

The power output of T_3 to the following IF stage is

$$0.125 \times \frac{7^2}{1000} \approx 6 \ \text{mW}$$

The power output of the oscillator is

$$0.5 \times \frac{7^2}{1000} = 24 \ \text{mW}$$

Therefore, the total power output of the stage is 30 mW. Allowing for variations, the transistor should be capable of dissipating 60 mW, at the highest ambient temperature, and should be capable of producing at least 30 mW, at the highest operating frequency (about 2000 kHz).

4-8. Detectors

RF voltage detectors are used to convert RF signal voltages into a dc voltage. If there is audio-frequency information present on the RF signal (in the form of amplitude modulation) the detector functions to convert the audio voltages into pulsating dc voltages. In either case, the detector uses a diode that acts as a rectifier.

Figure 4-59 shows the working schematics for two typical diode detectors, such as can be found in radio receivers. Usually, the detector stage follows one or more IF stages. The output of the detector stage (pulsating dc voltages at audio frequencies) is applied to the receiver audio section. In the circuit of Fig. 4-59a, the detector output is applied to a volume control in the audio circuits. In Fig. 4-59b, the detector load resistance acts as the volume control, eliminating the need for further controls in the audio circuits.

The AVC–AGC circuits of a receiver are associated with the detector circuit, since the AVC–AGC signal is obtained from the detector output. AVC–AGC circuits are discussed in Sec. 4-10. The following section will describe the design of the basic detector circuit.

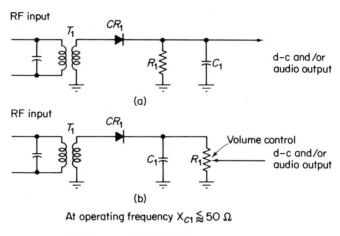

At operating frequency $X_{C1} \lessapprox 50\ \Omega$

R_1C_1 time constant $\lessapprox 250\ \mu\text{sec}$

Fig. 4-59. Basic detector.

4-8.1. Design Considerations

The basic design procedure is essentially the same for both circuits shown in Fig. 4-59. Selection of component values is a relatively simple matter, with wide variation in values producing little effect on operation. However, the design values are usually based on characteristics of other circuits. Here are some examples.

The impedance presented by the secondary of IF output transformer T_1 is the input impedance to the detector circuit. Ideally, the value of resistance R_1 should match the impedance of T_1's secondary output. However, R_1 is also the output load impedance of the detector and should match the input impedance of the following stage (usually an audio stage). Therefore, it is often necessary to trade off between these two impedance levels.

If the circuit of Fig. 4-59b is used, the volume control represents a variable input impedance to the audio stages. The effect of such a variable impedance is determined by the relative impedance of the following stage. For example, if the volume control circuit is coupled to an audio stage with a bypassed emitter, the input impedance of the audio stage will be quite low (about equal to the input impedance of the transistor, as discussed in Chapter 2). When the volume control is set to the high end (near the diode for maximum voltage output), the mismatch is the most severe, resulting in a signal loss. With the volume control at the low end for minimum output, the mismatch is less severe, and thus there is less signal loss.

If either circuit of Fig. 4-59 is used as part of an AVC–AGC system (Sec. 4-10), as is usually the case, the load resistance R_1 forms part of the bias network for the variable-gain stage. Since the bias of the variable-gain stage must be considered, this sets a maximum limit for the value of R_1. For example, if the bias required by the variable gain stage was 1 V, the bias drop across R_1 could never exceed 1 V.

Detector diode. Diodes used in detectors are rated as to their voltage and current capability, as are other diodes and rectifiers. Typically, the voltage at the secondary of T_1 is about 1 V and rarely exceeds 2 V. Therefore, the maximum reverse (or peak inverse) voltage rating of the diode need not exceed about 2 V. Current through the diode is determined by the value of R_1 and the voltage. Assuming a maximum voltage of 2 V and a minimum resistance value of 5 kΩ for R_1, current through the diode should never exceed about 1 mA. Because of the low signal voltage available, a low forward-voltage drop is desired for the diode.

Load resistance. As discussed, the value of load resistance R_1 is chosen on the basis of a tradeoff among input impedance, output impedance, and voltage drop in a bias network.

Bypass capacitor. Ideally, the bypass capacitor should be large enough to provide little or no reactance for RF signals but small enough to provide a short RC time constant (with the load resistance R_1) so that the audio signals will fully charge the capacitor. This may involve a tradeoff. As rules of thumb, the reactance at the radio frequency should be 50 Ω or less, and the RC time constant should be 250 μs or less.

4-8.2. Design Example

Assume that the circuit of Fig. 4-59a is to be used as the detector in a solid-state radio receiver such as a transistor portable. The value of load resistance R_1 must be something less than 5 kΩ, since R_1 is used as part of an AVC–AGC network. The nearest standard value is 4.7 kΩ.

Selection of the diode should present no problem, since the reverse voltage (peak inverse) is less than 2 V, and forward current is less than 1 mA. Any number of signal diodes would meet these requirements. Any conventional germanium diode could be used.

The time constant $R_1 C_1$ should be less than 250 μs. With R_1 at 5 kΩ (assuming that the 4.7-kΩ standard resistor was on the high side), the value of C_1 should be $(250 \times 10^{-6}) \div 5000$, or 0.05 μF.

The reactance of a 0.05-μF capacitor at 455 kHz (the usual broadcast receiver IF) is

$$\frac{1}{6.28 \times (455 \times 10^3) \times (0.05 \times 10^{-6})} \quad \text{or approximately 6 Ω}$$

4-9. Discriminator and Ratio Detector

Discriminators and ratio detectors are used with FM radio receivers. Both circuits function to remove audio-frequency information present on the RF signal (in the form of frequency modulation), and to convert the audio voltages into pulsating dc voltages. In either circuit, a diode is used as the basic detector. A ratio detector has an advantage in that it can be driven directly by an IF stage. A discriminator usually requires a limiter stage between the IF amplifier output and discriminator input.

Figures 4-60 and 4-61 show the working schematics of a discriminator and ratio detector, respectively. In both cases, the input transformer T_1 shows an extra pick-up winding in addition to the primary and secondary windings. A portion of the primary signal (from the IF amplifier or limiter stages) is sampled by the pick-up coil and injected into the electrical center of the circuit. Not all discriminator and ratio detector transformers are so constructed. An alternative method is to obtain the sampled portion by

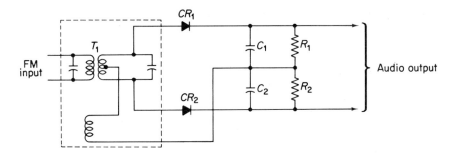

At operating frequency X_{C1} and $X_{C2} \lesssim 50\ \Omega$
$R_1 C_1$ and $R_2 C_2$ time constant $\lesssim 125\ \mu\mathrm{sec}$
$R_1 + R_2 \approx$ output impedance

Fig. 4-60. Basic FM discriminator.

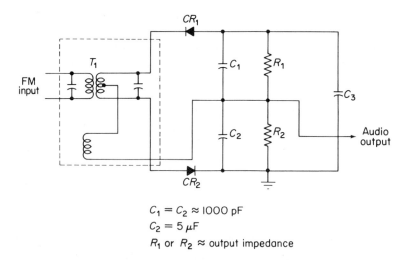

$C_1 = C_2 \approx 1000\ \mathrm{pF}$
$C_2 = 5\ \mu\mathrm{F}$
R_1 or $R_2 \approx$ output impedance

Fig. 4-61. Basic FM ratio detector.

eliminating the pick-up coil, and coupling this line to the primary by a capacitor.

Discriminator and ratio detectors used in broadcast receivers are tuned to the intermediate frequency of 10.7 MHz. Most discriminator and ratio detector transformers available from commercial sources are resonant at this frequency (complete with capacitors). Of course, special order transformers could be produced to cover other frequencies. In the following discussion, it is assumed that the discriminator or ratio detector is to be used at 10.7 MHz.

4-9.1. Design Considerations for Discriminator

One of the major problems in design of a discriminator is that both halves of the circuit are matched. That is, the characteristics of CR_1 should be identical to those of CR_2, the value of C_1 should match that of C_2, and thus the value of R_1 should match that of R_2. Operation of the circuit is based on the theory that both detectors will produce equal voltages, but of opposite polarity, across R_1 and R_2, when there is no frequency modulation. Slight variations in component values or diode characteristics can be picked up by tuning the input transformer. However, any major unbalance in the circuit will result in a distorted output.

Load resistances. The load impedance presented to the audio stages following the discriminator is the *total series resistance* of both R_1 and R_2. Usually, the values of R_1 and R_2 are the result of a tradeoff between a suitable impedance output and a realistic RC time constant with capacitors C_1 and C_2.

Bypass capacitors. Ideally, the bypass capacitors C_1 and C_2 should be large enough to provide little or no reactance for RF signals, but small enough to provide a short RC time constant (with R_1 and R_2) so that the audio signals will fully charge the capacitors. This usually involves a tradeoff. As rules of thumb, the reactance at the radio frequency should be 50 Ω or less, and the RC time constant should be 125 μs or less.

Detector diodes. Typically, voltage at the secondary of T_1 is about 1 V, and rarely exceeds 2 V. Therefore, the maximum reverse (or peak inverse) voltage rating of the diode need not exceed about 2 V. Current through the diode rarely exceeds 1 mA. Because of the low signal voltage available, a low forward-voltage drop, and minimum reverse current (leakage current) are always desirable. A diode will always have some capacitance. As the operating frequency increases, the reactance becomes quite small, and RF signals tend to bypass the diode. However, at the usual broadcast FM frequencies, the diode capacitance should present no problem.

4-9.1.1. Design Example for Discriminator

Assume that the circuit of Fig. 4-60 is to be used as the discriminator in a solid-state radio receiver, such as an FM stereo-broadcast receiver. The output of the discriminator must be fed to a 450-kΩ volume control. Therefore, the total series resistance of both R_1 and R_2 should be 450 kΩ. This would require a value of 225 kΩ for R_1 and R_2. The nearest standard value would be 220 kΩ. Therefore, both R_1 and R_2 should be 220 kΩ, with a total output impedance of 440 kΩ.

Selection of the diodes should present no problem, since the reverse voltage (peak inverse) is less than 2 V, and forward current is less than 1 mA. Any number of germanium diodes would meet these requirements. However, care should be taken to match the diode characteristics. Ideally, forward-voltage drop and reverse (leakage) current should be identical for both diodes.

The time constant of R_1C_1 and R_2C_2 should be less than 125 μs. With R_1 and R_2 at 220 kΩ, the values of C_1 and C_2 are each $(125 \times 10^{-6}) \div 220,000$, or approximately 550 pF.

The reactance of a 550-pF capacitor at 10.7 MHz is

$$\frac{1}{6.28 \times (10.7 \times 10^6) \times (550 \times 10^{-12})} \approx 27 \ \Omega$$

4-9.2. Design Considerations for Ratio Detector

As in the case of the discriminator, both halves of the radio detector must be matched. Operation is based on both detectors producing equal voltages when there is no frequency modulation. These voltages are of the same polarity on each half of the cycle, and across both R_1 and R_2. In the presence of frequency modulation, the voltage drops across R_1 and R_2 are not equal. Therefore, the voltage between the center tap and ground will vary in the presence of FM. Audio can be taken from across either resistor R_1 or R_2, but not from both, as with the discriminator. Any major unbalance in the circuit that cannot be corrected by tuning of T_1 will result in a distorted output.

Load resistance. The load impedance presented to the audio stages following the ratio detector is the resistance of R_1 or R_2, whichever is used as the audio load. Resistor R_2 is shown as the audio load in Fig. 4-61. Usually, the values of R_1 and R_2 are the result of a tradeoff between a suitable impedance output and a realistic RC time constant with capacitors C_1 and C_2.

Bypass capacitors. Ideally, bypass capacitors C_1 and C_2 should be large enough to provide little or no reactance for RF signals but small enough to provide a short RC time constant (with R_1 and R_2) so that the audio signals will fully charge the capacitors. This usually involves a tradeoff. The values of C_1 and C_2 are typically on the order of 1000 pF, and should not be less than about 300 pF to keep the reactance at a suitable level (assuming an operating frequency of 10.7 MHz).

Shunt diode. Shunt capacitor C_3 has a much larger value than C_1 or C_2. Typically, C_3 is about 5 μF. That value can be used as the first trial value for any ratio detector. Capacitor C_3 charges to the total voltage across R_1 and

R_2 (usually less than 4 or 5 V), and opposes any abrupt change of voltage. Thus, sharp bursts of static voltage or other AM signals riding on the FM carrier are effectively reduced by C_3.

Detector diodes. As in the case of the discriminator, the diode characteristics should be matched as nearly as possible. Maximum reverse voltage should not exceed 2 V, and forward current will usually be less than 1 mA. Low forward-voltage drop and minimum reverse (leakage) current are always desirable. Generally, diode capacitance can be ignored at the usual broadcast FM frequencies.

4-9.2.1. Design Example for Ratio Detector

Assume that the circuit of Fig. 4-61 is to be used as the ratio detector in a solid-state radio receiver such as the type found in communications work. The output of the ratio detector must be fed to an audio amplifier with an approximately 12-kΩ input. Therefore, the resistances of R_1 and R_2 should be 12 kΩ.

Using the rules of thumb, C_1 and C_2 should be approximately 1000 pF, while C_3 should be 5 μF. These values will provide suitable reactances. With the values of C_1 and C_2 established, find the RC time constants (R_1C_1 and R_2C_2). No problem should be encountered if the time constants are less than about 125 μs.

Selection of the diodes should present no problem, since the reverse voltage (peak inverse) is less than 2 V and forward current is less than 1 mA. Any number of germanium diodes would meet these requirements. However, care should be taken to match the diode characteristics. Ideally, forward-voltage drop and reverse (leakage) current should be identical for both diodes.

4-10. AVC–AGC Circuit

Most receivers have some form of AVC–AGC (automatic volume control-automatic gain control) circuit. The terms AVC and AGC are used interchangeably. AGC is a more accurate term since the circuits involved control the gain of an IF or RF stage rather than the volume of an audio signal in an AF stage. However, in a broadcast receiver, the net result is an automatic control of volume. Either way, the circuit functions to provide a constant output despite variations in signal strength. An increased signal will reduce stage gain, and vice versa.

Figure 4-62 shows the working schematic of two AGC systems that are common to broadcast and communications receivers. Diode CR_1 acts as a variable shunt resistance across the input of the IF stages. Diode CR_2 functions as the detector and AGC as the bias source.

Under no-signal conditions, or in the presence of a weak signal, diode CR_1 is reverse-biased and has no effect on the circuit. In the presence of a very

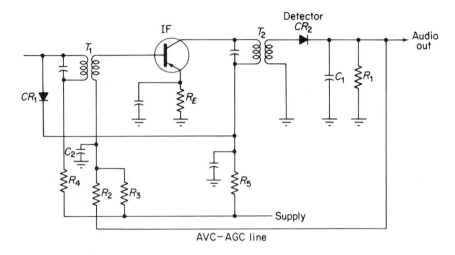

CR$_1$ is reverse biased with no signal
$C_2 \approx 10\,\mu F$
Drop across $R_1 \approx 0.5-1.0$ V
Drop across $R_1 + R_2 \approx 1.0-2.0$ V
Drop across $R_3 =$ supply $- (R_1 + R_2)$
$R_1 + R_2 \approx 10 \times R_E$

Fig. 4-62. Basic AVC–AGC circuit.

large signal, CR_1 is forward-biased and acts as a shunt resistance to reduce gain.

The output of CR_2 is developed across resistor R_1 and applied to the audio stages as described in Sec. 4-8. Resistor R_1 also forms part of the bias network for the IF stage transistor. Therefore, the combined fixed bias (from the network) and variable bias (from the detector) is applied to the IF stage base-emitter circuit. The detector bias varies with signal strength and is of a polarity that opposes variations in signal. That is, if the signal increases, the detector bias will be more positive (or less negative) for the base of a *PNP* transistor, and vice versa for an *NPN* transistor.

4-10.1. Design Considerations

Both AGC systems (shunt diode and variable bias) are often incorporated in the same receiver. The variable bias system handles normal variations in the signal, while the shunt diode handles large-signal variations.

Shunt diode. The shunt diode CR_1 should have a maximum reverse (peak inverse) voltage rating equal to the supply voltage. In most cases, the diode will never have a reverse voltage greater than 1 or 2 V. However, if the diode is capable of handling the full supply voltage, there will be no danger of breakdown. The forward-current capability of CR_1 should be such that the diode can pass the current if there were a full voltage drop across the collector resistor. The values of R_4 and R_5 must be such that CR_1 will be reverse-biased under no-signal conditions (with the IF stages at the Q-point).

Bias network. The values of the bias network (R_1, R_2, and R_3) should be chosen to provide the desired fixed bias for the IF stage as described in Sec. 4-6. The drop across R_1 and R_2 is the bias value applied to the base of the IF stage. The drop across R_1 is combined with the pulsating detector signal output. Typically, the drop across R_1 is on the order of 0.5 to 1 V. The drop across R_1 and R_2 is between 1 and 2 V. The value of C_2 is quite large in relation to other bypass capacitors, and is typically 10 μF, or larger. The value of C_1, CR_2, and all other components not part of the AGC systems should be chosen as described in Secs. 4-6 and 4-8.

4-10.2. Design Example

Assume that the circuits of Fig. 4-62 are to be used with a broadcast receiver such as a solid-state portable radio. The supply voltage is -9 V. The IF stage base current is 0.04 mA, with a collector current of 1.0 mA under no-signal conditions. The collector of the preceding stage (IF or RF) is at -7 V, under no-signal conditions. The desired operating-point base bias for the IF stage is 1 V. The value of the IF emitter resistance is 800 Ω.

The drop across R_1 and R_2 must be 1 V, with an 8-V drop across R_3. As discussed in Sec. 4-6, the value of the bias return resistance (the total of R_1 and R_2 in this case) should be approximately 10 times the resistance of R_E (the IF emitter resistance). With an R_E of 800 Ω, the total resistance of R_1 and R_2 should be 8 kΩ. R_1 and R_2 could be 4 kΩ each. However, 4-kΩ resistors are not standard. The nearest two standard values would be 4.7 kΩ and 3.3 kΩ, making a total of 8 kΩ. Make the larger value (4.7 kΩ) R_1 and the smaller value R_2.

With a 1-V drop across 8 kΩ, the current through R_1 and R_2 is approximately 0.12 mA. This combines with the 0.04-mA base current to produce a 0.16-mA flow through R_3.

With an 8-V drop across R_3, and a current of 0.16 mA, the resistance of R_3 is 50 kΩ. The nearest standard value is 51 kΩ.

The value of C_2 is arbitrarily set at 10 μF. The value of C_1 should be found as discussed in Sec. 4-8 (and will be on the order of 0.05 μF).

With 0.12 mA through R_1, and a resistance of 4.7 kΩ, the no-signal voltage drop across R_1 will be approximately 0.56 V (within the 0.5- to 1-V typical value).

Diode CR_1 should be capable of handling 9-V maximum reverse voltage (although the actual reverse voltage will probably never exceed about 2 V).

Since the collector of the preceding stage is at -7 V, the collector of the IF stage should be at -6 V, so that the cathode of CR_1 will be positive (or less negative) by 1 V in relation to the anode. This will keep CR_1 reverse-biased under no-signal conditions.

The drop across R_5 must be 3 V (9 V supply -6 V collector). With a 3-V drop, and a 1.0-mA collector current, the resistance of R_5 must be 3 kΩ. The nearest standard value would be 2.7 kΩ.

4-11. Radio-Frequency-Circuit Test Procedures

The following sections describe test procedures for radio-frequency circuits. The first sections are devoted to test and measurement procedures for the resonant circuits used at radio frequencies (resonant-frequency measurements, Q measurement, etc.). The remaining sections cover test procedures for the specific radio-frequency circuits described in this chapter (RF amplifiers, transmitters, receivers, detectors, etc.).

The procedures can be applied at any time during design. As a minimum, the tests should be made when the circuit is first completed in breadboard form. If the test results are not as desired, the component values should be changed as necessary to obtain the desired results.

Radio-frequency circuits should always be retested in final form (with all components soldered in place). This will show if there is any change in circuit characteristics due to the physical relocation of components. Such tests are especially important at the higher radio frequencies. Often, there will be capacitance or inductance between components, from components to wiring, and between wires. These stray "components" can add to the reactance and impedance of circuit components. When the physical locations of parts and wiring are changed, the stray reactances change, and alter circuit performance.

4-11.1. Basic RF Voltage Measurement

When the voltages to be measured are at radio frequencies and are beyond the frequency capabilities of the meter circuits or oscilloscope amplifiers, an RF probe is required. Such probes rectify the RF signals into a dc output which is almost equal to the peak RF voltage. The dc output of the probe is then applied to the meter or oscilloscope input and is displayed as a voltage readout in the normal manner.

If a probe is available as an accessory for a particular meter, that probe should be used in favor of any homemade probe. The manufacturer's probe will be matched to the meter in calibration, frequency compensation, etc. If a probe is not available for a particular meter or oscilloscope, the following notes discuss the fabrication of probes suitable for measurement and test of RF circuits during and after design.

4-11.1.1. Half-Wave Probe

The half-wave probe (Fig. 4-63) will provide an output to the meter (or oscilloscope) that is approximately equal to the peak value of the voltage being measured. Since most meters are calibrated to read in rms values, the probe output must be reduced to 0.707 of the peak value, by means of R_1. The value of R_1 could be found by calculation. But for practical purposes, a variable resistor should be used during calibration and then be replaced by a fixed resistor of the correct value. The following steps describe the calibration and fabrication procedure:

1. Connect the breadboard probe circuit to a signal generator and meter.
2. Set the meter to measure dc voltage. Either a VOM or electronic voltmeter can be used, but best results will be found with a high-input-impedance meter.
3. Adjust the signal-generator voltage *amplitude* to some precise value, such as 10 V rms, as measured on the generator's output meter.
4. Adjust the calibrating resistor R_1 until the meter indicates the same value (10 V rms).
5. As an alternative procedure, adjust the signal generator for a 10-V peak output, then adjust R_1 for a reading of 7.07 V on the meter being calibrated.
6. Remove the power, disconnect the circuit, measure the value of R_1, and replace the variable resistor with a fixed resistor of the same value.

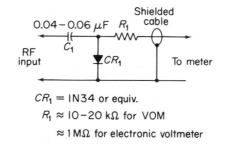

$CR_1 = $ IN34 or equiv.

$R_1 \approx $ I0–20 kΩ for VOM

$\approx $ 1 MΩ for electronic voltmeter

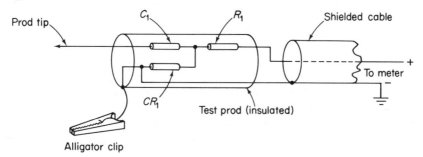

Fig. 4-63. Half-wave RF probe.

7. Repeat the test with the fixed resistance in place. If the reading is correct, mount the circuit in a suitable package, such as within a test prod. Repeat the test with the circuit in final form. Also repeat the test over the entire frequency range of circuits to be designed. This will check the frequency response of the probe. Generally, the probe will provide satisfactory response up to about 250 MHz.

8. Keep in mind that the meter must be set to measure direct current, since the probe output is dc.

4-11.1.2. Demodulator Probe

When RF signals contain modulation, a demodulator probe (Fig. 4-64) will be most effective for testing RF circuits during design. For example, an RF signal modulated by a fixed audio tone can be applied to an RF amplifier being tested. The demodulator probe will measure the RF amplifier response in terms of both RF signal and audio signal.

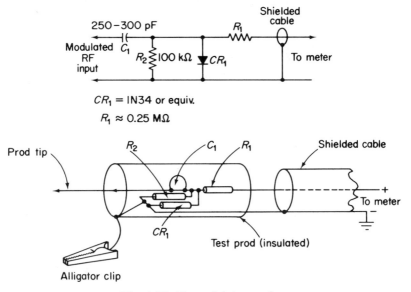

Fig. 4-64. Demodulator probe.

The demodulator probe is similar to the half-wave probe except for the low capacitance of C_1 and the parallel resistor R_2. These two components act as a filter. The demodulator probe produces both an ac and a dc output. The RF signal is converted into a dc voltage approximately equal to the peak value. The low-frequency modulation voltage on the RF signal appears as ac at the probe output.

In use, the meter is set to dc, and the RF signal is measured. Then the meter is set to ac, and the modulating voltage is measured. The calibrating resistor R_1 is adjusted so that the dc scale reads the rms value. The procedure for calibration and fabrication of the demodulator probe is the same as for the half-wave RF probe (Sec. 4-11.1.1), except that the schematic of Fig. 4-64 should be used. Also, keep in mind that R_1 should be adjusted on the basis of the RF signal (not the modulating signal) with the meter set to dc.

4-11.2. Measuring Resonant Frequency of LC Circuits

Once an *LC* circuit has been designed, using theoretical values such as described in Sec. 4-1, it is often convenient to measure the actual resonant frequency of the circuit. A meter can be used in conjunction with an RF signal generator to find the resonant frequency of either series or parallel *LC* circuits. The generator must be capable of producing a signal at the resonant frequency of the circuit, and the meter must be capable of measuring the frequency. If the resonant frequency is beyond the normal range of the meter, an RF probe must be used. The following steps describe the measurement procedure.

1. Connect the equipment as shown in Fig. 4-65. Use the connections of Fig. 4-65a for parallel resonant circuits, or the connections of Fig. 4-65b for series resonant circuits.
2. Adjust the generator output until a convenient midscale indication is obtained on the meter. Use an unmodulated signal output from the generator.

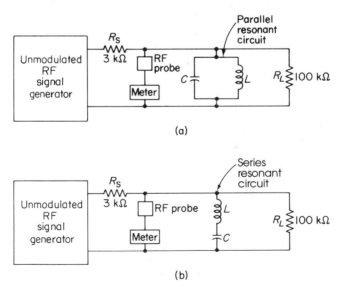

Fig. 4-65. Measuring resonant frequency of *LC* circuits.

3. Starting at a frequency well below the lowest possible frequency of the circuit under test, slowly increase the generator output frequency. If there is no way to judge the approximate resonant frequency, use the lowest generator frequency.

4. If the circuit being tested is parallel-resonant, watch the meter for a maximum, or peak, indication.

5. If the circuit being tested is series-resonant, watch the meter for a minimum, or dip, indication.

6. The resonant frequency of the circuit under test is the one at which there is a maximum (for parallel) or minimum (for series) indication on the meter.

7. There may be peak or dip indications at harmonics of the resonant frequency. Therefore, the test is most efficient when the approximate resonant frequency is known.

8. The value of load resistor R_L is not critical. The load is shunted across the LC circuit to flatten or broaden the resonant response (to lower the Q). Thus, the voltage maximum or minimum is approached more slowly. A suitable trial value for R_L is 100 kΩ. A lower value of R_L will sharpen the resonant response, and a higher value will flatten the curve.

4-11.3. Measuring Inductance of a Coil

Once a coil has been designed and wound, using theoretical values such as described in Sec. 4-1, it is often convenient to measure the actual inductance. A meter can be used in conjunction with an RF signal generator and a fixed capacitor, of known value and accuracy, to find the inductance of a coil. The generator must be capable of producing a signal at the resonant frequency of the test circuit, and the meter must be capable of measuring the frequency. If the resonant frequency is beyond the normal range of the meter, an RF probe must be used. The following steps describe the measurement procedure.

1. Connect the equipment as shown in Fig. 4-66. Use a capacitance value such as 10 μF, 100 pF, or some other even number to simplify the calculation.

2. Adjust the generator output until a convenient midscale indication is obtained on the meter. Use an unmodulated signal output from the generator.

3. Starting at a frequency well below the lowest possible resonant frequency of the inductance–capacitance combination under test, slowly increase the generator frequency. If there is no way to judge the approximate resonant frequency, use the lowest generator frequency.

4. Watch the meter for a maximum or peak indication. Note the frequency at which the peak indication occurs. This is the resonant frequency of the circuit.

5. Using this resonant frequency, and the known capacitance value, calculate the unknown inductance using the equation in Fig. 4-66.

6. Note that the procedure can be reversed to find an unknown capacitance value, when a known inductance value is available.

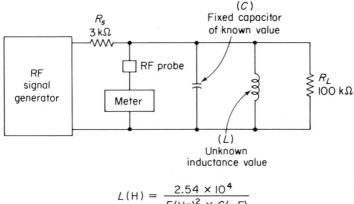

$$L(H) = \frac{2.54 \times 10^4}{F(Hz)^2 \times C(\mu F)}$$

$$C(\mu F) = \frac{2.54 \times 10^4}{F(Hz)^2 \times L(H)}$$

Fig. 4-66. Measuring inductance and capacitance in *LC* circuits.

4-11.4. *Measuring Self-Resonance and Distributed Capacitance of a Coil*

No matter what design or winding method is used, there will be some distributed capacitance in any coil. When the distributed capacitance combines with the coil's inductance, a resonant circuit is formed. The resonant frequency is usually quite high in relation to the frequency at which the coil will be used. However, since self-resonance may be at or near a harmonic of the frequency to be used, the self-resonant effect may limit the coil's usefulness in *LC* circuits. Some coils, particularly RF chokes, may have more than one self-resonant frequency.

A meter can be used in conjunction with an RF signal generator to find both the self-resonant frequency and distributed capacitance of a coil. The generator must be capable of producing a signal at the resonant frequency of the circuit, and the meter must be capable of measuring voltages at that frequency. Use an RF probe if required. The following steps describe the measurement procedure.

1. Connect the equipment as shown in Fig. 4-67.
2. Adjust the generator output until a convenient midscale indication is obtained on the meter. Use an unmodulated signal output from the generator.
3. Tune the signal generator over its entire frequency range, starting at the lowest frequency. Watch the meter for either peak or dip indications. Either a peak or dip indicates that the inductance is at a self-resonant point. The generator output frequency at that point is the self-resonant frequency.

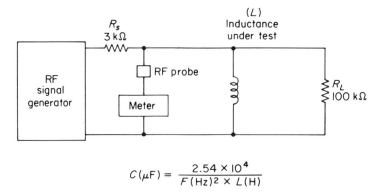

$$C\,(\mu F) = \frac{2.54 \times 10^4}{F\,(Hz)^2 \times L\,(H)}$$

Fig. 4-67. Measuring self-resonance and distributed capacitance of coil.

Make certain that peak or dip indications are not the result of changes in generator output level. Even the best laboratory generators may not produce a flat (constant level) output over the entire frequency range.

4. Since there may be more than one self-resonant point, tune through the entire signal generator range. Try to cover a frequency range up to at least the third harmonic of the highest frequency involved in circuit design.

5. Once the resonant frequency has been found, calculate the distributed capacitance using the equation of Fig. 4-67.

4-11.5. Measuring Q of Resonant Circuits

As discussed in Sec. 4-1, the Q of a resonant circuit is dependent upon the ratio of reactance and resistance. A high-Q circuit, where reactance is high in relation to resistance, will have a narrow bandwidth (sharp resonant peak or dip). A low Q produces a broad bandwidth. The Q of a resonant circuit is dependent upon the Q of the individual components (coil, capacitor, wiring, etc.).

Since the ultimate design goal of a resonant circuit is bandwidth (narrow or wide, depending upon design requirements), the most practical test of a completed resonant circuit is bandwidth at the resonant frequency.

The Q of a circuit can be measured using a signal generator and a meter with an RF probe. An electronic voltmeter will provide the least loading effect on the circuit and will therefore provide the most accurate indication.

Figure 4-68a shows the test circuit in which the signal generator is connected directly to the input of a complete stage, and Fig. 4-68b shows the indirect method of connecting the signal generator to the input.

When the stage or circuit has sufficient gain to provide a good reading on the meter with a nominal output from the generator, the indirect method (with isolating resistor) is preferred. Any signal generator has some output impedance (such as a 50-Ω output resistor). When this resistance is connected

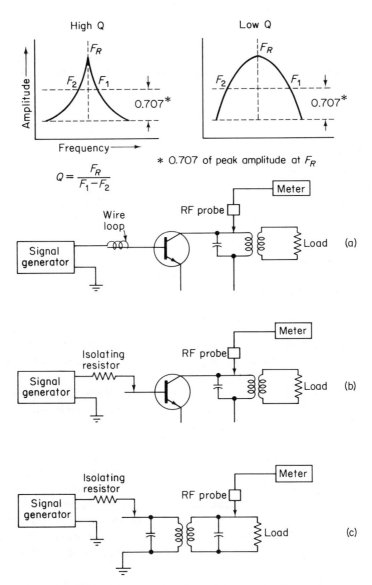

Fig. 4-68. Measuring Q of resonant circuits.

directly to the tuned circuit, the Q is lowered, and the response becomes broader. (In some cases, the generator output impedance will seriously detune the circuit.)

Figure 4-68c shows the test circuit for a single component (such as an IF transformer). The value of isolating resistance R is not critical, and is typically in the range of 100 kΩ.

The procedure for determing Q is as follows:

1. Connect the equipment as shown in Fig. 4-68a, b, or c, as applicable. Note that a load is shown in Fig. 4-68c. When a circuit is normally used with a load, the most realistic Q measurement is made with the circuit terminated in that load value. A fixed resistance can be used to simulate the load. The Q of a resonant circuit is often dependent upon the load value.
2. Tune the signal generator to the circuit resonance frequency. Operate the generator to produce an unmodulated output.
3. Tune the generator frequency for maximum reading on the meter. Note the generator frequency.
4. Tune the generator below resonance until the meter reading is 0.707 of the maximum reading. Note the generator frequency. To make the calculation more convenient, adjust the generator output *level* so that the meter reading is some even value, such as 1 V or 10 V, after the generator is tuned for maximum. This will make it easy to find the 0.707 mark.
5. Tune the generator above resonance until the meter reading is 0.707 of the maximum reading. Note the generator frequency.
6. Calculate the circuit Q using the equation of Fig. 4-68.

4-11.6. Measuring the Impedance of Resonant Circuits

Any resonant circuit will have some impedance at the resonant frequency. The impedance will change as frequency changes. This includes transformers (tuned and untuned), tank circuits, etc. In theory, a series-resonant circuit has zero impedance, while a parallel-resonant circuit has infinite impedance, at the resonant frequency. In practical circuits, this is impossible, since there will always be some resistance in the circuit.

In practical design, it is often convenient to find the actual impedance of a completed circuit, at a given frequency. Also, it may be necessary to find the impedance of a component so that circuit values can be designed around the impedance. For example, an IF transformer will present an impedance at both its primary and secondary. These values may not be specified.

The impedance of a resonant circuit or component can be measured using a signal generator and a meter with an RF probe. An electronic voltmeter will provide the least loading effect on the circuit and will therefore provide the most accurate indication.

The procedure for impedance measurement at radio frequencies is the same as for audio frequencies, as discussed in Sec. 2-12.7, except as follows. An RF signal generator must be used as the signal source. The meter must be provided with an RF probe. If the circuit or component under measurement has both an input and output (such as a transformer), the opposite side or winding must be terminated in its normal load. For example, if the input (primary) impedance of a transformer is to be measured, the output (secondary) impedance should be terminated in a load. A fixed resistance can be used to simulate the load resistance. If the impedance of a tuned circuit is to be measured, tune the circuit to peak or dip, then measure the impedance at resonance. Once the resonant impedance is found, the signal generator can be tuned to other frequencies to find the corresponding impedance.

4-11.7. Testing Transmitter RF Circuits

It is possible to test and adjust transmitter RF circuits (such as discussed in Sec. 4-3) using a meter with an RF probe. If an RF probe is not available (or as an alternative), it is possible to use a circuit such as shown in Fig. 4-69.

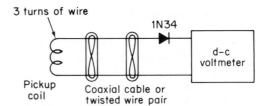

Fig. 4-69. Circuit for pick-up and measurement of RF signals.

This circuit is essentially a pick-up coil which is placed near the RF circuit inductance, and a rectifier that converts the RF into a dc voltage for measurement on a meter. The basic procedures are as follows:

1. Connect the equipment as shown in Fig. 4-70. If the circuit being measured is an amplifier, without an oscillator, a drive signal must be supplied by means of a signal generator. Use an unmodulated signal, at the correct operating frequency.
2. In turn, connect the meter (through an RF probe or the special circuit of Fig. 4-69) to each stage of RF circuit. Start with the first stage (this will

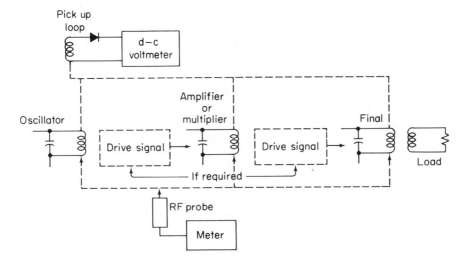

Fig. 4-70. Measuring transmitter resonant circuits.

be the oscillator if the circuit under test is a complete transmitter), and work toward the final (or output) stage.

3. A voltage indication should be obtained at each stage. Usually, the voltage indication will increase with each amplifier stage. Some stages may be frequency multipliers and provide no voltage amplification.

4. If a particular stage is to be tuned, adjust the tuning control for a maximum reading on the meter. If the stage is to be operated with a load (such as the final amplifier into an antenna), the load should be connected, or a simulated load should be used. A fixed resistance provides a good simulated load at frequencies up to about 250 MHz.

5. It should be noted that this tuning method or measurement technique will not guarantee each stage is at the desired operating frequency. It is possible to get maximum readings on harmonics. However, it is conventional to design RF transmitter circuits so that they will not tune to both the desired operating frequency and a harmonic. Generally, RF amplifier tank circuits will tune on either side of the desired frequency, but not to a harmonic (unless the circuit is seriously detuned, or the design calculations are hopelessly inaccurate).

4-11.8. Testing Receiver RF Circuits

It is possible to test and adjust receiver RF circuits (such as discussed in Secs. 4-6 through 4-10) using a meter and a signal generator. The same tests can be performed using an oscilloscope and sweep generator. [The oscilloscope/sweep generator methods are discussed fully in the author's *Handbook of Oscilloscopes: Theory and Application* (Prentice-Hall, Inc., Englewood Cliffs, N.J., 1968).]

The procedures for receiver circuit test using a meter and signal generator are as follows. Both AM and FM receivers require alignment of the IF and RF stages. An FM receiver also requires the alignment of the detector stage (discriminator or ratio detector). The normal sequence for alignment in a complete FM receiver is (1) detector, (2) IF amplifier and limiter stages, and (3) RF and local oscillator (mixer/converter). The alignment sequence for an AM receiver is (1) IF stages, and (2) RF and local oscillator. The following procedures can be applied to a complete receiver, or to individual stages, at any point during design.

If a complete receiver design is being tested, and the design includes an AVC–AGC circuit (Sec. 4-10), the AGC must be disabled. This is best accomplished by placing a fixed bias, of opposite polarity to the signal normally produced by the detector, on the AGC line. The fixed bias should be of sufficient amplitude to overcome the bias signal produced by the detector (usually on the order of 1 or 2 V). When such a bias is applied, the stage gain will be altered from the normal condition. Once alignment is complete, the bias should be removed.

If individual design circuits are to be tested, the precautions regarding AGC can be ignored.

4-11.8.1. FM-Detector Alignment

1. Connect the equipment as shown in Fig. 4-71 (for a discriminator) or Fig. 4-72 (for a ratio detector).

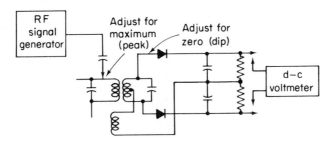

Fig. 4-71. FM-discriminator alignment.

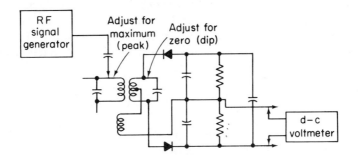

Fig. 4-72. FM-ratio-detector alignment.

2. Set the meter to measure dc voltage.
3. Adjust the signal generator frequency to the intermediate frequency (usually 10.7 MHz). Use an unmodulated output from the signal generator.
4. Adjust the secondary winding (either capacitor or tuning slug) of the discriminator transformer for zero reading on the meter. Adjust the transformer slightly each way and make sure the meter moves smoothly above and below the exact zero mark. (A meter with a zero-center scale is most helpful when adjusting FM detectors.)
5. Adjust the signal generator to some point below the intermediate frequency (to 10.625 MHz for an FM detector with a 10.7-MHz IF). Note the meter reading. If the meter reading goes down scale against the pin, reverse the meter polarity or test leads (the RF probe is not used for FM detector alignment).
6. Adjust the signal generator to some point above the intermediate frequency *exactly* equivalent to the amount set below the IF in step 5. For example, if the generator were set to 0.075 MHz below the IF (10.7 − 0.075 = 10.625), then the generator should be set to 10.775 (10.7 + 0.075 = 10.775).

7. The meter should read approximately the same in both steps 5 and 6, except the polarity will be reversed. For example, if the meter reads 7 scale divisions below zero for step 5 and 7 scale divisions above zero for step 6, the detector is balanced. If a detector circuit under design cannot be balanced under these conditions, the fault is usually a serious mismatch of diodes or other components.
8. Return the generator output to the intermediate frequency (10.7 MHz).
9. Adjust the primary winding of the detector transformer (either capacitor or tuning slug) for maximum reading on the meter. This sets the primary winding at the correct resonant frequency of the IF.
10. Repeat steps 4 through 8 to make sure that adjustment of the transformer primary has not disturbed the secondary setting. Invariably, the two settings will interact.

4-11.8.2. AM and FM Alignment

The alignment procedures for the IF stages of an AM receiver are essentially the same as those for an FM receiver. However, the meter must be connected at different points in the corresponding detector, as shown in Fig. 4-73. In either case the meter is set to measure direct current, and the RF probe is not

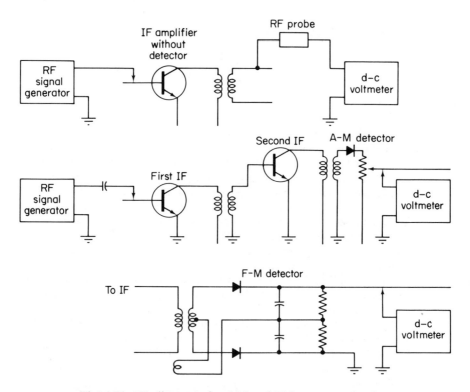

Fig. 4-73. IF alignment for AM and FM resonant circuits.

used. In those cases where IF stages are being tested without a detector (such as during design), an RF probe is required. As shown in Fig. 4-73, the RF probe is connected to the secondary of the final IF output transformer.

1. Connect the equipment as shown in Fig. 4-73.
2. Set the meter to measure dc, and connect it to the appropriate test point (with or without an RF probe as applicable).
3. Place the signal generator in operation and adjust the generator frequency to the receiver intermediate frequency (typically 10.7 MHz for FM and 455 kHz for AM). Use an unmodulated output from the signal generator.
4. Adjust the windings of the IF transformers (capacitor or tuning slug) in turn, starting with the last stage and working toward the first stage. Adjust each winding for maximum reading.
5. Repeat the procedure to make sure the tuning of one transformer has no effect on the remaining adjustments. Often, the adjustments will interact.

4-11.8.3. AM and FM RF Alignment

The alignment procedures for the RF stages (RF amplifier, local oscillator, mixer/converter) of an AM receiver are essentially the same as for an FM receiver. Again, it is a matter of connecting the meter to the appropriate test point. The same test points used for IF alignment can be used for aligning the RF stages as shown in Fig. 4-74. However, if an individual RF stage is to

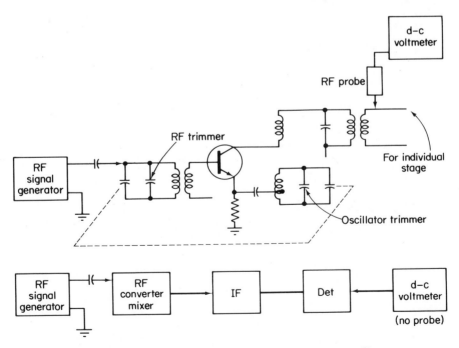

Fig. 4-74. Alignment of RF amplifier and local oscillator (converter/mixer).

be aligned, the meter must be connected to the secondary winding of the RF-stage output transformer, through an RF probe. The procedure is as follows:

1. Connect the equipment as shown in Fig. 4-74.
2. Set the meter to measure dc, and connect it to the appropriate test point (with or without an RF probe as applicable.)
3. Adjust the generator frequency to some point near the high end of the receiver operating frequency (typically 107 MHz for a broadcast-FM receiver, and 1400 kHz for an AM-broadcast receiver). Use an unmodulated output from the signal generator.
4. Adjust the RF-stage trimmer for maximum reading on the meter.
5. Adjust the generator frequency to the low end of the receiver operating frequency (typically 90 MHz for FM and 600 kHz for AM).
6. Adjust the oscillator-stage trimmer for maximum reading on the meter.
7. Repeat the procedure to make sure the resonant circuits "track" across the entire tuning range.

5

WAVEFORMING AND
WAVESHAPING CIRCUITS

5-1. Basic Design Considerations for Oscillators

All the design considerations discussed in Chapter 1 apply to oscillators (waveforming circuits). Of particular importance are how to interpret data sheets (Sec. 1-2), determining parameters at different frequencies (Sec. 1-3), temperature-related design problems (Sec. 1-4), and basic bias schemes (Sec. 1-6).

The main concern in any oscillator design is that the transistor will oscillate at the desired frequency and will produce the desired voltage or power. Most oscillator circuits operate with power outputs of less than 1 W. Many transistors will handle this power dissipation without heat sinks.

Another problem with oscillators is the class of operation. If an oscillator is biased class A (emitter–base forward-biased at all times), the output waveform will be free of distortion, but the circuit will not be efficient. That is, the power output will be low in relation to power input. Class A oscillators are typically 30 per cent efficient. Thus, a 3-W input is required for a 1-W output. For the purposes of calculation, input power for an oscillator can be considered as the product of collector voltage and collector direct current. The heat dissipation must be calculated on the basis of input power. For these reasons, class A oscillators are usually not used for RF, and are generally limited to those applications where a good waveform is the prime consideration.

A class C oscillator (emitter–base reverse-biased except in the presence of the feedback signal) is far more efficient (usually about 70 per cent). Thus, a 1.5-W input will produce a 1-W output. This cuts the heat-dissipation requirements in half. At radio frequencies, the waveform is usually not critical, so class C is in common use for RF circuits.

One drawback to class C is that the oscillator may not start in the reverse-bias condition. This can be overcome by forward biasing the emitter–base to start the oscillator (start collector current flow). The arrangement can be aided by an unbypassed-emitter resistor. Collector current flow will build up a reverse bias across the emitter resistor.

A particular problem with this bias scheme is that too much reverse bias may cause the transistor to cut off during the "on" half-cycle. To maintain the correct bias relationships, a variable bias charge is obtained by rectifying part of the oscillator signal and filtering the bias, using a large-value capacitor. In practice, the base–emitter diode serves as the rectifier, with the base (or emitter)-coupling (feedback) capacitor serving as the bias filter (to retain the correct bias charge during the "on" cycle).

Because the bias is variable (changes with the amplitude of the oscillator signal), the capacitor charge must also change. If the capacitor is too small, the oscillator may not start easily, or there will be distortion. If the capacitor is too large, the charge changes slowly, and the oscillator operates inter-mittently as a blocking oscillator.

To sum up, if the selected transistor is capable of producing the required power at the operating frequency, and the correct component values are selected for the resonant circuits, the only major problem in oscillator design is the correct bias point. Often, this must be found by trial-and-error test of the circuit in breadboard form.

Radio-frequency oscillators require resonant circuits for their operation. Therefore, all of the design considerations of Sec. 4-1 apply to the resonant circuits for RF oscillators. Likewise, the design considerations of Sec. 4-3 should be studied when RF oscillators are to be used with a frequency multi-plier and/or power amplifiers.

5-2. *LC* and Crystal-Controlled Oscillators

LC oscillators are those that use inductances (coils) and capacitors as the frequency-determining components. Typically, the coils and capacitors are connected in series- or parallel-resonant circuits, and adjusted to the desired operating frequency. Either the coil or capacitor can be variable. *LC* oscil-lators are used at higher frequencies (RF). Both the Hartley and Colpitts oscillators can be designed with transistors. (The Hartley and Colpitts are classic vacuum-tube oscillator circuits.) However, the Colpitts is generally the most popular.

Transistor *LC* oscillators can be crystal-controlled. That is, a quartz crystal can be used to set the frequency of operation, with an adjustable *LC* circuit used to "trim" the oscillator output to an exact frequency. In addition to the Hartley and Colpitts, there are a number of other crystal oscillator circuits suitable for transistors. These include the Pierce oscillator, harmonic or overtone oscillators, and oscillators that use two transistors to provide the necessary feedback required for oscillation.

5-2.1. Basic Solid-State LC Oscillators

Figure 5-1 shows two arrangements of the Hartley oscillator circuit. The circuit of Fig. 5-1a uses a bypassed emitter resistor to provide proper operating conditions. The circuit of Fig. 5-1b uses a base-leak resistor and biasing diode. The amount of feedback in either circuit depends upon the position of the coil (*L*) tap. Output from the Hartley circuits can be obtained through inductive coupling to the coil, or through capacitive coupling to the base.

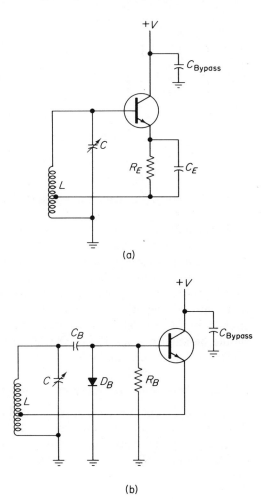

Fig. 5-1. Basic solid-state Hartley oscillator circuits.

Figure 5-2 shows two forms of the Colpitts oscillator circuit. Colpitts circuits are more commonly used in VHF and UHF equipment than the Hartley circuit because of the mechanical difficulty involved in making the tapped coils required at high frequencies. Feedback is controlled in the Colpitts oscillator by the ratio of capacitance C' to C''. Note that the LC circuits in the basic configurations of Fig. 5-2 are in the base circuit. Feedback is between emitter and base, through C' and C''. In practical Colpitts circuits, the LC components are in the collector, and feedback is between collector and emitter (determined by the ratio of capacitors). The base is held at a fixed dc level by bias resistors.

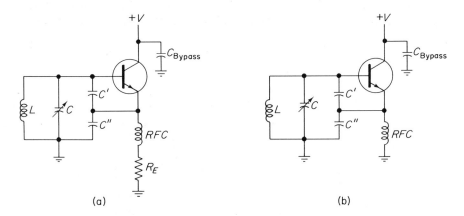

Fig. 5-2. Basic solid-state Colpitts oscillator circuits.

In the design examples described in later paragraphs of this section, the class of operation is *set by the amount of feedback* rather than the bias point. That is, the transistor is biased for an optimum operating point and then feedback is adjusted for the desired class of operation.

LC radio-frequency oscillators require resonant circuits for their operation. Thus, all the design considerations of Chapter 4 apply to the resonant circuits for LC oscillators. Likewise, the design considerations of Chapter 4 should be studied when RF oscillators are to be used with frequency multiplier and/or power amplifiers.

5-2.2. Basic Solid-State Crystal-Controlled Oscillator

Transistors operate efficiently in crystal oscillator circuits such as the Pierce-type oscillator shown in Fig. 5-3. This type of oscillator is very popular because of its simplicity and minimum number of components. No LC circuits are required for frequency control. Instead, the frequency is set by the crystal.

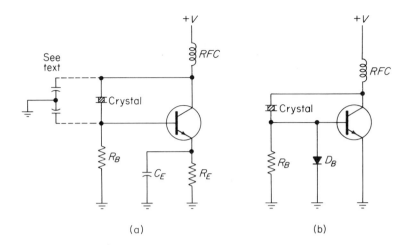

Fig. 5-3. Basic solid-state Pierce oscillator circuits.

At frequencies below 2 MHz, a capacitive voltage divider may be required across the crystal. The connection between the voltage-divider capacitors must be grounded so that the voltage developed across the capacitors is 180° phase-inverted.

It is frequently desirable to operate crystals in communications equipment at their *harmonic, or overtone, frequencies.* Figure 5-4 shows two circuits designed for overtone operation. Additional feedback is obtained for the overtone crystal by means of a capacitive divider, acting as the *LC* circuit bypass. Most third-overtone crystals operate satisfactorily without this additional feedback, but for the fifth and seventh harmonics, the extra feedback is required. The *LC* circuit in Fig. 5-4 is not fully bypassed and thus produces a voltage that aids oscillation. The crystals in both circuits are connected to the junction of the two capacitors C_D' and C_D''. The ratio of these capacitors should be approximately 1:3 to provide the required amount of additional feedback.

The circuit of Fig. 5-5 operates well at *low frequencies.* The crystal is located in the feedback circuit between the emitters of the two transistors and operates in the series mode. Capacitor C_2 is used for precise adjustment of the oscillator frequency. A reduction in C_2 capacitance increases the frequency slightly.

The practical limit of crystal fundamental resonance is about 25 MHz for transistor crystal oscillators. From 20 to 60 MHz, third-overtone crystals are used. Fifth-overtone crystals are used for frequencies between 60 and 120 MHz, and seventh-overtone crystals at frequencies above 120 MHz. Usually, 150 MHz is the top limit for a solid-state crystal oscillator. Multipliers are used above this frequency.

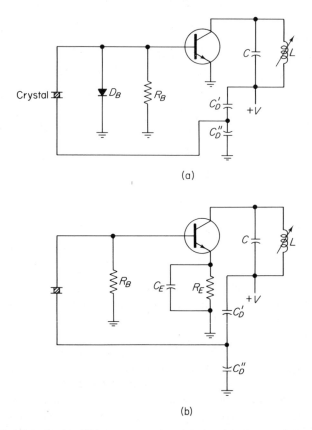

Fig. 5-4. Basic solid-state crystal oscillator circuits permitting operation at overtone or harmonic frequencies.

In most solid-state oscillator circuits, the crystal series-resonant frequency is most appropriate. At "overlapping" frequencies, where either a fifth or a seventh overtone can be used (such as 115 to 125 MHz), the power output is about the same for either crystal. Of course, the seventh overtone may present other problems, such as more critical adjustments and a tendency to oscillate at the fifth overtone.

In general, the oscillator LC circuit is tuned to the crystal overtone. A better viewpoint is that the reactive elements of the LC circuit and crystal form a coupling network between the transistor output and input, which produces the necessary phase shift. The function of the crystal is to introduce an additional reactive element capable of causing large phase-shift changes for a very small frequency change. To maintain the 360°-loop phase shift, changes in reactances in the circuit can be compensated for by only tiny frequency shifts, due to the extreme high crystal Q, and usually excellent tem-

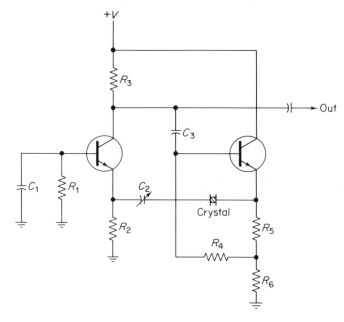

Fig. 5-5. Basic solid-state low-frequency crystal oscillator circuit.

perature stability. Where extreme stability is required, the crystal should be housed in a temperature-stable environment (crystal oven).

Crystals reduce oscillator efficiency as a result of losses in the crystal. These losses are represented by their series resistance. Typically, crystal series resistance is on the order of 20 to 100 Ω. Oscillator efficiency may be increased by reducing signal currents in all dissipative elements. Generally, this is done by including RF chokes (RFCs) in the transistor element leads. The values for such chokes are determined by frequency, as is discussed in the design examples.

5-2.3. Practical Solid-State Crystal Oscillator

Figure 5-6 is the working schematic of a crystal-controlled oscillator. This circuit is one of the many variations of the Colpitts oscillator. However, the output frequency is fixed and controlled by the crystal. The circuit can be used over a narrow range by L_1 (which is slug-tuned).

Many factors must be considered in the design of stable solid-state oscillators. In general, all the characteristics for vacuum-tube oscillators apply to solid-state oscillators. For example, the frequency-determining components must be temperature-stable, and mechanical movement of the individual components should not be possible.

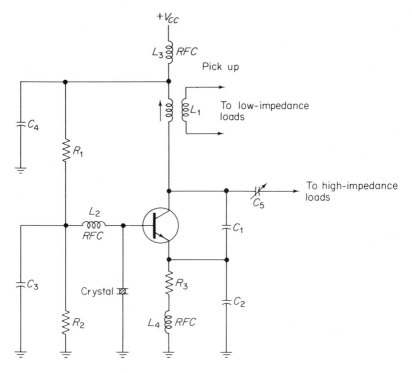

Fig. 5-6. Solid-state crystal-controlled oscillator circuit design.

Many factors affect oscillator frequency stability. For example, there is usually some optimum bias value and supply voltage that will produce maximum frequency stability over a given range of operating temperature. However, the one factor that can be controlled by the designer is percentage of feedback. (Note that this percentage refers to feedback versus output voltage.) The lowest practical feedback level is about 10 per cent. The best feedback level is about 15 per cent. Rarely is more than 25 to 30 per cent ever required, although some oscillators are operated at 40 per cent.

If the operating frequency of a solid-state oscillator is in the VHF and UHF regions, the percentage of feedback must be increased over that of a comparable oscillator operating at lower frequencies. Likewise, the percentage of feedback must be increased if the tuning circuits are made high-C through substantial increase in tuning capacitor value. The resonant frequency of oscillator circuits is set by the combination of L and C values. If the value of C is made quite high (with a corresponding lower value of L), the resonant circuit is said to be "high-C" and usually results in sharper resonant tuning. A large value of L (with a corresponding lower value of C)

introduces more resistance into the resonant circuit, thus lowering the circuit Q to produce broader resonant tuning. (Refer to Sec. 4-1.1.2.)

For maximum efficiency (maximum power output for a given supply voltage and current), the resonant circuit (C_1, C_2, L_1, and the transistor output capacitance of Fig. 5-6) should be at the same frequency as the crystal. If reduced efficiency is acceptable, the resonant circuit can be at a higher frequency (multiple) of the crystal frequency. However, the resonant circuit should not be operated at a frequency higher than the fourth harmonic of the crystal frequency.

Bias circuit. The bias-circuit components, R_1, R_2, and R_3, are selected to produce a given current flow under no-signal conditions. The bias circuit is calculated and tested on the basis of *normal operating point*, even though the circuit will never be at the operating point. A feedback signal is always present, and the transistor is always in a state of transition. The collector current should be set at a value to produce the required power output. With the correct bias-feedback relationship, the output power of the oscillator is about 0.3 times the input power.

Typically, the voltage drop across L_1 and L_3 is very small, so that the collector (or drain) voltage equals the supply voltage. Thus, to find a correct value of current for a given power output and supply voltage, divide the desired output by 0.3 to find the required input power. Then divide the input power by the supply voltage to find the collector-current flow.

Feedback signal. The signal output appears at the collector, and with the proper bias-feedback relationship, the output voltage is about 80 per cent of the supply voltage. The amount of feedback is set by the ratio of C_1 and C_2. For example, if C_1 and C_2 are of the same value, the feedback signal is one-half of the output signal. If C_2 is made about 3 times the value of C_1, the feedback signal is about 0.25 of the output signal voltage.

It may be necessary to change the value of C_1 in relation to C_2, in order to get a good bias-feedback relationship. For example, if C_2 is decreased in value, the feedback increases, and the oscillator operates nearer the class C region. An increase in C_2, with C_1 fixed, decreases the feedback and makes the oscillator operate as class A. Keep in mind that any change in C_2 (or C_1) will also affect frequency. Thus, if the C_2/C_1 values are changed, it will probably be necessary to change the value of L_1.

As a first trial value, the amount of feedback should be equal to, or greater than, cutoff. That is, the feedback voltage should be equal to or greater than the voltage necessary to cut off collector current flow. Under normal conditions, such a level of feedback should be sufficient to overcome the fixed bias (set by R_1 and R_2) and the variable bias set by R_3. As discussed, feedback is generally within the limits of 10 and 40 per cent, with the best stability in the 15 to 25 per cent range.

Frequency. Frequency of the circuit is determined by the resonant frequency of L_1, C_1, and C_2, and by the crystal frequency. Note that C_1 and C_2 are in series, so that the total capacitance must be found by the conventional series equation:

$$C = \frac{1}{1/C_1 + 1/C_2} \quad \text{or} \quad \frac{C_1 \times C_2}{C_1 + C_2}$$

Also note that the output capacitance of the transistor must be added to the value of C_1. At low frequencies, the output capacitance can be ignored since the value is usually quite low in relation to a typical value for C_1. At higher frequencies, the value of C_1 is lower, so the output capacitance becomes of greater importance.

For example, if the output capacitance is 5 pF at the frequency of interest, and the value of C_1 is 1000 pF, or larger, the effect of the output capacitance is small. (Transistor output capacitance can be considered as being in parallel with C_1.) If the value of C_1 is lowered to 5 pF, however, the parallel output capacitance will double the value. Thus, the output capacitance must be included in the resonant-frequency calculation.

As discussed, transistor output capacitance is not always listed on data sheets. The capacitance presented by the output of a transistor (collector-to-emitter) is composed of both output capacitance and reverse capacitance. However, reverse capacitance is usually small in relation to output capacitance, and can generally be ignored.

When output capacitance is not available on data sheets, it is possible to calculate an *approximate* value of output capacitance if output admittance is given. The imaginary part of output admittance (the *jb* part, such as jb_{22}) represents susceptance, which is the reciprocal of reactance. Thus, to find the reactance presented by the collector–emitter terminals of the transistor at the data-sheet frequency, divide the *jb* part into 1. Then find the capacitance that will produce such reactance at the data-sheet frequency using the equation

$$C = \frac{1}{6.28FX_C}$$

where C is the output capacitance, F is the frequency, and X_C is the capacitive reactance found as the reciprocal of the *jb* part of the output admittance.

Of course, this method assumes that the *jb* reactance is capacitive, and that the capacitance remains constant at all frequencies (at least that the data-sheet capacity is the same for the design frequency). Neither of these conditions is always true.

Capacitor C_1 can be made variable. However, it is generally easier to make L_1 variable, since the tuning range of a crystal-controlled oscillator is quite small.

Typically, the value of C_2 is about 3 times the value of C_1 (or the combined values of C_1 and the transistor output capacitance, where applicable). Thus, the signal voltage (fed back to the emitter terminal) is about 0.25 of the total output signal voltage (or about 0.2 of the supply voltage, when the usual bias-feedback relationship is established).

Resonant circuit. Any number of *L* and *C* combinations could be used to produce the desired frequency. That is, the coil can be made very large or very small, with corresponding capacitor values. Often, practical limitations are placed on the resonant circuit (such as available variable inductance values).

In the absence of some specific limitations, and as a starting point for resonant circuit values, the capacitance should be 2 pF/m. For example, if the frequency is 30 MHz, the wavelength is 10 m, and the capacitance should be 20 pF. Wavelength in meters is found by the equation

$$\text{wavelength (meters)} = \frac{300}{\text{frequency (MHz)}}$$

At frequencies below about 1 to 5 MHz, the 2 pF/m guideline may result in very large coils to produce the corresponding inductance. If so, the 2 pF/m can be raised to 20 pF/m.

As an alternative method to find realistic values for the resonant circuit, use an inductive reactance value (for L_1) between 80 and 100 Ω at the operating frequency. This guideline is particularly useful at low frequencies (below 1 MHz).

Output circuit. Output to the following stage can be taken from L_1 by means of a pick-up coil (for low-impedance loads) or coupling capacitor (for high-impedance loads). Generally, the most convenient output scheme is to use a coupling capacitor (C_5) and make the capacitor variable. This makes it possible to couple the oscillator to a variable load (a load that changes impedance with changes in frequency).

Crystal. The crystal must, of course, be resonant at the desired operating frequency (or a submultiple thereof, when the circuit is used as a multiplier). Note that the efficiency (power output in relation to power input) of the oscillator is reduced when the oscillator is also used as a multiplier. This is shown in Fig. 5-7, which also illustrates that efficiency is reduced when overtone crystals are used (instead of fundamental crystals).

The crystal must be capable of withstanding the combined dc and signal voltages at the transistor input (base). As a rule, the crystal should be capable of withstanding the full supply voltage, even though the crystal will never be operated at this level.

Bypass and coupling capacitors. The values of bypass capacitors C_3 and C_4 should be such that the reactance is 5 Ω or less at the crystal operating frequency. A higher reactance (200 Ω) could be tolerated. However, owing to the low crystal output, the lower reactance is preferred.

Crystal frequency of operation	RF Power output as a % of DC power input			
	Fund.	Second	Third	Fourth
		Harmonics		
Fundamental	30	15	10	5
Third overtone	25	15	10	5
Fifth overtone	20	12	7	3
Seventh overtone	20	12	7	3

Fig. 5-7. Typical RF power output of crystal oscillator circuits. Courtesy J. W. Streater, Mentor Radio Company.

The value of C_5 should be approximately equal to the combined parallel output capacitance of the transistor and C_1. Make this the midrange value of C_5 (if C_5 is variable).

Radio-frequency chokes. The values of RFCs L_2, L_3, and L_4 should be such that the reactance is between 1000 and 3000 Ω at the operating frequency. The minimum current capacity of the chokes should be greater (by at least 10 per cent) than the maximum anticipated direct current. Note that a high reactance is desired at the operating frequency. However, at high frequencies, this can result in very large chokes that produce a large voltage drop (or are too large physically).

5-2.4. Crystal Oscillator Design Example

Assume that the circuit of Fig. 5-6 is to provide an output at 50 MHz. The circuit is to be tuned by L_1. A 30-V supply is available. The crystal will not be damaged by 30 V and will operate at 50 MHz with the desired accuracy. The transistor has an output capacitance of 3 pF and will operate without damage with 30 V. The desired output power is 40 to 50 mW.

The collector is operated at 30 V (ignoring the small drop across L_1 and L_3). The values of R_1, R_2, and R_3 should be chosen to provide a current that will produce 40 to 50 mW output with 30 V at the collector. A 45-mW output divided by 0.3 is 150 mW. Thus, the input power (and total dissipation) is 150 mW. Make certain that the transistor will permit a 150-mW dissipation at maximum anticipated temperature.

For example, assume that the transistor has a 330-mW maximum dissipation at 25°C, a maximum temperature rating of 175°C, and a 2-mW/°C derating for temperatures above 25°C. If the transistor is operated at 100°C, or 75° above the 25°C level, the transistor must be derated by 150 mW (75 $\times$ 2 mW/°C), or $330 - 150$ mW $= 180$ mW. Under these conditions, the 150-mW input power dissipation is safe.

With 30 V at the collector and a desired 150-mW input power, the collector current must be 150 mW/30 V = 5 mA. Bias resistors R_1, R_2, and R_3 should be selected to produce this 5-mA collector current using the guidelines of Sec. 1-6. Then, with a 30-V supply, the output signal will be about 24 V (30 × 0.8 = 24). Of course, this is dependent upon the bias-feedback relationship. Also, the collector current will not remain at 5 mA when the circuit is oscillating, since the transistor is always in a state of transition.

As a starting point, make C_2 3 times the value of C_1 (plus the transistor output capacitance). With this ratio the feedback will be about 25 per cent of the output, or 6 V (24 × 0.25 = 6). Considering the amount of fixed and variable bias supplied by the bias network, a feedback of 6 V may be large. However, the 6-V value should serve as a good starting point.

For realistic values of L and C in the resonant circuit, let $C_1 = 2$ pF/m, or 12 pF (50 MHz = 6 m; 300/50 = 6). With C_1 at 12 pF, and the transistor output capacitance of 3 pF, the value of C_2 is 45 pF (12 + 3 = 15; 15 × 3 = 45).

The total capacitance across L_1 is

$$\frac{15 \times 45}{15 + 45} \approx 13 \text{ pF}$$

With a value of 13 pF across L_1, the value of L_1 for resonance at 50 MHz is

$$L\,(\mu\text{H}) = \frac{2.54 \times 10^4}{(50)^2 \times 13} \approx 0.8 \ \mu\text{H}$$

For convenience, L_1 should be tunable from about 0.5 to 1.6 μH.

Keep in mind that an incorrect bias-feedback relation results in distortion of the waveform, or low power, or both. The final test of correct operating point is a good waveform at the operating frequency, together with frequency stability at the desired output power. Also, feedback is set by the relationship of C_1 and C_2. Any change in this relationship requires a corresponding change in the value of L_1. As a guideline, if no realistic combination of C_1, C_2, and L_1 produces the desired waveform and power output, try a change in the bias.

The values of C_3 and C_4 should be:

$$\frac{1}{6.28 \times 50 \text{ MHz} \times 5} \approx 630 \text{ pF}$$

A slightly larger value (say 1000 pF) will assure a reactance of less than 5 at the operating frequency.

The values of L_2, L_3, and L_4 should be

$$\frac{2000}{6.28 \times 50 \text{ MHz}} \approx 6.3 \ \mu\text{H nominal}$$

Any value between about 3 and 9 μH should be satisfactory. The best test for the correct value of an RFC in an oscillator is to check for RF at the power-supply side of the line, with the oscillator operating. That is, check for

RF at the point where L_3 connects to the power supply (not at the L_1 side). There should be no RF (or RF should be no greater than a few microvolts for a typical solid-state oscillator) on the power-supply side of the RFC. If RF is removed from the power-supply line, the choke reactance is sufficiently high. Next, check for dc voltage drop across the choke. The drop should be a fraction of 1 V (typically in the microvolt range).

5-2.5. Practical Variable-Frequency Oscillator

Figure 5-8 is the working schematic of a variable-frequency oscillator. This circuit is also one of the many variations of the Colpitts oscillator. The circuit is chosen here for maximum stability at frequencies up to about 0.5 MHz. Oscillation is sustained by feedback to the emitter from the junction of C_1 and C_2 (as is the case for the crystal oscillator).

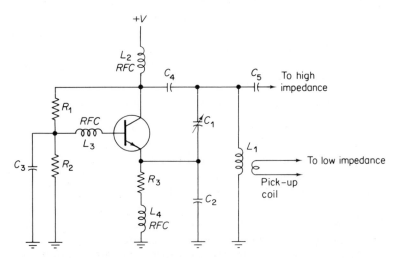

Fig. 5-8. Solid-state variable-frequency oscillator circuit design.

Design considerations. All the design considerations for the variable-frequency oscillator are the same as for the crystal-controlled oscillator (Sec. 5-2.3), with the following exceptions. Generally, C_1 is made variable to tune across a given frequency range. However, L_1 can be made variable if required.

The values of coupling and bypass capacitors C_3, C_4, and C_5 (if used) should be such that the reactance is 200 Ω at the lowest operating frequency (when variable capacitor C_1 is at full value). Note that capacitor C_4 and the output capacitance of the transistor may add to the C_1 capacitance. This tends to lower the resonant frequency of the L_1, C_1, C_2 circuit slightly from the calculated value. However, since C_1 is variable, there should be no problem in tuning to a desired frequency.

5-2.6. Variable-Frequency Oscillator Design Example

Assume that the circuit of Fig. 5-8 is to tune across a frequency range from about 10 kHz to 60 kHz. A 30-V supply is available. Thus, the transistor is operated at about 30 V (ignoring the small drop across L_2). The values of R_1, R_2, and R_3 are selected to produce the desired operating point current, using the guidelines of Sec. 1-6. Assume that the transistor has a negligible output capacitance (in relation to C_1) at the operating frequency. This is generally the case at these lower frequencies (10 to 60 kHz).

With the collector at 30 V, the power input is determined by the amount of current (set by the bias network), multiplied by 30 V. Assume that the transistor is operated at 1 mA of collector current. Under these conditions, the power input is 30 mW (30 V $\times$ 1 mA = 30 mW). Assuming a typical efficiency of 0.3, the output power is about 9 mW.

With a 30-V supply, the output signal should be about 24 V (30 $\times$ 0.8 = 24). Of course, this is dependent upon the bias-feedback relationship. When C_2 is made 3 times the value of C_1, the feedback signal will be 6 V (24 V $\times$ 0.25 = 6 V). Considering the amount of fixed and variable bias supplied by the bias network, a feedback of 6 V may be large. However, the 6-V value should serve as a good starting point.

For realistic L and C values in the resonant circuit, the inductive reactance of L_1 should be between 80 and 100 Ω at the operating frequency. Assume a value of 100 Ω as a first trial.

With an inductive reactance of 100 and a low-frequency limit of 10 kHz, the inductance of L_1 should be

$$\frac{100}{6.28 \times 10 \text{ kHz}} \approx 2 \text{ mH}$$

With a value of 2 mH for L_1, and a low-frequency limit of 10 kHz, the total capacitance of C_1 and C_2 (with the variable C_1 at its high limit) should be

$$\frac{2.54 \times 10^4}{(10)^2 \times 2000 \text{ } \mu\text{H}} \approx 0.12 \text{ } \mu\text{F}$$

With a 24-V output and a 6-V feedback, the value of C_1 is 0.12 (6 + 24)/ 24, or 0.15 μF. The value of C_2 is 0.15 $\times$ 3 = 0.45, which is rounded off to 0.5 μF.

Keep in mind that an incorrect bias-feedback relation will result in distortion of the waveform, or low power, or both. The final test of correct operating point is a good waveform at the operating frequency, together with the desired output power.

The values of C_3, C_4, and C_5 (if used) should be

$$\frac{1}{6.28 \times 10 \text{ kHz} \times 200} \approx 0.08 \; \mu\text{F (minimum)}$$

A slightly larger value (say 0.1 μF) will assure a reactance of less than 200 Ω at the lowest frequency (10 kHz).

The values of L_2 through L_4 should be

$$\frac{2000}{6.28 \times 10 \text{ kHz}} \approx 30 \text{ mH}$$

At the low currents involved, the 30-mH chokes should present little or no voltage drop.

5-3. *RC* Oscillator

Figure 5-9 is the working schematic of an *RC* (resistance–capacitance) oscillator. Such oscillators are used at audio frequencies instead of the *LC* (inductance–capacitance) oscillators described in Sec. 5-2. *RC* oscillators avoid the use of inductances, which are not practical in the audio-frequency

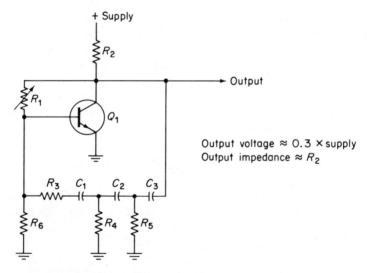

Frequency $\approx \frac{1}{18 \; RC}$, where $R = R_3 = R_4 = R_5$ and $C_1 = C_2 = C_3$

$$C_1 = C_2 = C_3 = \frac{1}{18 \; R \times \text{frequency}}$$
$R_1 = 500 \text{ k}\Omega$ variable
R_2 and R_6 = see text
$R_3 = R_4 = R_5 = 10 \text{ k}\Omega$ maximum

Fig. 5-9. *RC* oscillator.

range. *RC* oscillators are usually operated in class A, thus producing good waveforms. The feedback principle is used in *RC* oscillators. The collector signal is fed back through an *RC* network to the base. The time constants of the *RC* network determine the oscillator output frequency. In the circuit of Fig. 5-9, each of the three identical *RC* networks shift the phase by about 60°, resulting in a total phase shift of 180°.

5-3.1. Design Considerations

In addition to the general design consideration of Sec. 5-1, the following specific points should be considered.

Power output. *RC* oscillators do not have the efficiency of *LC* oscillators, since *RC* oscillators are operated class A. Also, there is considerable power loss in the *RC* network. Typically, *RC* oscillators are never more than 30 per cent efficient. Generally, power output is not a major design consideration in *RC* oscillators. Instead, voltage output is the prime factor. Typically, the voltage output is about 30 per cent of the supply voltage. Thus, the transistor–collector–voltage limits (or the available supply) set the maximum output voltage.

Output frequency. The oscillator frequency is determined by the *RC* time constant. To simplify design, the same values are used in all three *RC* networks. Thus, the output frequency is about one-third of the typical time constant-frequency relationships of $1/6.28RC$ or $1/18RC$. A more exact frequency calculation cannot be made in practical design, since the transistor capacitance and resistance values are added to the *RC* network. However, the $1/18RC$ relationship is satisfactory for trial values.

If a variable output frequency is required, either the *C* or *R* could be made variable. However, it is common practice to make the *C* variable, since three-section variable capacitors are readily available. Typically, the *RC* network resistors should not exceed about 10 kΩ. With this value, the lowest audio frequency (above 1 Hz) can be obtained with a *C* of less than 6 μF.

Bias network. The transistor should be biased class A. That is, there should be some collector current flowing at all times. The collector voltage should be one-half the supply voltage at the "operating point." A true operating point is never obtained since there is always some signal being fed back.

An emitter resistor need not be used since there will always be some reverse bias supplied by the feedback signal (on half of each cycle). The base is set to the correct voltage by R_1. Usually, R_1 is quite large (typically 0.25 to 0.5 MΩ), since base current is on the order of 0.1 mA or less.

Transistor selection. The transistor must be capable of oscillating at the desired frequency. This presents no particular problem since most *RC* oscillators are used at frequencies below 100 kHz (generally below 20 kHz).

However, the transistor must have a gain of about 60 at the operating frequency to overcome the power loss introduced by RC networks.

There is a design tradeoff relationship between transistor gain and the collector resistor (R_2) value. If the value of R_2 is equal to the value of the RC network resistors R_3, R_4, and R_5, a transistor gain of about 60 is required. If the value of R_2 is increased to about twice the value of R_3, R_4, and R_5, the required gain can be reduced to about 45. However, an increase in the value of R_2 lowers the operating-point collector voltage, thus lowering the available output-voltage swing.

If the transistor produces too much gain, resistor R_6 can be added, as shown in Fig. 5-9. As a first trial value, make R_6 the same value as R_2.

5-3.2. Design Example

Assume that the circuit of Fig. 5-9 is to provide a 6-V output at 60 Hz. The transistor is capable of oscillating over the entire audio-frequency range.

With a required output of 6 V, the supply voltage should be $6 \div 0.3$, or 20 V.

The values of R_3, R_4, and R_5 should be 10 kΩ maximum, and preferably nearer one-half that value. The nearest 10 per cent standard would be 4.7 kΩ.

With the RC network resistors at 4.7 kΩ, the values of the network capacitors should be $1 \div 18 \times 4700 \times 60$, or approximately 0.2 μF.

With the RC network resistors at 4.7 kΩ, the value of collector resistor R_2 should be between 4.7 and 14 kΩ. As a first trial, use 10 kΩ, which is about halfway between the two limits.

With R_2 at 10 kΩ, and a supply of 20 V, about 1-mA collector current will be required to drop the collector voltage to one-half ($20 \text{ V} \times 0.5 = 10 \text{ V}$; $10 \text{ V} \div 10,000 = 1 \text{ mA}$).

The value of R_1 should be adjusted to provide the 1-mA current flow, as indicated when the dc collector voltage is at 10 V. Then the output (ac) voltage can be measured. If the transistor gain is known, the approximate value of R_1 can be calculated. For example, if the gain is 50, the value of R_1 would be 100 kΩ. This is found as follows: 1 mA (collector current) $\div 50 = 20$ μA (base current); with 20 V (supply) $\div 20$ μA (base current), the value of R_1 is 100 kΩ.

As a practical matter, use a variable resistance for R_1 (arbitrarily 500 kΩ). Then adjust for the correct voltages, and waveforms, at the collector output.

It may be necessary to trade off R_1 and R_2 values. As a guide, if the waveform is poor, or oscillations are unstable, change the bias by changing R_1. If the waveform and oscillations are good, but the output voltage is low or high, change the value of R_2.

If the frequency is incorrect, change the values of the capacitors and/or resistors in the RC network. This problem usually does not arise when the circuit is used as a variable-frequency oscillator (where the capacitors are made variable). One exception is where a high- or low-frequency limit cannot be reached over the range of the variable capacitors. The problem does arise when the circuit is used to produce a fixed frequency. It is possible to modify the value of one RC network (say the value of R_3, R_4, or R_5) to change frequency. This may result in distortion, however. Likewise, a drastic change in only one RC network value may affect the overall phase shift, and result in unstable oscillation.

5-4. Twin-T *RC* Oscillator

Figure 5-10 is the working schematic of a twin-T RC (resistance–capacitance) oscillator. Such circuits are used at audio frequencies where a highly stable output at a fixed frequency is desired. This circuit is somewhat similar to the low-frequency oscillator described in Sec. 3-14. However, the circuit of Fig. 5-10 uses a two-stage amplifier made up of discrete components, rather than the IC amplifier described in Sec. 3-14.

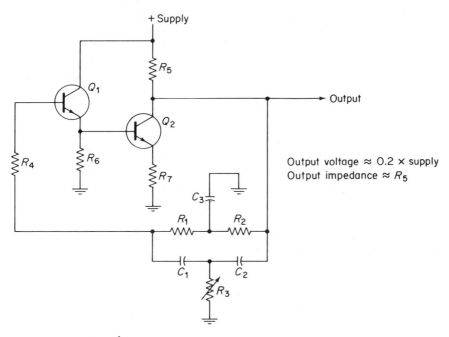

Frequency $\approx \dfrac{1}{5\,RC}$, where $R = R_1 = R_2$, $R_3 = 0.5R_1$ and $C = C_1 = C_2$, $C_3 = 2C$

$$R_4 \approx R_1 + R_2 \qquad R_5 \approx 60\,R_7$$
$$R_6 \approx 10\,R_7 \qquad R_7 \approx 50 - 100\,\Omega$$
$$R_1 = R_2 = 1000\,R_7 \qquad C_1 = C_2 = \dfrac{1}{5\,R_1 \times \text{frequency}}$$
$$R_3 = 0.5\,R_1 \qquad C_3 = 2C_1$$

Fig. 5-10. Twin-T *RC* oscillator.

5-4.1. Design Considerations

In addition to the general design considerations of Sec. 5-1, the following specific points should be considered.

Power output. Typically, twin-T RC oscillators are never more than about 30 per cent efficient. Likewise, the voltage output is about 30 per cent of the supply voltage. Thus, the transistor–collector–voltage limits (or the available supply) set the maximum output voltage.

Output frequency. The oscillator frequency is determined by the RC time constants of the twin-T network. This network is essentially a filter that has a sharp null, or balance, at the resonant frequency ($F = 1/6.28RC$). By decreasing the value of shunt resistor R_3 slightly from the balance point, the output of the twin-T network is a small, in-phase signal that is rapidly changing in phase at the balance frequency. When the network is adjusted off-balance (the normal oscillating condition) by R_3, the approximate frequency is found by $F \approx 1/5RC$.

Bias network. Both transistors should be forward-biased so that they operate class A. This will ensure a good waveform. The base bias for Q_1 (an emitter follower with no voltage gain) is supplied through R_4, network resistors R_1 and R_2, and collector resistor R_5. As a rule, R_4 should equal the total series resistance of R_1 and R_2. The collector of Q_1 is connected directly to the source, while the Q_1 emitter is returned through the base resistance of Q_2. Base resistor R_6 should be approximately 10 times the value of R_1. Transistor Q_2 should be biased similar to the class A amplifiers discussed in Chapter 2. However, considerable voltage gain is required to overcome the network loss. Therefore, R_5 (collector) should be approximately 60 times the value of R_7 (emitter) for maximum voltage gain. For a realistic R_5 value, the value of R_7 should be less than 100 Ω, typically 50 Ω (51 Ω standard).

Any number of RC combinations could be used to produce the desired operating frequency. However, since R_1 and R_2 also form part of the bias network, design of the RC network should start with these resistors. As first trial values, R_1 and R_2 should be 1000 times the value of R_7.

Transistor selection. The transistors must be capable of oscillating at the desired frequency. This presents no particular problem, since most RC oscillators are used at frequencies below 100 kHz (generally below 20 kHz). Both Q_1 and Q_2 can be the same transistor type, if convenient. However, Q_2 must have a gain of about 100 at the operating frequency to overcome the power loss.

5-4.2. Design Example

Assume that the circuit of Fig. 5-10 is to provide an output at 60 Hz. The transistors are capable of oscillating over the entire audio range, with the full supply voltage applied.

The key design value is the resistance of R_7. If the remaining resistor values are to be kept within reason, the value of R_7 should be less than 100 Ω. This will provide just enough reverse bias to stabilize both Q_1 and Q_2. A much larger value will usually provide so much reverse bias that the circuit cannot oscillate.

With R_7 at 51 Ω (the nearest standard value at approximately one-half of 100 Ω), the values of the remaining resistors are: R_1 and R_2 51 kΩ, R_3 25 kΩ (maximum), R_4 100 kΩ, R_5 3 to 3.3 kΩ, and R_6 510 Ω.

With R_1 and R_2 at 51 kΩ, the values of C_1 and C_2 should be

$$\frac{1}{5 \times (51 \times 10^3) \times 60} \quad \text{or} \quad 0.065 \ \mu F$$

With C_1 and C_2 at 0.065 μF, the value of C_3 should be 0.065×2, or 0.13 μF.

The critical element of this circuit is the adjustment of R_3. Resistor R_3 should always be made variable during the breadboard stage of design, even if the remaining resistance values are fixed. The value of R_3 should be increased until the circuit stops oscillating. Then reduce the value of R_3 slightly (detuning the *RC* filter) to allow an increase in the 60-Hz signal to sustain oscillation.

If it is necessary to detune the filter (change the value of R_3 from the nonoscillating point) by a large amount to get stable oscillations, the overall circuit gain is too low. This condition can be corrected by decreasing the value of R_4, increasing the value of R_5, or changing the transistor for higher gain.

With the correct value of R_3 chosen, the oscillations should remain stable despite changes in supply voltage (within reasonable limits). Using the values described, it should be possible to vary the supply voltage from about 12 V to 28 V, and still maintain stable oscillation. Of course, the output voltage will vary with changes in supply. Generally, the output voltage will be about 0.2 times the supply voltage. However, this ratio does not always remain true. A higher percentage can often be obtained by increasing the supply, all other factors being equal.

5-5. Blocking Oscillator

Figure 5-11 is the working schematic of a blocking oscillator. Note that two versions of the circuit are shown, one with a tapped transformer. The blocking oscillator is one of the simplest solid-state oscillators and operates on the relaxation principle. The transistor is initially forward-biased to conduct through the transformer primary winding. This causes a signal to be fed back to the base through C_1. The base is driven very hard in the forward direction, so that C_1 is rapidly charged through the forward-biased emitter–base junction. The output pulse at the T_1 secondary is generated by the rapid turn-on of the collector current, and the steepness of the pulse wavefront is limited

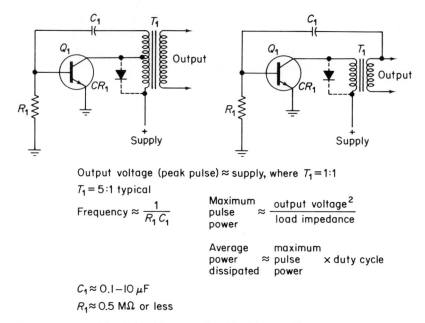

Output voltage (peak pulse) ≈ supply, where $T_1 = 1:1$

$T_1 = 5:1$ typical

Frequency ≈ $\dfrac{1}{R_1 C_1}$

Maximum pulse power ≈ $\dfrac{\text{output voltage}^2}{\text{load impedance}}$

Average power dissipated ≈ maximum pulse power × duty cycle

$C_1 ≈ 0.1 - 10\ \mu F$

$R_1 ≈ 0.5\ M\Omega$ or less

Fig. 5-11. Free-running blocking oscillator.

only by the leakage inductance of T_1. The top of the pulse is flattened by collector-current saturation. When saturation is reached, the feedback signal drops and is no longer able to forward bias the transistor. Capacitor C_1, in discharging, reverse-biases Q_1, and discharges slowly through R_1. The cycle is restarted by forward bias. The time constant $R_1 C_1$ determines the off-time between pulses and thus determines the oscillator frequency.

The output of a blocking oscillator is similar to that of Fig. 5-12. Blocking oscillators are not suited for radio frequencies, or an application where a sine wave (or anything approaching a sine wave) is needed. However, blocking oscillators are excellent sources of the steep wavefront pulses required in switching applications. A pulse rise time of 0.1 μs is not uncommon for blocking oscillator outputs.

Blocking oscillators can be used as driven oscillators where a single output pulse is produced in response to a trigger pulse. This requires that the transistor emitter–base be reverse-biased as shown in Fig. 5-13. With this arrangement, the transistor is initially cut off so that a trigger signal moves the operating point into the active region just long enough to start a pulse cycle.

One of the major advantages of a blocking oscillator is that very little current is drawn between pulses. Therefore, high pulse currents can be drawn from the output for short durations, without exceeding the power capability of the transistor.

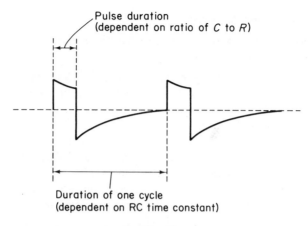

$$\text{Duty cycle} = \frac{\text{pulse duration}}{\text{duration of one cycle}}$$

Fig. 5-12. Typical blocking oscillator output waveform.

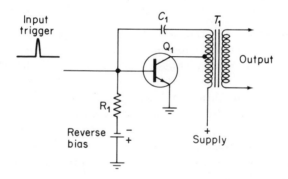

Output voltage (peak pulse) $\approx$ supply, where $T_1 = 1{:}1$

$$\text{Frequency} \approx \frac{1}{R_1 C_1} \qquad \frac{\text{Maximum}}{\text{pulse}} \approx \frac{\text{output voltage}^2}{\text{load impedance}}$$

$C_1 \approx 0.1 - 10 \ \mu\text{F}$

$R_1 \approx 0.5 \ \text{M}\Omega$ or less $\approx$ input impedance of oscillator

$$\frac{\text{Average}}{\text{power}} \approx \frac{\text{maximum}}{\text{pulse}} \times \text{duty cycle}$$
$$\text{dissipated} \quad \text{power}$$

Reverse bias $\approx$ peak trigger input $\times 0.5$

$$C_1 \approx \frac{1}{\text{freq.} \times R}$$

Fig. 5-13. Triggered or driven blocking oscillator.

5-5.1. Design Considerations

In addition to the general design considerations of Sec. 5-1, the following specific points should be considered.

Voltage output. If the primary and secondary windings of T_1 have the same number of turns (a 1:1 turns ratio), the output voltage (pulse peak) will be approximately equal to the supply voltage. Generally, the secondary winding of T_1 has fewer turns than the primary to provide a voltage step-down. When the untapped transformer version of the circuit (Fig. 5-5) is used (where C_1 is connected to the secondary), the full output voltage will be applied to the base during the brief pulse peak. In some cases, this could result in damage to the transistor. As a general rule, a 5:1 turns ratio is used, so that the output voltage is about $\frac{1}{5}$ of the supply voltage.

Transformer selection. Any transformer can be used with a blocking oscillator, provided the primary can withstand the voltage and current, and the secondary is at the desired output impedance. There is a momentary voltage surge across the primary winding equal to approximately twice the supply voltage. Often, special-purpose transformers are designed for use with blocking oscillators. Such transformers are supplied with design data for the blocking oscillator circuit. This information should be followed when available.

Transistor selection. The transistor must be capable of oscillating at the desired frequency and must be capable of withstanding the full supply voltage continuously. The momentary voltage buildup across the transformer primary is also applied to the transistor. Thus, the collector may be at about twice the supply voltage during the pulse peak. This condition can be corrected by the addition of diode CR_1 across the primary, as shown in dotted form on Fig. 5-11. Diode CR_1 must be capable of withstanding twice the supply voltage without breakdown. Since any diode will have some leakage, as well as some forward voltage drop, the addition of CR_1 to the circuit can result in a drop of the output voltage.

Current through the transistor, and the resultant power dissipation, is difficult to calculate in a blocking oscillator. As a rough approximation, divide the square of the anticipated output voltage (across the transformer secondary) by the anticipated load impedance to find the maximum pulse power. The average power dissipated by the transistor will then depend upon the duration of the pulse in relation to the spacing between pulses.

For example, assume that the output load impedance is 50 Ω, the output voltage is 10 V, and the pulse duration is 1 ms with a frequency of 100 Hz. The peak power dissipation is 10 $V^2 \div 50$, or 2 W. With 1-ms pulses spaced by 9 ms, the transistor is on 0.1 of the time. Thus, the average power dissipated by the transistor is 2 W $\times$ 0.1, or 0.2 W.

Section 1-4.3.6 should be studied concerning power dissipation and the use of heat sinks for pulse operation of transistors.

Bias requirements. The free-running version of the blocking oscillator (Fig. 5-11) is initially forward-biased through R_1. The amount of bias is not critical. Once the circuit begins to oscillate, the emitter–base junction is driven into full forward bias and full reverse bias by the charge and discharge of C_1.

The driven version of the blocking oscillator (Fig. 5-13) requires a fixed reverse bias on the emitter–base junction. This bias must be less than the available trigger source. As a first trial make the reverse bias voltage one half of the trigger voltage.

Operating frequency. The operating frequency is determined by the time constant of R_1C_1, and is approximately equal to the reciprocal of the time constant. The exact frequency is difficult to calculate, since both the transistor and transformer characteristics can affect the charge and discharge function. Also, the supply voltage can have some effect on frequency. However, blocking oscillators are fairly stable with respect to power-supply variations.

Various combinations of R and C could be used to produce a given time constant (and thus a given frequency). However, the following rules should be applied.

The value of C_1 should be between 0.1 and 10 μF, while the value of R_1 should be less than 0.5 MΩ, for a free-running blocking oscillator in the audio–frequency range.

When a larger value of C is used (with an R of corresponding low value) to produce a given RC time constant, the pulse duration will be longer in relation to the complete cycle. That is, the on-time (or duty cycle) will be longer. This increases the average power dissipation, as well as the average power output.

In practical design, the circuit should be tested in breadboard form, using the desired transformer, transistor, and supply voltage. The value of C should be fixed (at 10 μF for a first trial), and R_1 should be made variable (say a 0.5-MΩ potentiometer). Adjust R_1 to the approximate value required to produce the operating frequency. Apply power and observe the waveform for amplitude, pulse duration, and frequency.

If the frequency is incorrect, adjust R_1 until the desired frequency is obtained.

If the pulse duration is too long (with amplitude and frequency correct), decrease the value of C and increase the value of R by a corresponding amount.

If there is a sharp spike (or overshoot) on either edge of the pulse, connect diode CR_1 across the transformer primary.

5-5.2. Design Example (Free-Running Oscillator)

Assume that the circuit of Fig. 5-11 is to provide a pulse output of 4 V, across a 50-Ω load, using a transformer with an untapped primary. The operating frequency is 1000 Hz.

Although other turns ratios could be used, a 5:1 ratio is typical. That is, the transformer secondary should have a 50-Ω impedance to match the load, and the primary should have 5 times as many turns as the secondary. The primary impedance would be 1250 Ω. [Primary impedance ÷ secondary impedance = (primary turns ÷ secondary turns)², or $(5 \div 1)^2 = 25$; $25 \times 50 = 1250$.]

With a required 4 V at the secondary and a turns ratio of 5:1, the primary voltage is 20 V. Therefore, the supply voltage (and the collector voltage) would be 20 V.

With 4 V and 50 Ω at the secondary, the secondary current is 80 mA and the maximum power dissipation is 320 mW.

With a 5:1 turns ratio and a secondary current of 80 mA, the primary current (and collector current) is 16 mA.

If both the transformer and transistor are selected on the basis of the maximum voltage, current, and power dissipation calculations, there will be ample margin for safety. In practice, the transistor need not be capable of dissipating the full 320 mW. About one-half that value would still provide considerable safety, unless the pulse duration approached one-half the full cycle between pulses. Usually, blocking oscillators are operated at 0.1 to 0.2 duty cycles. Thus, the true dissipation would probably be more on the order of 32 to 64 mW.

The value of C_1 should be kept between the limits of 0.1 to 10 μF. Since the desired operating frequency is 1 kHz, the lower value (0.1 μF) should be used as the first trial. With a value of 0.1 μF for C_1, the value of R_1 should be $1/[1000 \times (0.1 \times 10^{-6})]$, or 10,000 Ω.

As discussed, R_1 must be made variable in the breadboard stage and then adjusted for the desired frequency. With the frequency established, check for proper waveform, pulse duration, and amplitude. Change the values of C and R if necessary.

5-5.3. Design Example (Driven Oscillator)

Assume that the circuit of Fig. 5-7 is to provide a pulse output of 10 V, across a 50-Ω load, using a transformer with an untapped primary. The circuit is to be driven at a rate of 3 kHz by a trigger pulse of 3 V. The trigger source impedance is 15 kΩ. The available supply voltage is 20 V.

With a 20-V supply and a 10-V output required, the transformer turns ratio must be 2:1. That is, the transformer secondary should have a 50-Ω impedance to match the load, and the primary should have twice as many turns as the secondary. The primary impedance would be approximately 200Ω.

With 10 V and 50 Ω at the secondary, the secondary current is 200 mA, and the maximum power dissipation is 2 W.

With a 2:1 turns ratio, and a secondary current of 200 mA, the primary current (and collector current) is 100 mA.

As in the case of the free-running oscillator, if both the transformer and transistor are selected on the basis of the maximum voltage, current, and power dissipation calculations, there will be an ample margin for safety. In practice, the transistor need not be capable of dissipating the full 2 W. About half that value would still provide considerable safety.

The value of R_1 should be approximately 15 kΩ, to match the trigger source impedance. With a value of 15 kΩ for R_1 and a trigger frequency of 3 kHz, the value of C_1 should be 1/(3000 $\times$ 15,000), or 0.022 μF. Note that this is lower than the nominal 0.1-μF low limit for free-running oscillators operating in the audio range. However, in a driven oscillator, the tradeoff should be such that the RC values match the driving frequency.

With a 3-V trigger pulse, the reverse bias should be 3 $\times$ 0.5, or 1.5 V.

The reverse bias should be made variable in the breadboard stage and then adjusted for the correct operating point. A high reverse bias may prevent the circuit from being triggered, A low reverse bias may cause the circuit to trigger at the wrong point on the trigger signal or to be triggered by undesired signals mixed with the trigger source.

With the reverse bias established at the correct point, check the output waveform. Adjust the values of C and R if necessary.

5-6. Multivibrator

Figure 5-14 is the working schematic of a multivibrator. The circuit is of the high-current type. That is, the emitter resistors are about half the collector resistor values, resulting in very high switching currents through the transistors. This requires transistors with a higher current capability (and higher power dissipation), but produces a circuit with high-frequency stability. That is, the circuit will maintain frequency to about 1 part in 10^4, in spite of large changes in supply voltage.

The same circuit could be used as a low-current type, where the collector resistors are about 10 times the emitter-resistance value. This will require a lower current (and power dissipation) capability for the transistors, but will result in frequency changes with variations of the supply voltage. This is not always an undesirable condition. For example, in telemetry circuits, a low-current-type multivibrator is used as a voltage-to-frequency converter to transmit voltage information. However, for stability, the high-current type is far superior.

In either type, the output is a symmetrical square wave. That is, both the positive and negative portions of the cycle are of the same duration and amplitude. The output can be taken from either half of the circuit. Likewise, either half of the circuit can be triggered, if desired. The circuit can be either free-running, where the frequency is determined by the RC time constant, or driven by a trigger source.

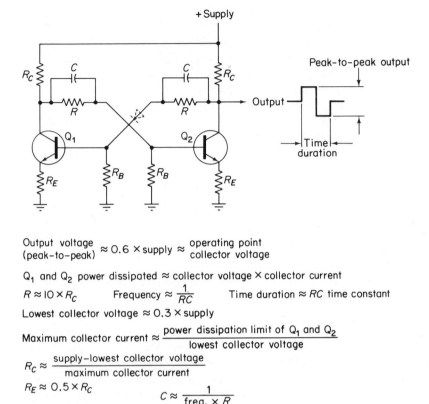

Output voltage (peak-to-peak) $\approx 0.6 \times$ supply $\approx$ operating point collector voltage

Q_1 and Q_2 power dissipated $\approx$ collector voltage $\times$ collector current

$R \approx 10 \times R_C$ Frequency $\approx \dfrac{1}{RC}$ Time duration $\approx RC$ time constant

Lowest collector voltage $\approx 0.3 \times$ supply

Maximum collector current $\approx \dfrac{\text{power dissipation limit of } Q_1 \text{ and } Q_2}{\text{lowest collector voltage}}$

$R_C \approx \dfrac{\text{supply} - \text{lowest collector voltage}}{\text{maximum collector current}}$

$R_E \approx 0.5 \times R_C$

$C \approx \dfrac{1}{\text{freq.} \times R}$

$R_B =$ see text

Fig. 5-14. High-current multivibrator.

5-6.1. Design Considerations

In addition to the general design considerations of Sec. 5-1, the following specific points should be considered. *Note that only the high-current-type circuit is discussed.*

Output voltage. The peak-to-peak output voltage (or square wave) will be about 0.6 times the supply voltage, if the collector resistance is about twice the emitter resistance. For example, if the supply is 10 V, each collector will vary between about 3.5 V and 9.5 V on alternate half-cycles.

Power dissipation. The power dissipated by each transistor can be approximated when the lowest collector voltage is multiplied by the collector current. For example, using the same 10-V supply, a 1-kΩ collector resistor, and a drop to 3 V, the current would be $10 - 3$, or 7; $7 \div 1000$, or 7 mA. With 7 mA and 3 V, the transistor power dissipation is 21 mW. Actually, this is minimum power dissipation (due to base current, etc.). To add a margin of safety, assume that the full supply voltage is applied with full current. In the case above, 7 mA $\times$ 10 V, or 70 mW, is the maximum power dissipation required.

Worst-case design. Since a multivibrator is a switching circuit, the principles of worst-case design should be applied. In simple terms, this means that the base current used should be about 3 times the theoretical base current. Generally, base current is calculated on the basis of required collector current divided by gain. In switching circuits, use 3 times the calculated base current.

When the worst-case design principle is applied, the transistor will always be overdriven. While this is undesirable in most circuits, it is required for switching circuits. Under normal operating conditions, a switching circuit is driven into full saturation, or full cutoff, in the shortest possible time.

Switching time. The total switching time of each transistor must be far less than the duration of one output cycle. Total switching time of a transistor is considered as the rise time, fall time, delay time, and storage time all added together. As a rule, switching time must be less than 0.1 of the pulse duration (for a complete cycle). For example, if a transistor has a total switching time of 1 μs, the pulse duration should be at least 10 μs. Thus, the maximum multivibrator operating frequency would be 100 kHz. Switching time information is found on the data sheets for switching transistors (or transistors that could be used in switching applications).

Symmetrical output. If both halves of the circuit are symmetrical, the output will be symmetrical. In some free-running circuits, a slight unbalance is introduced to ensure starting. A multivibrator is inherently stable, and may not start in the free-running configuration. An alternative method for starting a free-running multivibrator is to connect a diode in series with either of the feedback resistances, as shown in dotted form on Fig. 5-14. The circuit will then be asymmetrical until it starts and reaches full operation.

Operating frequency. The operating frequency is determined by the time constant of the feedback RC, and is approximately equal to the reciprocal of the time constant. The exact frequency is difficult to calculate, since the transistor characteristics can affect the charge and discharge function.

Various combinations of R and C could be used to produce a given time constant (and thus a given frequency). However, the value of R should be approximately equal to 10 times the value of the collector resistor. Capacitors of corresponding values should then be selected to produce the desired frequency.

Bias relationships. The bias circuit can be calculated, and tested, on the basis of normal operating point, even though the circuit will never be at the operating point. A feedback signal will always be present, and the transistors will always be in a state of transition between full saturation and full cutoff.

With the capacitors removed, both halves of the circuit should be forward-biased, and the collector should be approximately 0.6 of the supply voltage. Likewise, the emitter should be about 0.2 (or less) of the supply voltage, and the base should be about 0.5 V positive (for *NPN* transistors) or negative (for *PNP* transistors) in relation to the emitter. Since the values of the collector, emitter, and feedback resistances are set by other circuit considerations, the value of R_B (base resistance) must be chosen to provide the necessary bias.

In practice, the values should be calculated, and the circuit assembled in breadboard form, but with the capacitors omitted. The emitter, collector, and base should be checked for the desired voltage relationships. Then, if necessary, the value of R_B should be adjusted to produce the desired relationships. For a symmetrical output, the element voltages should be the same at both transistors.

5-6.2. Design Example

Assume that the circuit of Fig. 5-14 is to provide a 12-V output at 7 kHz, operating as a free-running multivibrator. The available transistor has a total switching time of 1 μs, and will dissipate 100 mW without a heat sink.

With a required output of 12 V (peak to peak), the supply must be $12 \div 0.6$, or 20 V. With a 20-V supply, each transistor will drop to about 7 V when fully saturated. This is a 13-V drop.

If the transistors are to be operated without heat sinks, the transistor power dissipation cannot be exceeded. To allow some safety margin, assume that the dissipation limit is 90 mW instead of the rated 100 mW. With the collector at 7 V and a maximum dissipation of 90 mW, the maximum allowable current is $0.090 \div 7$, or 13 mA.

With a 13-mA current and a 13-V drop (under maximum conditions), the collector resistors R_C should be $13 \div 0.013$, or 1 kΩ.

With 1 kΩ for the collector resistors, the emitter resistors R_E should be 500 Ω (510 Ω as the nearest standard), and the feedback resistors R_{feedback} should be 10 kΩ.

The value of the base resistors R_B should be calculated on the basis of operating point. Under the operating-point conditions (capacitors removed from the circuit), the collectors should be at about 0.6 of the supply voltage, or at about 12 V, while the emitter should be slightly less than 0.2 of the supply voltage, or about 4 V. Under these conditions, there must be about 8 mA of collector current flowing, producing an 8-V drop across R_C and a 4-V drop across R_E.

With the emitters at about 4 V, the base should be at about 4.5 V (4.5 V for the *NPN* shown). With 4.5 V at each base and 12 V at each collector, there must be a 7.5-V drop across each feedback resistor. Since the feedback resistors are 10 kΩ, the current must be 0.75 mA to produce a 7.5-V drop.

The 0.75-mA current through the feedback resistors is a combination of base current and the current through the base resistors R_B. The base current can be estimated on the basis of gain and required collector current. Assume that the rated gain is 120. Using half of this value for safety and a required 8-mA collector current, the base current is $0.008 \div 60$, or 0.13 mA.

By subtracting the 0.13-mA base current from the 0.75-mA feedback resistor current, the base resistor R_B current is $0.75 - 0.13$, or 0.62 mA.

With base voltage of 4.5 V and an R_B current of 0.62 mA, the R_B resistance values are $4.5 \div 0.00062$, or 7300 Ω (the nearest 10 per cent standard value would be 6800 Ω).

In practice, the circuit should be assembled in breadboard form, omitting the capacitors and using variable resistors for each R_B. Both R_B values should be adjusted as necessary for equal voltages at corresponding transistor elements. Then the capacitors can be connected and the output waveform checked for correct frequency, waveform, and amplitude. With the resistance ratios selected, the output amplitude should be approximately equal to the operating point collector voltage (about 12 V in this case).

With a value of 10 kΩ for the feedback resistors, and a frequency of 7 kHz, the value of the feedback capacitors is $1/(10{,}000 \times 7000)$, or 0.014 μF.

5-7. Modification of Square Waves

In the design of pulse circuits, it is often necessary to modify the square-wave output of multivibrators (or similar circuits) to achieve a given waveform. The most common waveforming circuits used with square waves are the high-pass and low-pass RC filters described in Sec. 2-10.

5-7.1. Design Considerations

When a square wave is applied to a low-pass RC filter, the output will be an integrated waveform, as shown in Fig. 5-15. In this case, the RC network produces an output voltage proportional to the algebraic sum of the instantaneous values of the input voltage. Not only will the amplitude be affected, but the waveform will change.

Any RC filter will have a cutoff frequency, at which the output will drop by 3 dB from the input value (a 10-V input will drop to about 7 V). The cutoff frequency can be approximated by the equation $1/6.28RC$, as shown in Fig. 5-15.

The amount of attenuation, and the amount of waveform modification, is determined by the relation of the cutoff frequency to the square-wave frequency. For example, if the square-wave frequency is much lower than the cutoff frequency, both attenuation and modification will be at a minimum. As the square-wave frequency increases, the output (an integral of a square wave) is a series of increasing and decreasing ramps. The ramps are linear,

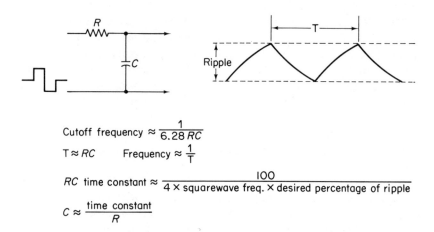

Cutoff frequency $\approx \dfrac{1}{6.28\,RC}$

$T \approx RC$ Frequency $\approx \dfrac{1}{T}$

RC time constant $\approx \dfrac{100}{4 \times \text{squarewave freq.} \times \text{desired percentage of ripple}}$

$C \approx \dfrac{\text{time constant}}{R}$

Fig. 5-15. Low-pass RC filter with square-wave input and integrated ramp output.

but only for a time that is shorter than about half the RC time constant. For very accurate integration of the square wave, the integration time should be about 0.1 of the RC time constant.

The relationship between frequency and waveform is shown in Fig. 5-16. In Fig. 5-16a, the frequency of the square waves is about one-third of the

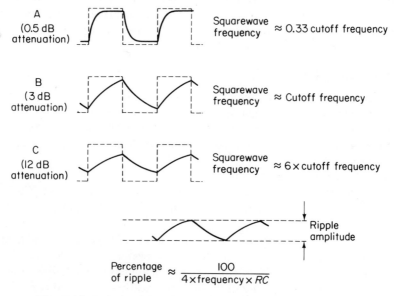

Fig. 5-16. Relationship of frequency and waveform in output of low-pass RC filter (with square waves at the input).

cutoff frequency. Note that there is little attenuation but slight modification of the waveform. When the square-wave frequency is increased to the cutoff frequency, as shown in Fig. 5-16b, the peak-to-peak waveform is attenuated by 3 dB, and the waveform has the shape of an almost completed exponential rise or fall. When the square-wave frequency is increased to about 6 times the cutoff frequency (Fig. 5-16c), the output is a series of (almost) linear ramps, with a peak-to-peak amplitude of about 12 dB below that of the input wave.

A more accurate prediction of the peak-to-peak amplitude relationships between input and output signals can be found by using the equation of Fig. 5-16. The peak-to-peak amplitude of the linear ramps, often known as the *ripple*, or *ripple amplitude*, is always less than 100 per cent of the square-wave input amplitude. However, the equation applies only when ripple amplitudes are less than 50 per cent of the input amplitude.

When a square wave is applied to a high-pass RC filter, the output will be a differentiated waveform, as shown in Fig. 5-17. In this case, the *RC* network produces an output voltage proportional to the time rate of change of the input voltage.

As in the case of the low-pass (integrator) circuit, the amount and nature of the waveform modification by a high-pass (differentiator) circuit is determined by the relationship of cutoff frequency to square-wave frequency. The relationship between frequency and waveform is shown in Fig. 5-18. In Fig. 5-18a, the square-wave frequency is about 10 times the cutoff frequency. Note that there is slight modification of the waveform. The tops of the waveform show some slope (or droop, as it is sometimes called). When the square-wave frequency is decreased to about 3 times the cutoff frequency,

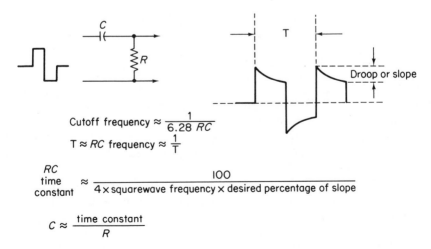

$$\text{Cutoff frequency} \approx \frac{1}{6.28 \, RC}$$

$$T \approx RC \quad \text{frequency} \approx \frac{1}{T}$$

$$\frac{RC}{\text{time constant}} \approx \frac{100}{4 \times \text{squarewave frequency} \times \text{desired percentage of slope}}$$

$$C \approx \frac{\text{time constant}}{R}$$

Fig. 5-17. High-pass *RC* filter with square-wave input and differential output.

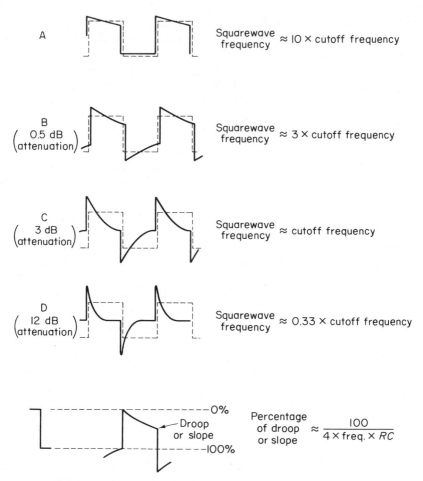

A

Squarewave frequency $\approx 10 \times$ cutoff frequency

B
(0.5 dB attenuation)

Squarewave frequency $\approx 3 \times$ cutoff frequency

C
(3 dB attenuation)

Squarewave frequency $\approx$ cutoff frequency

D
(12 dB attenuation)

Squarewave frequency $\approx 0.33 \times$ cutoff frequency

0%
Droop or slope
100%

Percentage of droop or slope $\approx \dfrac{100}{4 \times \text{freq.} \times RC}$

Fig. 5-18. Relationship of frequency and waveform in output of high-pass RC filter (with square waves at the input).

as shown in Fig. 5-18b, the amount of slope increases, and there is some attenuation of the average output. When the square-wave frequency is decreased to the cutoff frequency, the slope becomes quite severe. When the square-wave frequency drops below the cutoff frequency, the output becomes a series of "spikes."

When the percentage of slope is below about 50 per cent, the equation of Fig. 5-18 can be used to predict the slope. Note that the equation for slope in a differentiated wave is the same as for ripple in an integrated wave.

5-7.2. Design Example

Assume that the circuit of Fig. 5-15 is to convert a square-wave input to a series of linear ramps. The input is 10 V peak to peak, at 1 kHz. The ramps are to have a 3-V ripple amplitude. Calculate the cutoff frequency of the RC circuit and the approximate time during which the ramps will be linear. Assume that R must be 1 kΩ.

With a 10-V input and a 3-V output, the percentage of ripple amplitude is 30 per cent. Therefore, the RC time constant must be 100 $\div$ 4 × 1000 × 30, or 833 μs.

With R fixed at 1 kΩ, and a required RC time constant of 833 μs, the value of C must be 0.000833 $\div$ 1000, or 0.833 μF.

With an RC time constant of 833 μs, the cutoff frequency of the low-pass RC filter is 1/(6.28 × 0.000833), or 200 Hz. Thus, the square-wave frequency is about 5 times the cutoff frequency.

The integrating time where the ramp is linear (a time where the output is a true integral of the input) is about half of the total RC time constant, or 833 × 0.5 = 416 μs. However, for greatest accuracy, the integration time should be about 0.1 of the RC time constant, or 83 μs.

5-8. Modification of Sine Waves

There are many circuits used to modify and control sine waves. The most important of these circuits are clippers and clamps. The following sections describe the most commonly used versions of these circuits.

5-8.1. Design Considerations for Clippers

Figure 5-19 is the working schematic of a *basic diode clipper*, also known as a *peak limiter*. As shown, the output waveform is clipped (held at a peak limit) to a value determined by the bias voltage and the forward-voltage drop of the diode. For example, if the bias is 3 V, and the forward-voltage drop is 0.5 V, the output will be clipped to 3.5 V.

Either the positive or negative peaks can be clipped, depending on the polarity of the bias and diode. Both peaks can be clipped if two diodes and bias sources, with opposite polarities, are used.

The four basic diode characteristics must be considered in clipper circuits. However, maximum reverse voltage (peak inverse) is the prime consideration. The maximum reverse voltage rating of the diode must exceed the input voltage. The forward-voltage drop should always be minimum, and is typically 0.5 V for silicon diodes and 0.2 V for germanium diodes. Reverse current (leakage) should also be minimum. Any excessive leakage in the diode will clip or distort the waveform. Maximum forward current of the diode is not critical but does set the value of series resistance R_S.

The minimum value of R_S is set by the value of input voltage and maximum forward current rating of the diode. For example, if the input were 10 V and the diode were rated at 0.1 A, the minimum value of R_S would be 10 $\div$ 0.1, or 100 Ω. As a safety factor, the value of R_S should be at least twice the

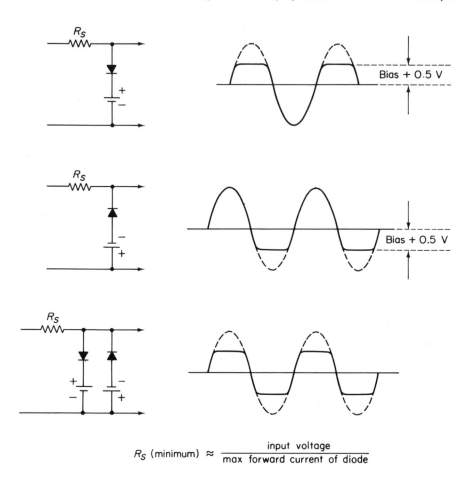

$$R_S \text{ (minimum)} \approx \frac{\text{input voltage}}{\text{max forward current of diode}}$$

Fig. 5-19. Biased diode clipper.

minimum value. This may result in a tradeoff. The output current is passed through R_S, resulting in some voltage drop across R_S. This means that the unclipped peak will be attenuated. A low value of R_S will produce a minimum of attenuation, assuming a given output current. However, the value of R_S must never be less than that required to produce a safe current level in the diode.

Figure 5-20 is the working schematic of an *unbiased clipper using a Zener diode*. The output waveform is clipped to a value determined by the Zener voltage and the forward voltage drop of the diode. The advantage of this circuit is that no bias voltage is required. The disadvantage is that Zener diodes do not have a sharp breakdown below about 4 V. Therefore, the circuit should be used at voltages greater than 4 V.

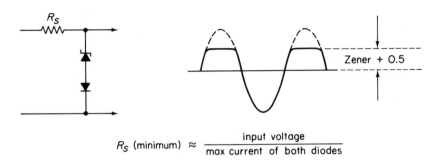

$$R_S \text{ (minimum)} \approx \frac{\text{input voltage}}{\text{max current of both diodes}}$$

Fig. 5-20. Zener-diode clipper (unbiased).

The value of R_S is calculated in the same way as for the basic diode clipper, except that the maximum current rating of both the Zener and clipper diodes must be considered. However, the current rating of the Zener is usually greater than that of the clipper, so only the diode current rating need be considered.

Figure 5-21 is the working schematic of *clipper circuit using only a Zener diode*. As shown, the output waveform is clipped in both polarities. One polarity is clipped at the Zener voltage, with the other polarity clipped at about 0.5 V (which is the typical drop across a Zener).

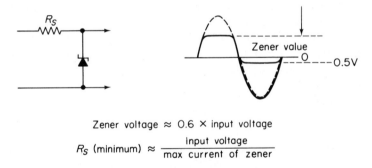

Zener voltage ≈ 0.6 × input voltage

$$R_S \text{ (minimum)} \approx \frac{\text{input voltage}}{\text{max current of zener}}$$

Fig. 5-21. Zener clipper (without clipping diode).

The value of R_S is calculated in the same way as the previous clipper circuits. However, the required Zener characteristics are somewhat different than those of the basic diode clipper. The Zener voltage rating should be about 0.6 times the input voltage. If a lower Zener voltage is used, there may be too much current flowing through the Zener, or the value of R_S will be excessive. Zener diodes are also rated as to power dissipation. This power is approximately equal to the product of the Zener voltage and the maximum current (as set by R_S).

Figure 5-22 is the working schematic of a *double-peak clipper using two Zener diodes*. This circuit is similar to that of Fig. 5-21, except that the two Zener diodes are connected back to back. As shown, the output waveform is clipped by an equal amount in both polarities. The peaks are clipped to the Zener voltage plus about 0.5 V. The value of R_S is calculated in the same way as the previous clipper circuits.

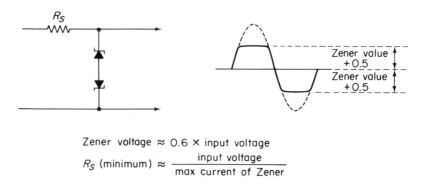

Zener voltage ≈ 0.6 × input voltage

$$R_S \text{ (minimum)} \approx \frac{\text{input voltage}}{\text{max current of Zener}}$$

Fig. 5-22. Double-peak clipper (with two Zener diodes).

Figure 5-23 is the working schematic of a *low-level noise suppression circuit using two diodes*. When the input voltage is below the usual diode voltage drop, there will be considerable attenuation. If the input voltage exceeds the diode voltage drop, there will be little attenuation. As a rule of thumb, if silicon diodes with a 0.5-V drop are used, any input signals below about 0.1 V will be attenuated at least 30 dB, and any input signals above about 1 V will be attenuated less than 3 dB.

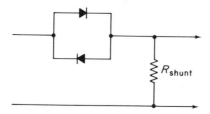

$R_{shunt} \approx 10\ k\Omega$ when minimum input is 0.1 V

Fig. 5-23. Low-level noise suppressor (series-diode type).

As in other clipper circuits, the diodes must be capable of withstanding the maximum input voltage and passing the maximum circuit current. The value of shunt resistor R depends upon the minimum input voltage level. As a rule of thumb, the value of R should be about 10 kΩ, when the minimum input signals are about 0.1 V.

Figure 5-24 is the working schematic of a *low-level clipper using two diodes.* The output voltage will be limited to the usual diode voltage drop (about 0.5 V for silicon diodes). Any input signals with a peak-to-peak amplitude of less than 0.5 V should pass without distortion or attenuation. As before, the diodes must be capable of withstanding the maximum input voltage. The value of R_S is set by the value of the input voltage and the maximum forward-current rating of the diode.

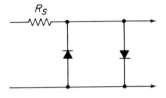

Fig. 5-24. Low-level suppressor (shunt-diode type).

$$R_S \text{(minimum)} \approx \frac{\text{input voltage}}{\text{max current of diodes}}$$

Figure 5-25 is the working schematic of a *high-level noise suppressor,* often known as a *base clipper.* The circuit is similar to that of the basic clipper (Fig. 5-19), but with the diode polarity reversed. The output waveform follows peak voltages in excess of the bias voltage. Whenever the peak voltage drops below the bias level or swings in the opposite polarity, the diode conducts, and the output voltage returns to the bias level. Therefore, any signals below the bias level are suppressed.

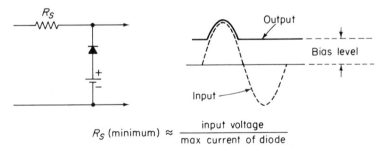

$$R_S \text{(minimum)} \approx \frac{\text{input voltage}}{\text{max current of diode}}$$

Fig. 5-25. Base clipper.

As in the case of other clipper circuits, the diode must be capable of withstanding the maximum input voltage. The minimum value of R_S is set by the maximum current rating of the diode.

5-8.2. Design Examples for Clippers

Assume that the circuit of Fig. 5-19 or 5-20 is to clip the positive peaks of a signal to a level of 3.5 V. The input signal is 10 V peak to peak. The available diode has a maximum reverse voltage rating of more than 10 V and a maximum forward current rating of 20 mA.

Since the desired peak limit is less than 4 V, the fixed bias circuit (Fig. 5-19) should be used instead of the Zener circuit (Fig. 5-20).

The minimum value of R_s should be 10 ÷ 0.020, or 500 Ω. The nearest standard is 510 Ω. A larger resistance could be used, but this would produce an unnecessary voltage drop to the output.

Assume that the circuit of Fig. 5-21 or 5-22 is to clip both the positive and negative peaks of a signal to a level of 5 V. The input signal is 10 V. It is desired to keep the power dissipation of the Zener diodes at 50 mW or less.

Since both peaks are to be clipped, the circuit of Fig. 5-22 should be used. To provide a 5-V peak, the Zener voltage should be 4.5 V, which is added to the nominal 0.5-V drop across the opposite diode. However, the nearest standard Zener voltage is 4.7 V. Both Zener diodes should be 4.7 V to provide an approximate 5-V peak in both polarities. With a 4.7-V Zener drop, and a maximum 50-mW power dissipation for the Zener diodes, the value of R_s should be 4.7^2 ÷ 0.050, or 440 Ω. The next largest standard value would be 510 Ω.

5-8.3. Design Considerations for Clamps

Figure 5-26 is the working schematic for a clamp. As shown, the circuit is essentially a high-pass RC filter with a diode across the shunt resistance. The time constant must be long to prevent distortion. As a rule, the time constant

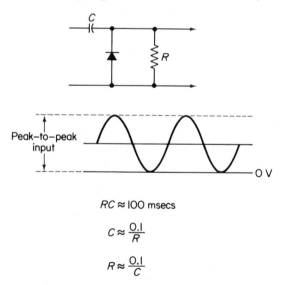

$$RC \approx 100 \text{ msecs}$$

$$C \approx \frac{0.1}{R}$$

$$R \approx \frac{0.1}{C}$$

Fig. 5-26. Basic clamp.

should be on the order of 100 ms, for any frequency from 10 Hz up to 1MHz. Often, it is necessary to select *RC* component values on the basis of external circuit considerations. For example, the resistance may have to be a certain value to match impedance. In that case, the capacitance should be of a corresponding value to produce the 100-ms time constant. Preferably, the resistance should be as large as possible, thus eliminating the need for frequent recharging.

With the circuit as shown, the negative peaks of an input signal will be clamped to ground (at 0 V). If the diode polarity were reversed, the positive peaks would be at zero, and negative peaks would be below zero. If a fixed bias source (battery) is connected in series with the diode, the negative peaks would be clamped to the bias value. For example, if the bias were 1 V, and the input were 3 V peak to peak, the output negative peak would be at 1 V with the positive peak at 4 V.

5-8.4. Design Example for Clamps

Assume that the circuit of Fig. 5-26 is used to clamp a 10-V input signal to ground. The resistance value should be 1 MΩ to match an external impedance.

With an *R* of 1 MΩ, the value of *C* should be 0.1 μF (for a 100-ms time constant). The diode must be capable of withstanding the 10-V input. Current requirements for a clamping diode are usually negligible.

5-9. Waveforming and Waveshaping Test Procedures

The following sections describe test procedures for waveforming and waveshaping circuits.

The only practical test for a waveshaping circuit is display of the circuit output on an oscilloscope. The same applies to a waveforming circuit such as an oscillator. In either case, the exact waveform, amplitude, and duration can be measured directly. The frequency can be found if the duration can be measured directly. The frequency can be found if the duration (or period) of one complete cycle is measured, since frequency is the reciprocal of period.

When the oscillator frequency is very high (in the radio-frequency range), it is not practical to measure the frequency with an oscilloscope. The oscilloscope amplifiers may not pass the frequency. Even if the frequency can be passed without severe attenuation or distortion, it is usually very difficult to measure the duration of a single pulse at high frequencies. Therefore, the only practical method is to measure radio frequencies by the zero-beat method, or to use a digital frequency counter. The amplitude of radio frequencies can be measured by means of an RF probe, or RF pick-up coil, as described in Sec. 4-11.

5-9.1. Oscilloscope Measurement of Waveforms

Figure 5-27 shows the basic connections required for measurement of waveforms on an oscilloscope. A low-capacitance probe should be used for all measurements, except radio frequencies. It is impossible to describe the procedures for checking each type of oscillator, multivibrator, audio genera-

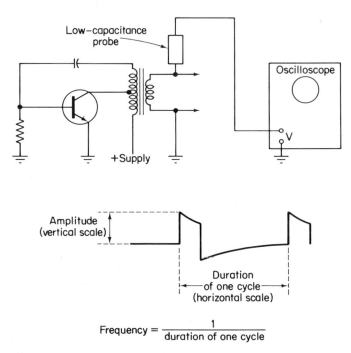

Fig. 5-27. Measuring waveform, amplitude, and frequency with an oscilloscope display.

tor, etc. However, the following procedures are applicable to most self-generating circuits, and can be adapted to meet the test needs of specific circuits.

1. Connect the equipment as shown in Fig. 5-27. The oscilloscope should have a voltage-calibrated vertical axis, and a time-calibrated horizontal axis.
2. Switch on the internal recurrent sweep. Set the sweep selector and "sync" selector to "Internal."
3. Apply power to the circuit under design. Touch the low-capacitance probe to the circuit test point.
4. Adjust the sweep frequency and "sync" controls to produce two or three stationary cycles of each waveform on the oscilloscope screen.
5. Measure the waveform amplitude of interest on the voltage-calibrated vertical axis. For example, if the oscilloscope vertical axis is 1 V per division,

and the waveform occupies 3 vertical divisions, the peak-to-peak amplitude of the waveform is 3 V.

6. Measure the waveform duration of interest on the time-calibrated horizontal axis. If it is desired to find the pulse repetition rate, or output frequency, measure the duration of *one complete cycle* (not one pulse), then convert this time duration into frequency. For example, if the oscilloscope horizontal axis is 0.1 ms per division, and one complete cycle of the waveform occupies 8 horizontal divisions, the time duration is 0.8 ms, and the frequency is $1 \div 0.0008$, or 1250 Hz.

7. If the horizontal axis is frequency calibrated, and it is desired to find the pulse repetition rate or output frequency, set the oscilloscope sweep and "sync" controls to produce one complete cycle that occupies the entire sweep length. Then read the frequency from the oscilloscope dials.

5-9.2. Zero-Beat Frequency Measurement

The basic components required for any form of zero-beat frequency measurement are a signal or voltage source of known accuracy that can be varied over the frequency range of the unknown signal or voltage, and a detector that will show when the two signals (standard signal and design-circuit signal) are at the same frequency, or some exact multiple of each other.

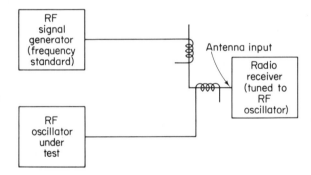

Fig. 5-28. Zero-beat frequency measurement for *RF* circuits.

A radio receiver is a common detector for zero-beat frequency measurement. Figure 5-28 shows how a radio-frequency oscillator can be checked against a frequency standard such as a precision signal generator. Both signals are applied to the radio-receiver input. When the standard frequency generator is adjusted close to the frequency of the oscillator, a tone, or "beat note," is heard on the receiver. When the generator is adjusted to exactly the same frequency as the oscillator, the tone can no longer be heard. The oscillator frequency can then be read from the generator frequency dial, or from a digital frequency counter connected at the generator output.

Once the frequency is established, the amplitude of the radio-frequency signal can be measured using an RF probe (Sec. 4-11).

5-9.3. *Measuring Power Output of Oscillators*

The power output of an oscillator is found by noting the output voltage E_O across a load resistance R_L, as shown in Fig. 5-29. Power output is $E_O^2 \div R_L$.

The load resistance should be equivalent to the load impedance of the circuit with which the oscillator will operate.

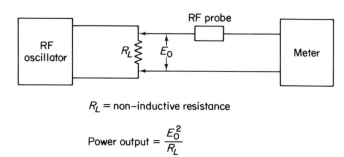

R_L = non-inductive resistance

$$\text{Power output} = \frac{E_O^2}{R_L}$$

Fig. 5-29. Measuring power output of *RF* oscillators.

6

POWER-SUPPLY CIRCUITS

6-1. Basic Solid-State Power Supply

Figure 6-1 is the working schematic of a basic solid-state power supply. Note the simplicity of the circuit. Unlike vacuum-tube equipment, the need for elaborate filters in solid-state power supplies is not critical. In fact, the added weight and bulk produced by large filter chokes is usually not worth the reduc-

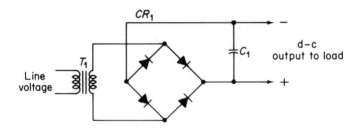

d-c output ≈ 1.3 × secondary voltage
Transformer power (or VA) rating > 1.3 × d-c output power
Load resistance = output voltage/output current
Power supply resistance = (no-load V - full load V)/current
C_1 working voltage > 1.3 × output voltage
C_1 capacitance (in μF) ≈ 200,000/load resistance × max ripple
Secondary voltage ≈ required d-c output/1.3
d-c power output = output voltage × load current
CR_1 ratings = see text

Fig. 6-1. Basic solid-state power supply.

tion in output ripple voltage. Generally, solid-state equipment can tolerate larger ripple voltages than equivalent vacuum-tube circuits, as a result of the large operating voltage differential.

The circuit of Fig. 6-1 is a full-wave bridge, with a capacitor filter. This is the most efficient circuit for solid-state applications.

6-1.1. Design Considerations

In addition to the general design considerations of Chapter 1, the following specific points should be considered.

Power transformer. The power transformer T_1 in a solid-state circuit is usually a step-down transformer. Typically, the primary voltage is 110 to 120 V, while the secondary is between 12 and 50 V. If the transformer is used with a full-wave bridge rectifier and capacitor filter, the dc output of the supply will be about 1.3 times the transformer secondary voltage. Thus, a T_1 secondary of 10 V will produce about 13-V dc output.

In addition to voltage, transformers are rated as to power capability. Usually, this is a volt–ampere (VA) rating rather than a true ac power rating. As a rule, the transformer power rating should be at least 1.3 times the dc output power of the complete circuit. Thus, if 10 W of dc power is required, the transformer should be capable of handling at least 13 W.

Load resistance and power-supply resistance. These two terms are often confused. Neither value can be measured directly but must be calculated on the basis of voltage and current. Load resistance (or effective load impedance) is found when the output voltage across the load is divided by the load current. For example, with a dc output of 10 V and a load current of 2 A, the load impedance is 5 Ω.

Power-supply resistance (or internal impedance) is found when the change in output voltage is divided by a change in load current.

There are two methods for calculating power-supply resistance. In the first method, the output voltage is measured with no load, then with full load. The difference in voltage is divided by the full-load current. For example, if the no-load output voltage is 30 V, and the full-load output is 25 V, with a 2-A full-load current, the power-supply resistance is $(30 - 25)/2$, or 2.5 Ω.

In the second method, the output voltage is measured with a small load, then with a full load. The difference in voltage is divided by the difference in load current. For example, if the small-load voltage is 28 V, with 0.5 A, and the full-load output is 25 V, with 2 A, the power-supply resistance is $(28 - 25)/(2 - 0.5) = 3/1.5$, or 2 Ω.

Diode characteristics. There are four basic characteristics that must be considered for any type of diode. These are: maximum reverse voltage, forward voltage, reverse current, and forward current.

Maximum reverse voltage (also known as *peak inverse voltage*) is the amount of reverse voltage that a diode can withstand without breakdown. Data sheets often list two values for reverse voltage: average, or nominal, reverse voltage, and peak, or maximum, reverse voltage. As a rule for full-wave bridge rectifiers, the average reverse voltage rating should be about twice the dc output voltage. In any event, the reverse voltage rating must be greater than either the secondary output voltage or the dc output voltage.

Forward voltage (or forward-voltage drop) is the amount of drop across the diode in the forward-bias condition. Ideally, the forward-voltage drop should be zero. Typically, the forward voltage drop is on the order of 0.5 to 1 V. In practice, the diodes will drop the secondary output voltage, but this will be offset by the voltage buildup across the capacitor.

Reverse current (or leakage current) is the amount of current flow when the diode is reverse-biased. Ideally, reverse current should be zero. Typically, reverse current is on the order of a few microamperes (at most, a few milliamperes).

Forward current is the maximum current capacity of the diode in the forward-bias condition. As a rule for full-wave bridge rectifiers, the forward current rating of each diode must be greater than the dc output current, and should be about twice the dc output current.

Capacitor characteristics. Capacitors used in full-wave bridge rectifiers are of the electrolytic type, since a high capacitance value is required. The voltage rating of the capacitor should be at least 1.3 times the dc output voltage. The capacitance rating should be selected on the basis of allowable ripple and the load resistance, as shown by the equation in Fig. 6-1. For example, assuming a load resistance of 50 Ω and a maximum permissible ripple of 5 per cent, the capacitance would be 200,000 ÷ 50 × 5, or 800 μF.

6-1.2. Design Example

Assume that the circuit of Fig. 6-1 is to provide a dc output of 28 V at 4 A, with a ripple of not more than 3 per cent. A line supply of 110 to 120 V is available.

With a required 28-V output, the transformer T_1 secondary should be 28 ÷ 1.3, or 21.5 V.

With a dc power output of 112 W (28 V × 4 A), the volt–ampere (or power) rating of transformer T should be 112 W × 1.3, or 145.6 W, which is rounded off to 150 W.

The load resistance is 28 ÷ 4, or 7 Ω.

The diode characteristics should be as follows:

Maximum reverse voltage is 28 × 2, or 56 V. In practical design, any reverse voltage rating above about 30 V should be satisfactory.

Forward voltage drop should not exceed 1 V.

Reverse current should not exceed 100 μA.

Forward current is 4×2, or 8 A.

Note that these characteristics are for individual diodes. It is possible to obtain four diodes (connected in a bridge configuration) as a package (or module). In that case, the data-sheet rating for the package must be considered as the ratings for individual diodes.

The capacitor C_1 should have a voltage rating of 28×1.3, or 35.4 V. Any working voltage about 35 V should be satisfactory.

With a load resistance of 7 Ω, and a maximum ripple of 3 per cent, the capacitance value of C_1 should be $200,000 \div 7 \times 3$, or 9523 μF.

6-2. Zener-Diode Regulation

The basic solid-state power supply described in Sec. 6-1 is not provided with any regulation. That is, the dc output will vary with changes in load resistance and with changes in input voltage. While this condition can be tolerated in some equipment, there are a number of solid-state circuits where the supply voltage must remain fixed (within a tolerance) despite changes in load and/or input voltage.

Figure 6-2 is the working schematic of a basic Zener-diode voltage regulator. A Zener diode provides the simplest method for voltage regulation. However, the Zener diode has certain limitations which must be considered in design.

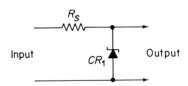

Output voltage = Zener voltage

Zener power dissipation = Zener voltage × Zener current

Minimum input voltage > 1.4 × Zener voltage

Zener voltage ≈ 0.7 × lowest input voltage

Safe Zener power ≈ 3 × load power

R_S (in ohms) $\approx \dfrac{\text{(max input voltage} - \text{Zener voltage)}^2}{\text{safe power dissipation}}$

Input voltage variation < 30 % of max input voltage

Load current = Zener voltage/load resistance

Load power = Zener voltage × load current

Safe Zener current ≈ 3 × load current

Fig. 6-2. Basic Zener-diode regulator circuit.

6-2.1. Design Considerations

In addition to the general design considerations of Chapter 1, the following specific design points should be considered.

Zener characteristics. Data sheets do not always show the same characteristics for Zener diodes. In practical design, the critical characteristics are Zener (or avalanche) voltage, maximum current, and power capability.

Power-dissipation capability of a Zener diode can be considered as the Zener voltage times the maximum current rating. Zener diodes are rarely rated for greater than a few watts' power dissipation. This is a room-temperature (typically 25°C) power dissipation and must be derated for higher temperature.

The Zener voltage is typically in the range between about 4 V and 27 V, with maximum currents up to about 200 mA (typically less than 100 mA). (For special applications, there are high-voltage Zeners with low-current ratings.)

In theory, the Zener diode will maintain the output at the avalanche voltage (within a tolerance from about 0.1 to 0.9 V) despite any changes in input and load.

In practice, if the input voltage falls below about 1.4 times the Zener voltage, the diode will not go into the avalanche condition, and there will be no regulation. On the other hand, if the input voltage rises to a point where the maximum current is exceeded, the Zener will overheat (and possibly be destroyed).

Current through the Zener diode is limited by series resistor R_S.

Trial values. The equations for trial values in Zener-diode regulators are shown in Fig. 6-2. The following summarizes the use of these equations. The Zener voltage should be about 0.7 times the lowest input voltage. Thus, if the input voltage varies between 10 and 15 V, the Zener voltage should be 10×0.7, or 7 V. If the design problem is stated in reverse, the minimum input voltage must be 1.4 times the desired Zener voltage. Thus, if the required Zener (or regulated output) voltage is 10 V, the minimum input voltage from the basic power supply must be 10×1.4, or 14 V.

A safe power-dissipation rating for the Zener diode is 3 times the load power. Thus, with a 250-mW load (say 10 V across 400 Ω) a Zener capable of dissipating 750 mW (or more) should be satisfactory.

The value of R_S is found using the maximum input voltage, the desired Zener voltage, and the safe power dissipation. Using the same 10- to 15-V input, a desired Zener voltage of 7 V, and a safe power dissipation of 600 mW, the first trial value of R_S would be $(15 - 7)^2 \div 0.6$, or 106 Ω (use the next highest standard of 110 Ω).

Note that if the input voltage varies by more than about 30 per cent (of the highest input voltage), it will be difficult (if not impractical) to provide regulation with a Zener diode.

Although the circuit of Fig. 6-2 could be assembled in breadboard form using the trial values just calculated, it is convenient to check the design on paper, using maximum and minimum input voltage values. If the trial value for R_S has been properly calculated, the maximum input voltage will produce a current that can be safely dissipated by the Zener diode, while the minimum input voltage will provide more than enough current to meet the load requirements. The on-paper calculations for maximum and minimum input voltage are included in the following design example.

6-2.2. Design Example

Assume that the circuit of Fig. 6-2 is to provide a regulated 7.5-V output to a 250-Ω load. Assume that the maximum input voltage is 15 V. Assume that 7.5-V Zener diodes are available. Find the minimum input voltage that would be practical, the current (or power) rating for the Zener, and the optimum value for series resistor R_S.

With a 7.5-V Zener, the minimum input voltage would be 7.5 $\times$ 1.4, or 10.5 V.

With a 7.5-V output across a 250-Ω load, the load current would be 7.5 $\div$ 250, or 30 mA. The load power would be 7.5 $\times$ 0.030, or 225 mW.

A safe power-dissipation rating for the Zener is 225 $\times$ 3, or 675 mW. The nearest standard value would probably be a 1-W Zener.

With a safe power-dissipation rating of 675 mW, a Zener voltage of 7.5, and a maximum input voltage of 15, the first trial value for R_S should be $(15 - 7.5)^2 \div 0.675$, or 83.3 Ω. The nearest 10 per cent standard value would be 82 Ω.

With a load current of 30 mA, the current rating of the Zener should be 30 $\times$ 3, or 90 mA. A 100-mA rating would provide an added measure of protection.

The design values (82 Ω for R_S and a Zener of 7.5 V with a 100-mA rating) should be checked against the maximum and minimum input voltages, as follows. The maximum input current (combined Zener current and load current) will be equal to the drop across R_S (15 $-$ 7.5 $=$ 7.5 V) divided by the resistance (82 Ω), or 91 mA. Therefore, with the usual load current of 30 mA, the Zener will normally pass only 61 mA (91 $-$ 30 $=$ 61 mA). If the load current drops to zero, the full 91 mA will flow through the Zener, so the 100-mA rating will be adequate.

Maximum power to the Zener will be 7.5 $\times$ 0.090, or 675 mW, as calculated. A 1-W rating will be adequate.

Maximum power dissipated by R_S will also be 675 mW (7.5-V drop, 0.090 A). Therefore, R_S should also have a 1-W dissipation rating.

As a final calculation before going to the breadboard stage, check that the minimum input voltage will provide more than enough current to meet the load requirements. Minimum input voltage was calculated as 10.5 V. Using this value and a 7.5-V Zener, the drop across R_S is 10.5 — 7.5, or 3 V.

With a 3-V drop across 82 Ω, the minimum current would be 3 ÷ 82, or 36 mA. Under normal conditions, the load would draw 30 mA, with the Zener drawing the remaining 6 mA.

If the current had been less than 30 mA, the value of R_S would have to be decreased.

6-3. Extending the Zener Voltage Range

On rare occasions, it may be necessary to regulate a voltage not normally obtainable from Zener diodes. For example, it may be necessary to regulate a higher voltage with high current, beyond the capability of existing commercial Zeners. Generally, high-voltage Zeners have correspondingly low current ratings. In that case, it is possible to series-connect several Zener diodes to get a desired regulation level, as shown in Fig. 6-3.

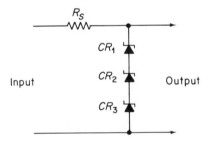

Output voltage = total series Zener voltage
Individual Zener power dissipation = individual Zener voltage × Zener current
Minimum input voltage > 1.4 × total series Zener voltage
Total series Zener voltage ≈ 0.7 × minimum input voltage
Safe power for each Zener ≈ 3 × individual Zener power dissipation
Total load current = total series Zener voltage/load resistance
Total load power = total series Zener voltage × total load current
Safe Zener current ≈ 3 × total load current

$$R_S \text{ (in ohms)} \approx \frac{(\text{max input voltage} - \text{total series Zener voltage})^2}{\text{safe power dissipation}}$$

Fig. 6-3. Extending Zener voltage ramp.

6-3.1. Design Considerations

The design considerations for the circuit of Fig. 6-3 are similar to those of Fig. 6-2, as discussed in Sec. 6-2.1. The diodes need not have equal breakdown voltages since the arrangement is self-equalizing. However, the power-handling ability of each diode should be the same. In addition, the current ranges should be similar, or the loads so arranged to avoid damaging any of the diodes.

In practice, if the input voltage falls below about 1.4 times the total Zener voltage of the series string, some or all of the diodes will not go into the avalanche condition, and there will be no regulation. On the other hand, if the input voltage rises to a point where the maximum current of any one diode is exceeded, that Zener will overheat (and possibly be destroyed).

Trial values. The total Zener voltage should be about 0.7 times the lowest input voltage. In the reverse, the minimum input voltage must be 1.4 times the total Zener voltage. Thus, if each of the three Zeners was 7 V, the total Zener voltage would be 21 V, and the minimum input voltage should be $21 \times 1.4 = 29.4$ V.

A safe power-dissipation rating for each Zener diode is 3 times the load power. Thus, with a 100-mW load, each of the Zeners should be capable of dissipating 300 mW.

The value of R_S is found using the maximum input voltage, total series Zener voltage, and the safe power dissipation.

As in the case of the basic Zener circuit, if the input voltage varies by more than about 30 per cent of the highest input voltage, it will be difficult to design a practical regulation circuit (with one or more Zeners). Also, before going to the breadboard stage, check the design on paper. With a correct value for R_S, the maximum input voltage will produce a current that can be safely dissipated by the Zener with the lowest current rating, while the minimum input voltage will provide more than enough current to meet the load requirements.

6-3.2. Design Example

Assume that the circuit of Fig. 6-3 is to provide a regulated 22.5-V output to a 75-Ω load. Assume that the highest current rating of an available 22.5-V Zener is 100 mA, that 7.5-V Zener diodes are available, and that the maximum input voltage is 40 V. Find the minimum input voltage that would be practical, the current (or power) rating for each Zener, and the optimum value for series resistor R_S.

With a required 22.5 V (to be provided by three 7.5-V Zeners), the minimum input voltage would be 22.5×1.4, or 31.5 V.

With a 22.5-V output across a 75-Ω load, the load current would be $22.5 \div 75$, or 300 mA. The load power would be 22.5×0.3, or 3.75 W. The power dissipated by each Zener would be 7.5×0.3, or 2.25 W.

A safe power-dissipation rating for each Zener is 3 times the individual power, or 6.75 W. However, the nearest standard value would probably be a 10-W Zener.

With a load current of 300 mA, the current rating of each Zener should be $300 \times 3 = 900$ mA.

With a safe power-dissipation rating of 6.75 W, a total Zener voltage of 22.5 V, and a maximum input voltage of 40 V, the first trial value for R_S should be $(40 - 22.5)^2 \div 6.75$, or 45 Ω. The nearest 10 per cent standard is 47 Ω.

The design values (47 Ω for R_S, and three Zeners of 7.5 V with 900-mA ratings) should be checked against the maximum and minimum input voltages as follows. The maximum input current (combined Zener current and load current) will be equal to the drop across R_S ($40 - 22.5 = 17.5$ V) divided by the resistance (47 Ω), or 307 mA. This would provide the full 300 mA for the load, plus 7 mA through the Zener. However, with a minimum input voltage calculated at 31.5 V and a Zener point of 22.5 V, the drop across R_S is 9 V. With a 9-V drop and a 47-Ω resistance, the current would be slightly less than 200 mA, not enough to supply the 300-mA load. Therefore, the 47-Ω value for R is too high. In that event, cut the calculated value in half (to 23.5 Ω), or the nearest standard value of 27 Ω.

With a minimum drop of 9 V across 27 Ω, the current is 333 mA, or more than enough to meet the maximum load of 300 mA. Going back to the maximum input voltage of 40 V, the input current would be $40 - 22.5 = 17.5$ V; 17.5 V $\div$ 27 = 650 mA. Therefore, the 900-mA rating would be adequate.

6-4. Extending Zener Current Range

Although there are some Zener diodes that will handle heavy currents and dissipate large amounts of power (50 W), it is usually less expensive to extend the current rating by means of a power transistor. There are two basic circuits: the shunt circuit and the series circuit. In either circuit, the power transistor is mounted on a heat sink to dissipate maximum power. The emitter–base junction of the power transistor is held at some fixed value by means of a low-power Zener. Thus, the power transistor (silicon) emitter remains within 0.5 V of the Zener voltage (or about 0.2 V in the case of a germanium transistor), despite changes in input voltage or load current.

Although popular at one time, the shunt circuit has now been replaced by the series circuit. The shunt regulator circuit has two disadvantages; it consumes more power (which is lost to the load), and it continuously loads the basic power supply. For these reasons we will concentrate on the series regulator, such as the one shown in Fig. 6-4.

6-4.1. Design Considerations

In addition to the general design considerations of Chapter 1, the following specific design points should be considered.

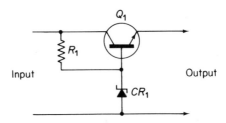

Output voltage : p-n-p silicon ≈ Zener − 0.5 V
                p-n-p germanium ≈ Zener − 0.2 V
                n-p-n silicon ≈ Zener + 0.5 V
                n-p-n germanium ≈ Zener + 0.2 V

Maximum voltage, power, and current ≈ Q_1 ratings; see text

Minimum input voltage ≈ 1- 2 V + Zener

Current thru R_1 ≈ 5 % of total load current (at minimum input value)

Zener power rating ≈ 0.1 × total power output

$$R_1 \text{(in ohms)} \approx \frac{\text{(minimum input voltage} - \text{Zener voltage)}}{\text{(maximum output current} \times 0.5)}$$

$$\text{Maximum current thru } R_1 \approx \frac{\text{(maximum input voltage} - \text{Zener voltage)}}{R_1 \text{ resistance}}$$

R_1 (in watts) ≈ (maximum current thru R_1)^2 × R_1 resistance

Zener maximum current rating ≈ maximum current thru R_1

Fig. 6-4. Extending Zener current range (series transistor regulator).

Maximum voltage, current, and power dissipation. The input voltage, current, and power-dissipation characteristics of the regulator circuit are set by the series transistor Q_1 and not by the Zener diode. For example, maximum input voltage to the circuit is set by the maximum collector voltage of Q_1. The maximum current and power ratings of Q_1 fix the maximum load current and power-dissipation capability of the circuit. Power dissipation of the transistor can be considered as the voltage drop from collector to emitter, or the difference between supply voltage and output (load) voltage. It is assumed that the transistor will be operated on a heat sink to obtain maximum current and power capabilities (see Sec. 1-4).

Output voltage. The output voltage of the regulator is set by the Zener diode, and is within 0.5 V (for silicon transistors) or 0.2 V (for germanium transistors) of the Zener voltage. If a *PNP* is used, the output voltage will be 0.2 or 0.5 V below the Zener value. If an *NPN* is used, the output voltage will be 0.2 or 0.5 V above the Zener value.

Minimum input voltage. The minimum voltage capability of the regulator is also set by the Zener voltage. As a rule, the minimum voltage (at the collector) must be at least 1 V, and preferably 2 V, greater than the Zener voltage (at the base).

Base resistor. The value of base resistor R_1 is selected to provide the optimum base current. As a rule, R_1 should provide approximately 5 per cent of the total load current, when the input voltage is at minimum.

Zener diode. The Zener diode should be selected on the basis of Zener voltage, and current or power capability. Since the output voltage is close to the Zener voltage, this sets the Zener value. Although the output current and power dissipation of the regulator are primarily dependent upon the transistor capabilities, it is usually not practical to design any regulator circuit that will provide satisfactory operation with currents greater than 10 times the maximum rated Zener current.

6-4.2. Design Example

Assume that the circuit of Fig. 6-4 is to provide 32 W of power (8 V at 4 A maximum). The input voltage varies from 11 to 15 V. A germanium *PNP* transistor capable of handling 35 W is available (with heat sink).

Verify that the transistor is capable of handling the full voltage, current, and power. The maximum voltage applied will be 15 V. Any transistor capable of 35 W dissipation should operate satisfactorily with 15 V at the collector. However, this should be verified by reference to the data sheet. The same is true of maximum current, or 4 A in this case. In operation, the transistor will not have to dissipate the full power, since the maximum emitter–collector differential will be 7 V (15 V input − 8 V output). In theory, the transistor should never dissipate more than 28 W (7 V × 4 A). So the 35-W capability should provide a good margin of safety.

Since the output voltage must be 8 V, and the transistor is germanium *PNP*, the Zener rating should be 8 + 0.2, or 8.2 V. (If the transistor had been silicon *NPN*, the Zener voltage would be 8 − 0.5, or 7.5 V.)

The minimum input voltage should be 8.2 + 1, or 9.2 V or, preferably, 8.2 + 2, or 10.2 V, or greater. Therefore, the 11-V minimum should present no problem.

As a first trial for base resistor R_1, use a resistance that will produce a current at the base equal to 5 per cent of the maximum output current, when the input voltage is minimum, or $(11 - 8.2)/(4 \times 0.05) = 14\ \Omega$. The nearest standard would be 15 or 18 Ω. The wattage rating for R_1 should be based on maximum current when the input voltage is maximum, or

$$I = \frac{15 - 8.2}{15} \approx 0.5\ \text{A}$$

$$P = (0.5)^2 \times 15 = 3.75\ \text{W}$$

The nearest standard would be 5 W.

The current through R_1 will be divided between the base–emitter junction, and the Zener diode. Since this current will be a maximum of 0.5 A, the Zener diode should also be capable of 0.5 A, or 4.1 W (8.2 × 0.5). The

nearest standard Zener would be 10 W. (In operation, the Zener would probably never pass the full 0.5 A, since part of the current would pass through
the emitter–base junction. However, the Zener rating should be on the basis
of maximum possible current.)

6-5. Overload Protection

The series regulator is subject to damage caused by overloads. This can be
in the form of a high input voltage or an excessive output load. Either way,
the series transistor will burn out. A high input voltage is not likely. Any
prolonged period of high input voltage will probably burn out the basic
power-supply component first (unless the series-regulator design was on the
borderline of safety). On the other hand, an excessive output load (low load
resistance and high output current) is quite common. (A screwdriver shorted
across the regulator output terminals will do the job nicely.) Therefore, a
series regulator should be provided with some form of protection from overloads due to excessive output current.

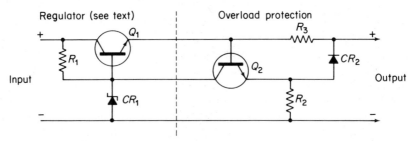

With Q_2 and CR_1 silicon, drop across $R_3 \approx 1.1$ V, at maximum current output
Output voltage $\approx$ output from regulator $- 1$ V
Output current and power $\approx$ same as regulator
$R_2 \approx 10 \times$ load resistance, at maximum current
Load resistance $\approx$ output voltage / load current
R_3 (in ohms) $\approx 1.1/$maximum load current
R_3 (in watts) $\approx$ (maximum load current)$^2 \times R_3$ resistance

Fig. 6-5. Overload protection for series regulator.

Figure 6-5 is the working schematic of an overload-protection circuit for
series regulators. Operation of the regulator portion of the circuit (Q_1, R_1,
CR_1) is the same as described for Fig. 6-4. Operation of the overload-protection circuit is controlled by the voltage across R_3. All of the load current
passes through R_3 and develops a corresponding voltage. When the load
current is below a certain value (at a safe level below the maximum rating
of the series regulator), the voltage drop across R_3 is not sufficient to forward-
bias Q_2. Thus, Q_2 remains nonconducting as long as the load current is at a

safe level. If diode CR_2 is silicon, there will be about a 0.5-V drop from the output terminal to the emitter of Q_2. Likewise, if Q_2 is silicon, another 0.5-V drop will be required before Q_2 is turned on. Therefore, the drop across R_3 must be about 1 V or more (typically 1.1 V) before Q_2 turns on. The value of R_3 is chosen so that the drop will be about 1.1 V when the output load current is at the maximum safe level. With Q_2 turned on, part of the current through R_1 is passed through Q_2, thus drawing current away from the base of Q_1. Transistor Q_1 is cut off (or partially cut off), thus preventing (or limiting) current flow to the output load.

When the load is returned to normal, the drop across R_3 is below 1.1 V, and transistor Q_2 is turned off.

6-5.1. Design Considerations

In addition to the general design considerations of Chapter 1, the following specific design point should be considered.

Regulator circuit. Design considerations for the regulator portion of the circuit are the same as described in Sec. 6-4.1.

Output voltage. The output voltage from the overload-protection circuit will be about 1 V less than the output from the regulator. This is due to the drop across R_3. The current and power capacity of the overall circuit will be approximately the same as for the regulator circuit.

Transistor characteristics. Overload-control transistor Q_2 must be capable of withstanding the full output voltage. However, the current (or power) requirements for Q_2 are not high in comparison to those of Q_1. Transistor Q_2 is off and dissipates no power except in case of an overload. When overload occurs, current through Q_2 is limited by emitter resistor R_2.

Emitter resistor. The first trial value of R_2 should be 10 times the load resistance, at maximum current. For example, if the output voltage is 8 V and the maximum load is 400 mA, the load resistance is $8 \div 0.4$, or 20 Ω. With a 20-Ω load, the value of R_2 should be 20×10, or 200 Ω. Since the nominal current through R_1 is approximately 5 per cent of the total load current, the current set by R_2 should be sufficient to drive Q_1 into cutoff. For example, assuming a load current of 400 mA and an output of 8 V, the nominal current through R_1 would be 400×0.05, or 20 mA. With the output at 8 V, and a 0.5-V drop across CR_2, the drop across R_2 would be 7.5 V. With R_2 at 200 Ω and a 7.5-V drop, the current through R_2 (and the emitter–collector of Q_2) would be approximately 37 mA.

The final test for the correct value of R_2 is that Q_1 will approach cutoff when the maximum current limit is reached.

Base resistor. The first trial value of R_3 should be such that the voltage drop is 1.1 V when maximum load current is reached. For example, using the same 400-mA load, the value of R_3 should be 1.1 ÷ 0.4, or 2.75 Ω. The nearest standard value would be 2.7 or 3.0 Ω. The base resistor must be capable of passing the full-load current without overheating. In this example, the power requirement should be $(0.4)^2 \times 3$, or 0.48 W. A 0.5 W would be the nearest standard.

6-5.2. Design Example

Assume that the circuit of Fig. 6-5 is to provide approximately 2.5 W of power (8.5 V at 300 mA). The input voltage varies from 11 to 15 V. A silicon *PNP* transistor capable of handling 5 W is available (with heat sink). The output must be maintained at 8.5 V, at any input voltage between 11 and 15 V, and at any current up to 300 mA. If the output becomes shorted, or the load resistance drops, the output current must be limited to a safe level.

Verify that transistor Q_1 is capable of handling the full voltage, current, and power. The maximum voltage applied will be 15 V, with a maximum current of 0.3 A. This results in a maximum power dissipation of 15×0.3, or 4.5 W. The 5-W transistor should provide a satisfactory margin of safety, since the maximum emitter–collector differential will never reach 15 V.

Since the output must be 8.5 V, and there is a drop of approximately 1 V across R_3, the output from Q_1 must be 8.5 + 1, or 9.5 V. Since Q_1 is silicon, the Zener CR_1 rating should be 9.5 + 0.5, or 10 V.

The minimum input voltage should be 10 + 1, or 11 V or greater. Therefore, the low input voltage is on the borderline. A minimum input voltage of 12 V would be preferable. If the circuit does not provide proper regulation with minimum (11-V) input, it may be necessary to raise the minimum input, or lower the output capability, depending upon other circuit requirements.

As a first trial for base resistor R_1, use a resistance that will produce a current at the base equal to 5 per cent of the maximum output current, when the input voltage is minimum, or: $(11 - 10)/(0.3 \times 0.05) = 66.6$ Ω. The nearest standard would be 68 Ω. The wattage rating for R_1 should be based on maximum current when the input voltage is maximum, or

$$I = \frac{15 - 10}{68} \approx 73 \text{ mA}$$

$$P = (0.073)^3 \times 68 = 0.005 \text{ W}$$

The nearest standard would be 0.25 W.

The current through R_1 will be divided between the base–emitter junction of Q_1 and Zener diode CR_1. Since this current will be a maximum of 73 mA, the Zener diode should also be capable of 73 mA, or $10 \times 0.073 \times 0.73$ W. The nearest standard Zener would be 1 W. (In operation, the Zener would probably never pass the full 73 mA, since part of the current would always pass through the emitter–base junction of Q_1. However, the Zener rating should be on the basis of maximum current.)

The first trial value of R_2 should be 10 times the load resistance, at maximum current. With an output voltage of 8.5 V, and a maximum load of 300 mA, the load resistance is 8.5 ÷ 0.3, or approximately 28 Ω. With a 28-Ω load, the value of R_2 should be 28 × 10, or 280 Ω. The next highest standard would be 300 Ω.

With R_2 at 300 Ω, the output at 8.5 V, and an approximate 0.5-V drop across CR_2, the drop across R_2 should be 8 V. This would produce a current flow through R_2 and Q_2 of approximately 27 mA. Since the nominal current through R_1 was calculated at 73 mA, it may be necessary to decrease the value of R_2 so that more current can be drawn by Q_2 (away from the base of Q_1) when Q_2 conducts in response to an overload. If breadboard tests prove it is necessary to lower the value of R_2, try a value of 100 Ω. This will result in an available 80 mA through Q_2.

With a possible 80 mA through Q_2 and 10 V at the collector, the maximum dissipation required of Q_2 is 10 × 0.08, or 0.8 W. However, since the emitter of Q_2 will be at an approximately 8 V, the differential between emitter and collector is 2 V. Thus, the probable power-dissipation capability of Q_2 is 0.2 W. As a rule, the power dissipation and current rating for Q_2 should be based on the maximum possible current.

The first trial value of R_3 should be such that the voltage drop is 1.1 V with 300 mA, or 1.1 ÷ 0.3 = 3.7 Ω (3.6 and 3.9 Ω are the nearest standard values). Resistor R_3 must be capable of passing the full 300 mA without overheating. Thus, the power requirement would be $(0.3)^2$ × 3.7, or 0.333 W. A 0.5-W resistor would be the nearest standard.

6-6. Feedback Regulator

Figure 6-6 is the working schematic of a feedback regulator and rectifier circuit. Note that the rectifier section is similar to the basic rectifier circuit of Fig. 6-1. Also note that the output is adjustable. In operation, the output voltage will remain within about 0.5 V of the voltage at potentiometer R_3 (at the base of Q_3). Any variations in load (which might result in an undesired change in voltage across the load) cause corresponding changes in current through the base–emitter of Q_3. Transistors Q_1 and Q_2 are connected as a Darlington pair or compound (Sec. 2-5.1). Any variations in output current (due to load changes) change the base current of Q_2. These changes are amplified through Q_1 and Q_2. The change in emitter–collector current of Q_1 opposes any change in output current.

6-6.1. Design Considerations

Transistor characteristics. All three transistors should be capable of operating at the highest output voltage (from the rectifier circuit) without breakdown. Transistor Q_1 handles the heavy current load. The power-

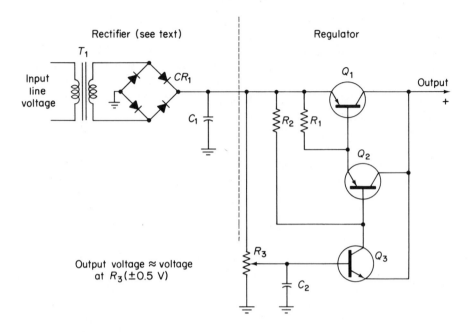

Rectifier (see text)   Regulator

Input
line
voltage

Output voltage ≈ voltage
at $R_3(\pm 0.5$ V)

Q_1, Q_2, Q_3, ratings ≈ see text

T_1, CR_1, C_1 ≈ see text
  Current thru R_3 < 5% max. load current

R_1 ≈ 0.01 × R_3 C_2 ≈ 0.5 × C_1

R_2 ≈ 0.1 × R_3

R_3 ≈ max output voltage/(max load current × 0.05)

Fig. 6-6. Solid-state rectifier with feedback regulator.

dissipation rating of Q_1 should be based on the full input voltage (from the rectifier) and the full-load current. For example, if the output from the rectifier is 25 V, and the output current is 1 A, Q_1 should be capable of handling at least 25 W. Transistor Q_2 will be subjected to about 0.3 times the current and/or power dissipation of Q_1. As a safety factor, select Q_2 on the basis of 0.5 times the Q_1 current and power dissipation. Transistor Q_3 will handle approximately 0.1 times the current and/or power dissipation of Q_1. As a safety factor, select Q_3 on the basis of 0.2 times the Q_1 current and power dissipation.

Rectifier circuit. Design considerations for the rectifier portion of the circuit are the same as described in Sec. 6-1.1.

Resistance values. The value of resistor R_3 should be such that current through R_3 is less than 5 per cent of the maximum output load current. With resistor R_3 properly calculated, the first trial value for R_1 is 0.01 times the value of R_3, and the first trial value for R_2 is 0.1 times the value of R_3. These ratios should provide sufficient current gain through the Darlington compound, even with transistors of low gain (beta).

Capacitor values. The value of C_1 is selected as part of the rectifier circuit, as described in Sec. 6-1. As a first trial value, capacitor C_2 should be half the capacitance value of C_1. However, the value of C_2 should be no less than 500 μF, for a typical solid-state power supply.

6-6.2. Design Example

Assume that the circuit of Fig. 6-6 is to provide a regulated output up to 32 V at 0.5 A. Assume that the rectifier output must have a ripple of less than 2 per cent. Under these conditions, the regulator output ripple will be less than 0.02 per cent. (As a rule, the ripple will be decreased by a factor of 0.01. That is, the regulator output ripple is approximately 1 per cent of the rectifier ripple.) A line supply of 110 to 120 V is available.

With a required 32-V output, input to the regulator from the rectifier must be about 32.5 to 33 V. This will allow for a 0.5- to 1-V drop across the regulator.

With a required 32.5-V output (from the rectifier), the transformer T_1 secondary should be 32.5 $\div$ 1.3, or 25 V.

With a maximum dc power output of 16 W (32 $\div$ 0.5), the volt–ampere (or power) rating of transformer T_1 should be 16 $\times$ 1.3, or 20.8 W, rounded off to 25 W.

The load resistance is 32 $\div$ 0.5, or 64 Ω.

The diode characteristics should be as follows:

Maximum reverse voltage should be 25 $\times$ 2, or 50 V. In practical design, any reverse voltage rating above about 28 V should be satisfactory.

Forward voltage drop should not exceed 1 V.

Reverse current should not exceed 100 μA.

Forward current should be 0.5 $\times$ 2, or 1 A.

Note that these characteristics are for individual diodes. It is possible to obtain four diodes (connected in a bridge configuration) as a package (or module). In that case, the data-sheet rating for the package must be considered as the ratings for individual diodes.

The capacitor C_1 should have a voltage rating of 32.5 $\times$ 1.3, or 42.25 V, rounded off to 50 V.

With a load resistance of 64 Ω, and a maximum ripple of 2 per cent, the capacitance value of C_1 should be 200,000 $\div$ 64 $\times$ 2, or 1560 μF, rounded off to 1600 μF.

All three transistors must be capable of withstanding at least 32.5 V without breakdown; 35 to 40 V would be a safer figure.

Transistor Q_1 should be capable of handling at least 0.5 A and 16.25 W (32.5×0.5). Safer figures would be 0.6 A and 20 W.

Transistor Q_2 should be capable of handling 0.3 A and 10 W ($0.6 A \times 0.5$; 20 W $\times$ 0.5). Transistor Q_3 should be capable of handling 0.12 A and 4 W ($0.6 A \times 0.2$; 20 W $\times$ 0.2).

The minimum value of R_3 should be $32.5 \div (0.5 \times 0.05) = 1300\ \Omega$. The next largest standard value would be 1500 Ω.

With R_3 at 1500 Ω, the value of R_2 should be 1500×0.1, or 150 Ω.

With R_3 at 1500 Ω, the value of R_1 should be 1500×0.01, or 15 Ω.

With C_1 at 1600 μF, the value of C_2 should be 1600×0.5, or 800 μF.

6-7. Feedback Regulator for Input-Voltage Variations

One problem with the circuit of Fig. 6-6 is that the output voltage may vary with changes in input voltage. While large input-voltage variations are not usually found in practical applications, there may be some solid-state circuits that require a more constant output voltage.

Figure 6-7 is the working schematic of a solid-state power supply that will provide a constant output voltage, despite changes in input voltage and/or output load. Note that the circuit is almost identical to that of Fig. 6-6, except for the addition of Zener diode CR_2 and resistor R_4.

The purpose of CR_2 is to provide a constant voltage source across potentiometer R_3. This circuit has one disadvantage, in that the output voltage is limited by the Zener voltage rather than by the available output from the rectifier.

6-7.1. Design Considerations

Design considerations for the circuit of Fig. 6-7 are the same as for the circuit of Fig. 6-6, except as follows. The Zener voltage should be 2 or 3 V below the minimum output voltage from the rectifier portion of the circuit. For example, if the normal rectifier output is 20 V, and the line input voltage to the power supply may vary by 10 per cent, the rectifier output could vary from 18 to 22 V. Under these conditions, the Zener voltage should be 15 to 16 V (maximum). (A lower Zener voltage could be selected, but this would be a waste of the available output voltage.)

As in the previous example, the current through R_3 should be less than 5 per cent of the total output current. The voltage across R_3 is set by the Zener value. Therefore, the resistance value calculated for R_3 (as described in Sec. 6-6) should be checked against the selected Zener voltage to make sure that

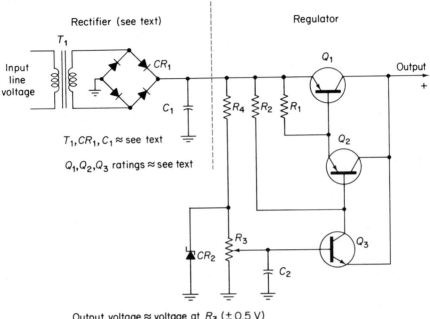

Output voltage ≈ voltage at R_3 (±0.5 V)

Max output voltage ≈ Zener voltage of CR_2 ≈ minimum output
voltage from
rectifier − 2 or 3 V

R_3 ≈ max output voltage /(max load current × 0.05)

Current thru R_3 < 5% max load current

R_4 (in ohms) ≈ $\dfrac{\text{minimum drop}}{\text{across } R_4}$ /1.2 × current thru R_3

Fig. 6-7. Feedback regulator with protection for input-voltage
variations.

the 5 per cent limit is not exceeded. For example, if the value is 1500 Ω and
the selected Zener voltage is 15 V, the current would be 10 mA. If the output
current is greater than 200 mA, the 1500-Ω value would stand.

The current through R_3 combines with the current through the Zener and
passes through R_4. The value of R_4 should be such that the current through
R_4 is 1.2 times the current through R_3, when the voltage drop across R_4 is at
minimum (minimum input voltage). For example, if the minimum input
voltage was 18 V and the Zener voltage was 15 V, the minimum drop across
R_4 would be 3 V. If the current through R_3 was 10 mA (1500 Ω, 15 V), the
current through R_4 should be 12 mA (minimum). With a 3-V drop and
12 mA, the resistance of R_4 should be 3 ÷ 0.012, or 250 Ω.

If the input voltage rises to 22 V, the Zener will remain at 15 V, and there will be a 7-V drop across R_4. The current through R_4 would be 7 ÷ 250, or 28 mA. Of this, 10 mA would pass through R_3, and the remaining 18 mA would pass through the Zener diode. Thus, the maximum current (and power) rating for CR_2 should be based on maximum input voltage.

6-7.2. Design Example

Assume that the circuit of Fig. 6-7 is to provide the same (approximate) output as that of Fig. 6-6 (as described in Sec. 6-6.2), except that the input voltage may vary as much as 10 per cent. Use a Zener that will provide the maximum output voltage.

With a nominal 32.5- 33-V output from the rectifier, a 10 per cent input-voltage variation would produce a rectifier output of 30 to 36 V.

The maximum Zener voltage must be 2 or 3 V below the minimum voltage, or 27 V.

With 27 V across a 1500-Ω potentiometer R_3, the current through R_3 is 27 ÷ 1500, or 18 mA. Since the output load current is 0.5 A maximum, 18 mA is less than 5 per cent of the maximum current (0.5 × 0.05 = 25 mA). Therefore, the 1500-Ω value for R_3 is acceptable for a first trial.

With 18 mA through R_3, the desired current through R_4 is 21.6 mA (18 × 1.2 = 21.6 mA) when the voltage drop across R_4 is minimum. With the Zener at 27 V and the minimum at 30 V, the minimum drop across R_4 is 3 V (30 − 27 = 3).

With a 3-V drop and 22 mA (rounded off from 21.6), the value of R_4 should be 3 ÷ 0.022, or 136 Ω. The nearest standard value would be 150 Ω.

With these values, the current through R_3 would remain at 18 mA, with about 4 mA through the Zener, at minimum input voltage.

If the input voltage rises to 36 V, the drop across R_4 will be 36 − 27, or 9 V. With a 9-V drop and 150 Ω, the current will be 9 ÷ 150, or 60 mA. The current through R_3 should remain at 18 mA, with about 42 mA through the Zener. Thus, the power-dissipation rating for the Zener should be 27 × 0.042, or 1.134 W. A 1-W Zener might get by, but a higher rating would provide a greater measure of safety.

6-8. Switching Regulator

This section describes the design of a switching regulator. The design parameters and special techniques described here have been developed by the Industrial Applications group of Motorola Semiconductor Products, Inc.

The function of a switching regulator is to convert an unregulated dc input to a regulated dc output. However, the method used to accomplish the function differs from the conventional series pass regulators described in Secs. 6-6 and 6-7. In a switching regulator, the power transistor is used in a switching mode rather than a linear mode. Regulator efficiencies are usually 70 per cent or better, which is about double that of the series pass regulator. High-frequency switching regulators offer considerable weight and size reductions, and better efficiency at high power, over conventional 60-Hz transformer-coupled, series-regulated power supplies.

Switching regulators are not without special problems. In addition to requiring more complex circuits, switching regulators produce electromagnetic interference (EMI). However, with proper design, EMI can be reduced to acceptable levels. Such design techniques involve the use of low-loss ferrite cores for transformers and chokes, the use of high-permeability magnetic alloys for shielding, and the use of miniature semiconductor and IC devices for switching and regulation circuitry.

Figure 6-8 shows the block diagram of a basic switching regulator. The circuit regulates by switching the series transistor (power switch) to either the ON or OFF condition. The duty cycle of the series transistor determines the average dc output. In turn, duty cycle is adjusted in accordance with a feedback that is proportional to the difference between the dc output and a reference voltage.

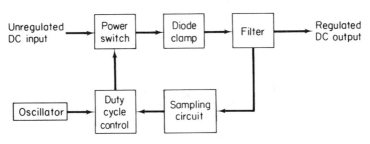

Fig. 6-8. Basic switching-regulator functions.

Switching is usually a constant frequency just above the audible range (20 kHz is typical), although some switching regulators use a variable frequency with changing line and load. Higher frequencies are generally less efficient since transistor switching losses and ferrite core losses increase.

Low-frequency unibase transistors (those with f_T of about 200 kHz, refer to Sec. 1-3) are sufficient for series pass regulators. Switching regulators must use transistors with f_T of about 4 MHz to operate efficiently. For best efficiency, higher-frequency transistors (annular and double-diffused) with f_T greater than 30 MHz should be used. Darlington transistors (Sec. 2-5.1) are also used in switching regulators. A fast-recovery rectifier, or a Schottky barrier diode, is used as a free-wheeling clamp diode (Sec. 5-8.3) to keep the switching transistor load line within safe operating limits and to increase efficiency. Other solid-state devices used in some switching regulators include gates, flip-flops, op-amp comparators, timers, and rectifiers.

6-8.1. Typical Switching-Regulator Circuits

Figure 6-9 shows four typical switching-regulator circuits. All the circuits have the following common elements:

1. Switching transistor.
2. Clamp diode.
3. *LC* filter.
4. Logic or control block.

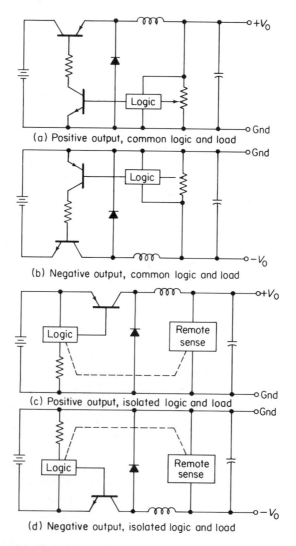

Fig. 6-9. Switching-regulator-circuit variations. Courtesy Motorola.

None of the Fig. 6-9 circuits provide isolation between the line and load. However, the one-transistor design results in simplicity and economy. It is usually desirable to have at least one line in common with the input and output to reduce ground loops. The one-line approach also determines whether the output voltage is considered positive or negative. However, most circuits will operate from either supply since the input and output grounds are usually isolated. This arrangement is suitable for the most general use of switching regulators.

In the circuits of Fig. 6-9a and b, the logic operates from the load voltage. These circuits are not self-starting, and provisions must be made to operate from the line during starting, and in the event of short circuits. In the circuits of Fig. 6-9c and d, the logic operates continuously from the line and is isolated from the load voltage. The sense and feedback elements must be electrically isolated for this reason. An optoelectronic coupler or photocoupler is well suited for this purpose.

The circuits of Fig. 6-9b and d are generally used in line-operated supplies because economical high-voltage *NPN* transistors are available, whereas *PNP* types are not. Of these two, the circuit of Fig. 6-9d is most popular, because the logic is tied directly to the series switch and switching can be much more efficient.

Driver transformers are also used in many designs to interface between the logic and switching transistor. In such a case, the switching transistor may be either *PNP* or *NPN*.

6-8.2. Switching-Regulator Theory of Operation

The high efficiency of switching regulators is a result of operating the series transistor in a switching mode. When the transistor is switched ON, full input voltage is applied to the *LC* filter. When the transistor is switched OFF, the input voltage is zero. With the transistor turned ON and OFF for equal amounts of time (50 per cent duty cycle), the dc load voltage will be half the input voltage. The output voltage V_o is always equal to the input voltage V_{in} times the duty cycle D as follows:

$$V_o = DV_{in}$$

Varying the duty cycle compensates for changes in the input voltage. This technique is used to obtain a regulated output voltage as follows.

Repetitive operation of the switching transistor at a fixed duty cycle produces the steady-state waveforms shown in Fig. 6-10. With the switch closed, inductor current I_L flows from the input voltage V_{in} to the load. The difference between the input and output voltage ($V_{in} - V_o$) is applied across the inductor. This causes I_L to increase during the "switch-closed" period.

With the switch open, stored energy in the inductor forces I_L to continue to flow to the load and return through the diode. The inductor voltage is now

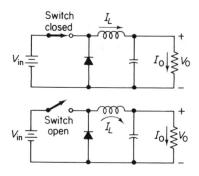

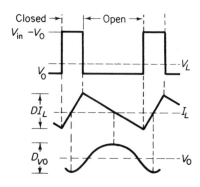

Fig. 6-10. Theoretical switching-regulator waveforms. Courtesy Motorola.

reversed and is approximately equal to V_o. I_L decreases during the "switch-open" period.

The average current through the inductor equals the load current. Since the capacitor keeps V_o constant, the load current (I_o) will also be constant. When I_L increases above I_o, the capacitor will charge, and when I_L drops below I_o, the capacitor will discharge. The theoretical waveforms are shown in Fig. 6-10. The end results of steady-state operation are as follows:

1. The average inductor voltage is zero, but a wide variation from $(V_{in} - V_o)$ to V_o is experienced.
2. The direct-current flowing through the inductor will equal the load current. A small amount of sawtooth ripple is also present.
3. The dc voltage on the capacitor is equal to the load voltage. A small amount of ripple is also present at the capacitor.

To be effective, a regulator must provide compensation for changes in both V_{in} and I_o. In a switching regulator, V_{in}, or input voltage, changes are automatically compensated for by appropriate duty-cycle variation in a closed-loop system. Input regulation and ripple rejection are dependent upon loop gain.

Changes in I_o or output current are more difficult to compensate for, and load transient response is generally poor. Changes in I_o are compensated for with temporary duty-cycle changes. For example, a change in load from half to full will result in the following:

1. Duty cycle increases to its maximum (transistor may just stay ON).
2. The inductor current takes many cycles to increase to its new dc level.
3. Duty cycle returns to its original value.

6-8.3. Design Example

The following paragraphs describe design of a solid-state power supply with a switching regulator. The circuit will supply a regulated 24-V dc output from a 120-V ac line. Typical load variations are from 1.5 to 3 A. Line variations are from 100 to 140 V. The circuit also includes short-circuit or overload protection (Sec. 6-5).

Figure 6-11 is a functional block diagram of the basic switching regulator circuit. Since the circuit is line-operated, an input rectifier and filter are required to convert the incoming ac to dc. The power switch uses a high-voltage transistor operating at a switching frequency of 20 kHz to control

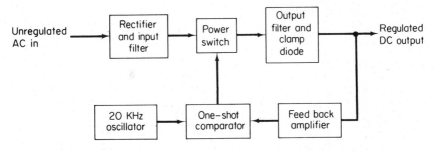

Fig. 6-11. Switching-regulator block diagram.

the pulse width of the voltage applied to the *LC* output filter.

The control portion of the switching regulator uses an oscillator, a one-shot comparator, and a feedback block which also includes an optocoupler. The oscillator generates the fixed frequency of operation. The resulting output forms a clock pulse for the comparator. The second input to the comparator is derived from the feedback amplifier, which, in turn, senses the output voltage and controls the duty cycle of the one-shot comparator. The output of the comparator is then fed directly to the power switch to complete the loop. The following paragraphs describe design of the individual circuits that make up the complete switching regulator.

Figures 6-12 and 6-13 are schematic diagrams for the complete switching regulator (power supply, logic, regulator, feedback, etc.). These illustrations provide information for ordering component parts. Note that the four comparator blocks shown in Fig. 6-13 are part of a single quad comparator (the Motorola MC3302).

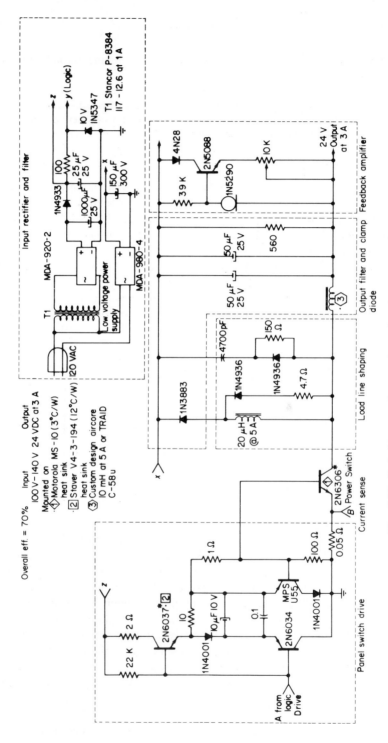

Fig. 6-12. Switching-regulator schematic diagram. Courtesy Motorola.

392

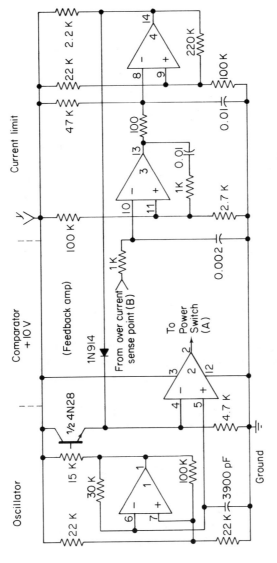

Fig. 6-13. Control circuit with **MC3302** quad comparator. Courtesy Motorola.

6-8.4. Input Rectifier and Filter

The input rectifier and filter (Fig. 6-12) consists of a bridge rectifier (MDA-980-4) and filter capacitor. Note that a series current-limiting resistance is not used. This improves efficiency but results in high starting current surges. Thus, a larger rectifier is used, even though the output current is not to exceed 3 A. The 150-μF capacitor is used to limit the input ripple to under 20 V peak to peak, with a full load.

6-8.5. Output Filter and Clamp Diode

The output filter (Fig. 6-12) consists of a 10-mH choke and two 50-μF capacitors. Note that the clamp diode 1N3883 is capable of handling 6 A, even though the nominal output current is 3 A maximum. This allows for current surges up to about 5 A without damage.

The minimum inductance value for the choke can be found using the equation

$$L = \frac{(V_{in} - V_o)V_o}{0.05 I_{o(max)} V_{in} F}$$

where V_{in} = input voltage
V_o = output voltage (nominal 24 V)
$I_{o(max)}$ = maximum output current
F = frequency (20 kHz)
0.05 restricts the current in choke L to variations no greater than 5 per cent of the maximum load current

Using the design values results in a calculated value for L of about 6 mH. The nearest standard choke available in this design case was a 10-mH air-core choke. A standard Triad C 58U, or an equivalent custom-designed ferrite core, may be used if a small air gap is provided to prevent saturation at currents up to 5 A (the short-circuit limit of the regulator).

The minimum capacitance value for the filter can be found using the equation

$$C = \frac{V_{in} - V_o}{(2L)F^2 V_{in}(\text{max ripple})}$$

where V_{in} = input voltage
V_o = output voltage
L = inductance of the choke
F = frequency (20 kHz)
max ripple is arbitrarily set at 50 mV

Using the design values results in a calculated value for C of about 50 μF. The dissipation factor of most electrolytic capacitors is too high for this application. Because of this, and to provide an additional safety margin in filtering, two 50-μF capacitors are used.

A conventional bleeder resistor (560 Ω) is used across the filter. However, note that high-frequency bypass capacitors and/or additional *LC* filters found in similar regulators are not used in this design. Such circuits are used to prevent switching spikes from reaching the load. However, such circuits are not required in this design because of the load-line shaping network described in Sec. 6-8.9.

6-8.6. Power Switch and Drive

The power switch and drive (Fig. 6-12) includes a push-pull driver (2N6034 and 2N6037) which provides the interface between the IC logic drive signal and the actual power switch (2N6306). Power for this circuit, as well as the IC logic, is provided by the low-voltage power supply shown in Fig. 6-12.

Note that the IC logic is in the position indicated by the arrangement of Fig. 6-9d and as discussed in Sec. 6-8.1. The main advantage of this layout is that the logic ground is common to the emitter of the power switch. Because of this, conventional speed-up capacitors can be used to couple the drive signal to the base terminal of the power switch to improve switching speeds. This is particularly important in this design. Because of the large step-down in voltage, and the high switching frequency, on-times of the power switch are very short.

The minimum switching time is found using the peak V_{in} of 200 V. As discussed in Sec. 6-8.2, this results in a duty cycle of

$$D = \frac{V_o}{V_{in}} = \frac{24}{200} = 12 \text{ per cent}$$

One complete cycle at 20 kHz is 50 μs (1/20,000). Thus, at 20 kHz, the ON time will be 12 per cent of 50 μs, or about 6 μs. With simple resistive termination of the power-switch base–emitter junction, the storage time alone could be as much as 10 μs. For this reason, a special drive circuit and a high-frequency 250-V, 8-A, 2N6306 power transistor were selected for design. The drive circuit shown in Fig. 6-12 does the job of switching the 2N6306 at 3 A and 20 kHz from the 120-V line. To do this, two unique design innovations are used.

1. An artificial negative bias supply is created from the single positive supply to improve fall time of the switching pulse.
2. Current limiting is added to the base current to limit overdrive and to reduce storage time.

Turn-off. Turn-off of the power switch is accomplished when the IC logic is moved to a logic low. (Operation of the IC logic is described in Sec. 6-8.8.) With point A of Fig. 6-12 at a logic low, the 2N6034 is turned ON to create a path for reverse base current. The 10-μF capacitor, which is used as a reverse bias supply, then removes the stored base charge from the 2N6306 and forces the power switch to turn off. Reverse base current flows for about 2 μs. The diodes are added to prevent the capacitor from discharging into the 10-Ω resistor once the base is cleared. This minimizes the amount of charge that must be replaced during each alternation.

Turn-on. When the logic goes high, the 2N6037 turns on and forward base current flows through the 10-Ω resistor, turning on the 2N6306 power switch. The 2N6034 and 2N6037 are both plastic Darlington power transistors. However, the 2N6037 should be used with an appropriate heat sink (as indicated in Fig. 6-12), since the 2N6037 supplies both forward base drive for the 2N6306 and recharge current for the 10-μF capacitor. Recharge takes approximately 2 μs. Most of the recharge current is bypassed around the 2N6306 by the MPSU55, which limits forward base drive to about 1 A. This particular drive level was chosen to keep storage time low, and at the same time to ensure that the 2N6306 will remain saturated, even under short-circuit conditions.

6-8.7. Feedback Amplifier

Because the load ground and logic ground are isolated, it is necessary to use a 4N28 optocoupler for the feedback amplifier. A simplified schematic of the 4N28 circuit is shown in Fig. 6-14. Note that the 4N28 optocoupler appears in both the feedback amplifier of Fig. 6-12 and the comparator of Fig. 6-13.

The 1N5290 shown in Figs. 6-12 and 6-14 is a 0.5-mA current-limiting diode, and causes the voltage drop across the 39-kΩ sense resistor to remain fixed at about 20 V. This unique design causes changes in load voltage to appear directly at the base of the 2N5088 and results in sensing without attenuation. Load voltage changes produce a linear change in LED current

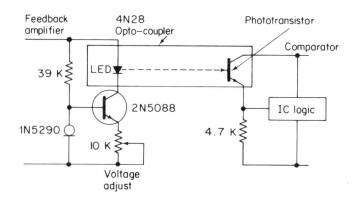

Fig. 6-14. Feedback amplifier with optocoupler.

and a proportional voltage change at the sense terminal of the logic circuitry (at the emitter of the phototransistor portion of the 4N28).

The 1N5290 diode is considered to be a zero-temperature coefficient unit. Also, the output level of the 4N28 of 1 mA is chosen to minimize its temperature dependence. A photo Darlington coupler was tried for increased gain. However, the Darlington coupler introduced low-frequency oscillation because of its slow speed.

The 10-kΩ potentiometer in the emitter circuit of the 2N5088 permits the output voltage to be set over a narrow range.

6-8.8. Oscillator, Comparator, and Current-Limiter Circuits

In addition to the feedback amplifier, the control circuit for the switching regulator also requires an oscillator, comparator, and current limiter. These three functions are all made possible by the use of a single IC, the MC3302 quad comparator, as shown in Fig. 6-13.

Comparator 1 is used as a 20-kHz oscillator which supplies a sawtooth output operating between the voltage limits defined by the 100-kΩ positive feedback resistor and the logic supply voltage. Comparator 2 takes the sawtooth output and compares it to the feedback signal to produce a variable duty cycle output pulse for the power switch (Sec. 6-8.6).

The timing diagram of Fig. 6-15 illustrates how the feedback signal and oscillator output are combined to produce the comparator output. In normal operation, the feedback signal is a constant dc voltage which is between the limits of the oscillator sawtooth. When the sawtooth exceeds the feedback threshold, comparator 2 switches to a high output level. The comparator is reset when the sawtooth drops back below the feedback signal. In actual practice, the comparator output pulse is delayed about 2 μs. However, this has no effect on regulation. Variations in the output and feedback signals will still produce compensated changes in the power-switch pulse width.

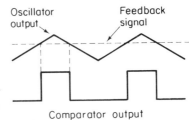

Fig. 6-15. Timing diagram showing relationship of oscillator output, feedback signal, and comparator output.

The remaining two comparators in Fig. 6-13 are used to initiate current-limiting action. Comparator 3 senses the overcurrent and triggers comparator 4, which is used as a one-shot multivibrator. The zero-input voltage threshold is derived from the reference voltage of 0.24 V from comparator 3. Power dissipation in the 0.05-Ω sense resistor (Fig. 6-12) is therefore kept at a minimum. With this combination of reference voltage and current resistor, current limiting is at about 4 A. When this level is reached (due to overload or short circuit), the one-shot capacitor is discharged, and a 1-ms time interval is initiated. Positive ac feedback is used to ensure that this capacitor does not completely reset. An *RC* filter is also provided to prevent inductive transients from accidentally shutting the regulator down.

While the one-shot is recharging, the output of comparator 4 remains high to simulate excessive output voltage at the feedback sense point (comparator 2). Drive pulses to the power switch are completely inhibited during this downtime. Comparator 4 is reset, and drive pulses continue when the one-shot reaches an 8-V threshold. Positive feedback to this reference prevents circuit oscillations as the threshold is approached. If the short is removed, the switching regulator will automatically reset to a full load. If the short remains, operation (with a 30-μs 4-A pulse) continues at an audible 1-kHz rate set by the one-shot multivibrator.

6-8.9. Load-Line Shaping

Load-line shaping is used to improve reliability and reduce EMI. Shaping is basically done by using reactive elements shown in Fig. 6-12 to absorb what would normally be switching losses. The loss-elimination network is shown in simplified form in Fig. 6-16, along with design equations.

During turn-on, the inductor L_X must block the input voltage V_{in} during the transistor turn-on time t_r. Diode current I_o, which has been flowing in

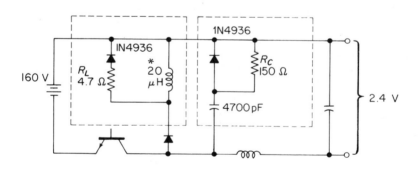

$$L_X = \frac{V_{in}\, t_r}{I_o} \qquad C_X = \frac{I_o\, t_f}{V_{in}}$$

*Core: Indiana General Ferrite Pot Core CF–214H
*Turns: 6T of #18 AWG

Fig. 6-16. Loss-elimination networks. Courtesy Motorola.

the clamp diode and inductor prior to this time, will decrease to zero. Measured current rise and fall times on the transistor are 0.3 and 0.4 μs, respectively. Thus,

$$L_X = \frac{V_{in}t_r}{I_o} = \frac{160(0.3\ \mu s)}{3} = 16\ \mu H$$

The value used is 20 μH. The inductor should not saturate at rated load to be effective, and current must build back up to I_o. An inductance/resistance time constant of 4 μs ensures this condition. Thus, $R_L = L_x/4 = 20/4 = 5\ \Omega$.

Using a 4.7-Ω resistor is practical since the collector-voltage overshoot is limited to about 15 V.

During turn-off, the capacitor C_X must supply load current I_o while the transistor is turning off (t_f). The capacitor will have charged to the input voltage and will now discharge to zero. Thus,

$$C_X = \frac{I_o t_f}{V_{in}} = \frac{3(0.4\ \mu s)}{160} = 7500\ pF$$

Because of parasitic circuit capacitance, only 4700 pF is required. The capacitor must recharge during the on-time of the transistor to 10 μs. The *RC* time constant of 1 μs is chosen to ensure this condition. Thus,

$$R_C = \frac{1\ \mu s}{4700\ pF} = 210\ \Omega$$

Using a 150-Ω resistor is practical because the collector current overshoot is limited to about 1 A. Thus, parasitic capacitance has no effect on this recharge time. In actual practice, the values of L_X and C_X may be adjusted to obtain the desired load-line effect.

6-8.10. Operating Characteristics

The output-voltage variations due to line and load changes are as described below.

With the output load current varied between 1.5 and 3 A and the input voltage held at 120 V, the output voltage varies no more than 0.2 V, or 0.8 per cent regulation.

With the input voltage varied between 100 and 140 V, and the output load current held at 3 A, the output voltage varies no more than 0.7 V, or 3 per cent regulation.

The ripple content of the output voltage (24 V) is as follows. The 60-Hz ripple is 200 mV peak to peak maximum, representing a ripple factor of approximately 10 mV/V.

The 20-kHz ripple is 50 mV peak to peak maximum and is less than 0.2 per cent of the output voltage.

Ripple is measured with a 24-V output into a 3-A load.

6-9. DC–DC Converter

Figures 6-17, 6-18, and 6-19 are working schematics for three dc–dc converters. These converters operate with dc inputs of 12 V and 28 V and produce dc output voltages from 100 to 500 V. Output current ratings are from 1 to 25 A. Output voltages can be increased or decreased by changing the number of secondary turns.

Note that components are assigned values, type numbers, or winding data. The values for these components depend upon the desired output. Proper selection of these values, based on the design considerations and examples of this section, will provide the desired power output. Other power outputs could be produced by means of intermediate values. However, the tabulated outputs should provide the designer with sufficient choice for most applications. The values tabulated are the nearest standard parts available that will meet or exceed the minimum specifications required for the particular converter.

Note that the part numbers given in the illustrations may or may not be in stock, particularly the transformer part numbers. For that reason, the *winding information* for all transformers is given for each power output. If the exact part numbers cannot be located, the practical solution is to order a transformer that has close to the same winding configuration. It may even be possible to order a custom transformer wound to exact specifications. However, that usually involves considerable extra expense, delay in manufacturing time, etc. Keep in mind that transformer T_2 (when used) is a simple, nonsaturable output transformer, but transformer T_1 (in all cases) requires a *saturable core*.

Following are brief circuit descriptions of the various converters to show the relationship of circuit function to design problem. Note that the circuits are essentially multivibrators that convert dc into ac, followed by a step-up transformer, followed by a bridge rectifier that converts the stepped-up ac into dc.

When the required power outputs are between 15 and 55 W, the circuit of Fig. 6-17 is used. Note that only one transformer is required. Any voltage imbalance in the circuit will cause one of the transistors, for example, Q_1, to conduct a small amount of current. Regeneration will turn Q_2 off, and Q_1 will be driven into saturation. The collector current will increase. As the core becomes saturated, the collector current increases rapidly, and is limited only by the resistance in the collector circuit and the transistor characteristics. When the core is saturated, the induced voltage in the winding is zero. The resulting lack of base drive causes Q_1 to turn off, and the collector current drops to zero. The drop in collector current causes the polarity in all windings to reverse, thereby biasing Q_1 off and Q_2 on. When the core approaches

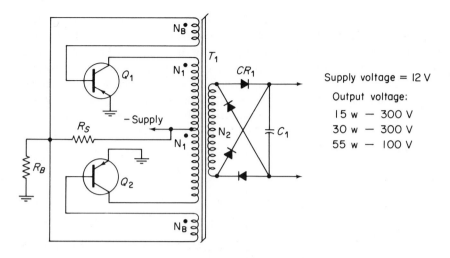

Supply voltage = 12 V

Output voltage:

15 w — 300 V

30 w — 300 V

55 w — 100 V

Power rating (watts)	Q_1 and Q_2	R_B (Ω)	R_S (Ω)	CR_1	C_1 (μF)	N_1 turns AWG	N_2 turns AWG	N_B turns AWG	T_1 type
15	2N1038	15	1200	1N2071	2	70 #18	1800 #30	20 #30	Arnold #5772D2
30	2N1042	15	1500	1N2071	4	78 #16	2000 #29	30 #29	Magnetics #500172A
55	2N456	5	180	1N2071	6	29 #17	275 #24	6 #24	Magnetics #500352A

Fig. 6-17. Dc–dc converter (15 to 55 watts). Courtesy Texas Instruments.

negative saturation, Q_2 is turned off, and the current drops to zero, turning Q_1 on.

Resistors R_B and R_S are added to place a bias on the bases of both Q_1 and Q_2. This bias provides a starting current and reduces the effect of base–emitter voltage variations. The frequency of oscillation is determined by the design of transformer T_1.

The secondary output voltage is rectified by the bridge-connected diodes CR_1 and filtered by capacitor C_1.

When the required power output is in excess of about 55 W, two transformers are used, as shown in Figs. 6-18 and 6-19. However, only transformer T_1 saturates. Thus, the extra current necessary at saturation is small compared to the load current. This allows use of a small drive transformer T_1 to drive a much larger, relatively inexpensive, power transformer T_2 which steps up the output voltage to the required value.

When one of the transistors (for example, Q_1) conducts, the collector swings from the supply voltage to near zero (saturation). The voltage builds up across the primary of T_2 and is applied to the primary of T_1 through feedback resistor R_F. Transistor Q_2 is biased off, while Q_1 is biased into conduction. As soon as the core of T_1 reaches saturation, the increasing current causes additional voltage drop across R_F, thus aiding the regenerative action (Q_2 off; Q_1 on). Transistor Q_2 continues in this state until reverse saturation of the transformer is reached. The circuit then switches back to the initial state, and the cycle is completed.

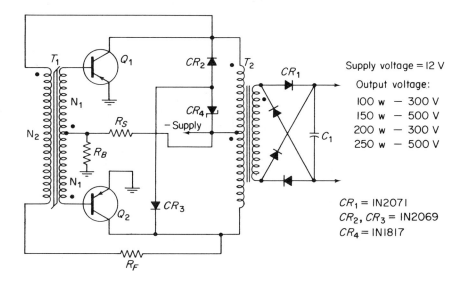

Power rating (watts)	Q_1 and Q_2	R_B (Ω)	R_S (Ω)	R_F (Ω)	C_1 μF	N_1 turns AWG	N_2 turns AWG	T_1 type	T_2 part no. (see text)
100	2N511	2	100	5	10	48 #24	185 #28	Magnetics #500942A	440402−1
150	2N512	2	75	10	20	48 #22	185 #26	Magnetics #501812A	440404−1
200	2N513	1	75	5	20	35 #20	140 #26	Magnetics #500262A	440406−1
250	2N514	1	75	5	30	35 #20	140 #24	Magnetics #500262A	440408−1

Fig. 6-18. Dc–dc converter (100 to 250 watts). Courtesy Texas Instruments.

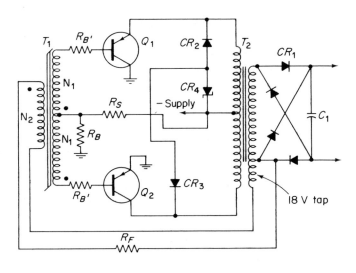

Supply voltage = 28 V   Output voltage = 300 V
                                 Output power = 500 w

CR_1 = IN1126	Q_1, Q_2 = 2N514A	R_S = 75 Ω
CR_2, CR_3 = IN2069	R_B = 0.5 Ω	R_F = 10 Ω
CR_4 = IN1825	$R_{B'}$ = 0.5 Ω	C_1 = 40 μF

T_1 = *Arnold* #523302
N_1 = 35 turns #20 AWG
N_2 = 140 turns #24 AWG
T_2 = 440413 −1 (see text)

Fig. 6-19. Dc–dc converter (500 watts). Courtesy Texas Instruments.

The collector current of the conducting transistor rises to the value of the load current, plus the magnetizing current of T_2, and the feedback current needed to produce the drive. The magnetizing current of T_2 is never more than a fraction of the rated load current, since T_2 is not allowed to saturate.

Resistors R_B and R_S are added to place a bias on the bases of Q_1 and Q_2. This bias provides a starting current and reduces the effect of base–emitter voltage variations. The frequency of oscillation is determined by the design of transformer T_1 and the value of feedback resistor R_F.

6-9.1. Design Considerations

The prime design consideration in any dc–dc converter is that the circuit will produce the desired output voltage, at a given current, or into a given load resistance. The next consideration is the efficiency of the converter (output power versus input power). Frequency of oscillation (that is, the multi-vibrator frequency) may also be a consideration in some applications.

Figures 6-20 through 6-27 show graphs of output voltage, output power, frequency, and efficiency, versus load current. As shown, the output voltage

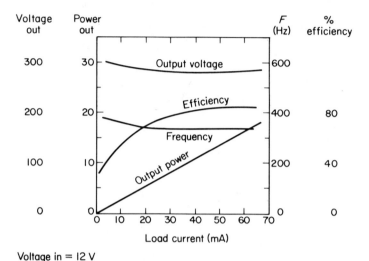

Voltage in = 12 V

Fig. 6-20. 15-W-converter characteristics. Courtesy Texas Instruments.

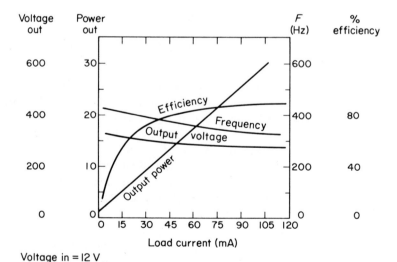

Voltage in = 12 V

Fig. 6-21. 30-W-converter characteristics. Courtesy Texas Instruments.

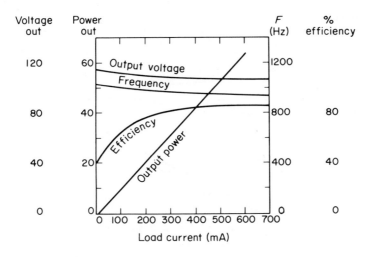

Voltage in = 12 V

Fig. 6-22. 55-W-converter characteristics. Courtesy Texas
Instruments.

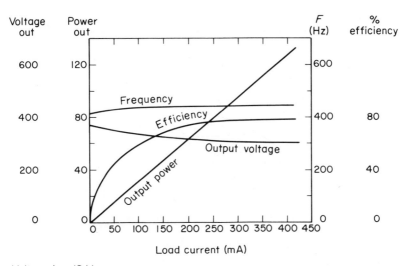

Voltage in = 12 V

Fig. 6-23. 100-W-converter characteristics. Courtesy Texas
Instruments.

regulation is less than 7 per cent from half- to full-rated load. The output
power is almost a straight-line function of load current. The frequency of
the circuit in Fig. 6-17 decreases slightly under load. The frequency of the

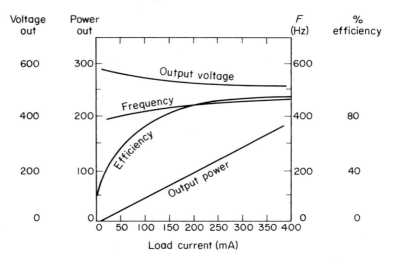

Fig. 6-24. 150-W-converter characteristics. Courtesy Texas Instruments.

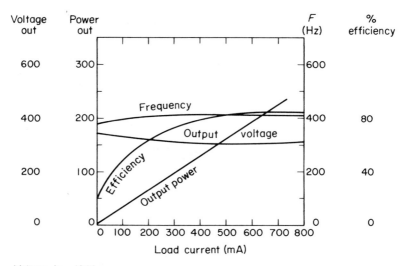

Fig. 6-25. 200-W-converter characteristics. Courtesy Texas Instruments.

circuits in Figs. 6-18 and 6-19 remains almost constant under load. The efficiency of all converters is greater than 80 per cent at rated load.

Use of these graphs, in conjunction with the schematic and tabulated values, is best understood by reference to the following design example.

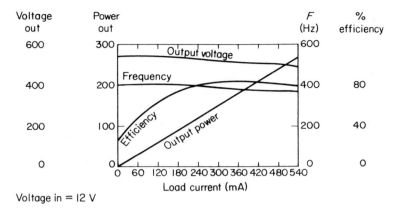

Fig. 6-26. 250-W-converter characteristics. Courtesy Texas
Instruments.

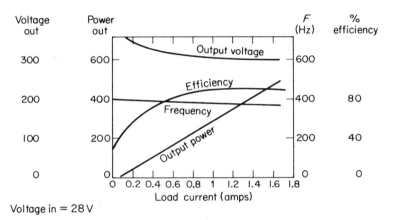

Fig. 6-27. 500-W-converter characteristics. Courtesy Texas
Instruments.

6-9.2. Design Example

Assume that the circuit of Fig. 6-18 is to provide 200 W of output power.
The output voltage must be 300 V, with 12-V input. The frequency is 400 Hz.

Converter efficiency. A converter efficiency of 80 per cent is assumed
(from the graph of Fig. 6-25). The necessary input power is:

$$\text{power input} = \frac{\text{power output}}{\text{efficiency}} = \frac{200}{0.8} = 250 \text{ W}$$

Collector current. The collector current of each transistor is given by:

$$\text{collector current} = \frac{\text{power input}}{\text{supply voltage}} = \frac{250}{12} = 20.8 \text{ A}$$

Transistor characteristics. The transistors must be selected on the basis of maximum collector breakdown voltage and maximum collector current (20.8 A, or greater, for safety). Each transistor, during its off time, is subjected to approximately twice the supply voltage, or (2×12) 24 V. The recommended transistor type is 2N513, as shown in Fig. 6-18.

Base current. The base current of each transistor is:

$$\text{base current} = \frac{\text{collector current}}{\text{minimum beta}} = \frac{20.8}{20} = 1.04 \text{ A}$$

Note that minimum beta (at the rated collector current) is obtained from the 2N513 data sheet.

Base drive voltage. The base–emitter drive voltage is made equal to twice the maximum base–emitter voltage shown on the transistor data sheet. This reduces the effect of variation in base–emitter voltage among transistors. The maximum base–emitter voltage for a 2N513 is 2 V. Therefore, the base–emitter drive voltage is 4 V.

Base drive power. The power required for base drive is given by:

$$\text{base drive power} = 4 \times 1.04 = 4.16 \text{ W}$$

Drive transformer T_1 efficiency and power input. A 90 per cent efficiency is assumed for the drive transformer T_1. Therefore, the power supplied to the primary of T_1 is:

$$\text{primary drive power} = \frac{\text{base drive power}}{\text{efficiency}} = \frac{4.16}{0.9}$$

$$= 4.6 \text{ W}$$

Turns ratio. The turns ratio of the drive transformer is chosen as $4:1$. Since the required base drive voltage is 4 V (at the secondary of T_1), the primary voltage of T_1 is 16 V.

Drive transformer T_1 primary current. With a primary power input (primary drive power) of 4.6 W and a primary voltage of 16 V, the primary current of T_1 is $4.6 \div 16$, or 287 mA.

The design considerations discussed thus far establish the characteristics necessary to order (or construct) drive transformer T_1. These characteristics are available with the transformer tabulated in Fig. 6-18.

Power transformer T_2 characteristics. The characteristics for ordering (or constructing) transformer T_2 are given in Fig. 6-28. The following notes should also be considered. The input impedance of T_2 (at the operating frequency) should be approximately 40 times the load resistance. The core loss at the rated load is approximately 5 per cent of the total power output. The primary and secondary winding losses are approximately 1 per cent of the total power output.

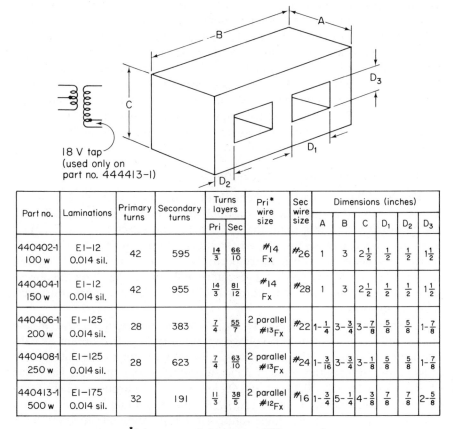

18 V tap (used only on part no. 444413-1)

Part no.	Laminations	Primary turns	Secondary turns	Turns layers		Pri* wire size	Sec wire size	Dimensions (inches)					
				Pri	Sec			A	B	C	D_1	D_2	D_3
440402-1 100 w	EI-12 0.014 sil.	42	595	$\frac{14}{3}$	$\frac{66}{10}$	#14 Fx	#26	1	3	$2\frac{1}{2}$	$\frac{1}{2}$	$\frac{1}{2}$	$1\frac{1}{2}$
440404-1 150 w	EI-12 0.014 sil.	42	955	$\frac{14}{3}$	$\frac{81}{12}$	#14 Fx	#28	1	3	$2\frac{1}{2}$	$\frac{1}{2}$	$\frac{1}{2}$	$1\frac{1}{2}$
440406-1 200 w	EI-125 0.014 sil.	28	383	$\frac{7}{4}$	$\frac{55}{7}$	2 parallel #13Fx	#22	$1-\frac{1}{4}$	$3-\frac{3}{4}$	$3-\frac{7}{8}$	$\frac{5}{8}$	$\frac{5}{8}$	$1-\frac{7}{8}$
440408-1 250 w	EI-125 0.014 sil.	28	623	$\frac{7}{4}$	$\frac{63}{10}$	2 parallel #13Fx	#24	$1-\frac{3}{16}$	$3-\frac{3}{4}$	$3-\frac{1}{8}$	$\frac{5}{8}$	$\frac{5}{8}$	$1-\frac{7}{8}$
440413-1 500 w	EI-175 0.014 sil.	32	191	$\frac{11}{3}$	$\frac{38}{5}$	2 parallel #12Fx	#16	$1-\frac{3}{4}$	$5-\frac{1}{4}$	$4-\frac{3}{8}$	$\frac{7}{8}$	$\frac{7}{8}$	$2-\frac{5}{8}$

*Primary parallel wires are bifilar wound

Fig. 6-28. Schematic and transformer core for T_2.

Base resistors. The value of R_B is chosen to drop approximately half of the 4-V drive voltage. The value of R_S is made as large as possible to provide some starting current but yet keep losses at a minimum.

Feedback resistor. The value of R_F is chosen to provide or drop the necessary voltage to give 16 V across the primary of T_1, at the rated load.

6-10. Power-Supply-Circuit Test Procedures

The following sections describe test procedures for power-supply circuits.

The basic function of a power supply is to convert alternating current into direct current. In the case of a dc–dc converter, direct current is converted to direct current at a different voltage (usually higher). Either way, the power-supply function can be checked quite simply by measuring the output voltage. For a thorough test of a power supply, the output voltage should be measured without load, with a full load, and possibly with a partial load. If a power supply will deliver the full-rated output voltage into a full-rated load, the basic power-supply function is met. In addition, it is often helpful to measure the regulating effect of a power supply, the power-supply internal resistance, and the amplitude of any ripple at the power-supply output. Also, since the operation of most power-supply circuits is dependent upon characteristics of a transformer, it is convenient (particularly at the breadboard stage) to test the transformer as a separate component.

6-10.1. Measuring Power-Supply Output

Figure 6-29 is the working schematic of a basic power-supply test circuit. This arrangement permits the power supply to be tested at no-load, half-load, and full-load, depending upon the position of load selector switch S_1. With S_1 in position 1, there is no load on the power supply. At positions 2 and 3, there is half-load and full load, respectively. The load resistors R_1 and R_2 should be noninductive so that a reactance will not be placed on the power-supply output. (Generally, the output load is considered as pure resistance, for test purposes.)

As shown by the equations in Fig. 6-29, the values of R_1 and R_2 are chosen on the basis of output voltage and load current (maximum or half-load). For example, if the power supply is designed for an output of 25 V at 500-mA full load, the value of R_2 should be $25 \div 0.5$, or 50 Ω. The value of R_1 should be $25 \div 0.25$, or 100 Ω. For practical purposes, where more than one power supply is to be tested, R_1 and R_2 should be variable, and adjusted to the correct resistance value before test (using an ohmmeter with power removed). If only one supply is to be tested, or as a temporary test setup, fixed resistors can be used. Either way, the resistors should be composition (not wire-wound), and must be capable of dissipating the rated power without overheating. For example, using the previous values for R_1 and R_2, the power dissipation of R_1 is 25×0.5, or 12.5 W (use at least 15 W), and the dissipation for R_2 is 25×0.25, or 6.25 W (use at least 10 W). To measure power-supply output, use the following procedure:

1. Connect the equipment as shown in Fig. 6-29.
2. If R_1 and R_2 are adjustable, set them to the correct values, using the equations.
3. Apply power. Measure the output voltage at each position of S_1.
4. Calculate the current at positions 2 and 3 of S_1, using the equation for actual load current. For example, assume that R_1 is 100 Ω, and the meter indicates 22 V at position 2 of S_1. Then, actual load current is $22 \div 100$, or 220 mA. If the power-supply output was 25 V at position 1, and dropped to 22 V at position 2, this means that the power supply will not produce full

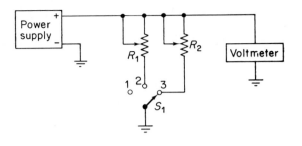

$$R_1 \text{ (in ohms)} = \text{half} - \text{load} = \frac{\text{rated output voltage}}{\frac{1}{2} \text{ max rated current}}$$

$$R_2 \text{ (in ohms)} = \text{full} - \text{load} = \frac{\text{rated output voltage}}{\text{max rated current}}$$

$$\text{Actual load current} = \frac{\text{voltage readout}}{\text{value of } R_1 \text{ or } R_2}$$

$$\% \text{ regulation} = \frac{\text{no} - \text{load voltage} - \text{full} - \text{load voltage}}{\text{full} - \text{load voltage}} \times 100$$

$$\begin{array}{l}\text{Power supply} \\ \text{internal} \\ \text{resistance}\end{array} = \frac{\text{no} - \text{load voltage} - \text{full} - \text{load voltage}}{\text{current (amperes)}}$$

Fig. 6-29. Basic power-supply test circuit.

output with a load. This is an indication of poor regulation (possibly result-ing from poor design).

6-10.2. Measuring Power-Supply Regulation

Power-supply regulation is usually expressed as a percentage and is deter-mined with the following equation:

$$\text{percentage of regulation} = \frac{\text{no-load voltage} - \text{full-load voltage}}{\text{full-load voltage}} \times 100$$

A low percentage-of-regulation figure is desired since it indicates that the output voltage changes very little with load. The following steps are used when measuring power-supply regulation. The same procedure can be used for power supplies with or without a regulation circuit.

1. Connect the equipment as shown in Fig. 6-29.
2. If R_2 is adjustable, set it to the correct value, using the equation.
3. Apply power. Measure the output voltage at position 1 (no-load) and position 3 (full-load).
4. Using the equation, calculate the percentage of regulation. For example, if the no-load voltage was 25 V, and the full-load voltage was 20 V, the

percentage of regulation would be $(25 - 20) \div 20$, or 25 per cent (very poor regulation).

5. Note that power-supply regulation is usually poor (high percentage) when the internal resistance is high.

6-10.3. Measuring Power-Supply Internal Resistance

Power-supply internal resistance is determined with the following equation:

$$\text{internal resistance} = \frac{\text{no-load voltage} - \text{full-load voltage}}{\text{current}} \times 100$$

A low internal resistance is most desirable since it indicates that the output voltage will change very little with load. To measure power-supply internal resistance, use the following procedure:

1. Connect the equipment as shown in Fig. 6-29.
2. If R_2 is adjustable, set it to the correct value, using the equation.
3. Apply power. Measure the output voltage at position 1 (no-load) and position 3 (full-load).
4. Calculate the actual load current at position 3 (full-load). For example, if R_2 was adjusted to 50 Ω, and the output voltage at position 3 was 20 V (as in the previous example), the actual load current would be $20 \div 50$, or 400 mA.
5. With the no-load voltage, full-load voltage, and actual current established, find the internal resistance using the equation given. For example, with no-load voltage of 25 V, a full-load voltage of 20 V, and a current of 400 mA (0.4 A), the internal resistance is $(25 - 20) \div 0.4$, or 12.5 Ω.

6-10.4. Measuring Power-Supply Ripple

Any power supply, no matter how well regulated or filtered, will have some ripple. This ripple can be measured with a meter or oscilloscope. Usually, the factor of most concern is the ratio between ripple and full dc output voltage. For example, if 3 V of ripple is measured, together with a 100-V dc output, the ratio would be 3 : 100 (or 3 per cent ripple).

6-10.4.1. Measuring Ripple with a Meter

1. Connect the equipment as shown in Fig. 6-29.
2. If R_2 is adjustable, set it to the correct value, using the equation. Ripple is usually measured under full-load power.
3. Apply power. Measure the dc output voltage at position 3 (full-load).
4. Set the meter to measure alternating current. Any voltage measured under these conditions will be ac ripple.
5. Find the percentage of ripple, as a ratio between the two voltages (ac/dc).
6. One problem often overlooked in measuring ripple is that any ripple voltage (single-phase, three-phase, full-wave, half-wave, etc.) is not a pure sine wave.

Most meters provide accurate ac voltage indications only for pure sine waves. A more satisfactory method of measuring ripple is with an oscilloscope, where the peak value can be measured directly.

6-10.4.2. Measuring Ripple with an Oscilloscope

An oscilloscope will display the ripple waveform from which the amplitude, frequency, and nature of the ripple voltage can be determined. Usually, the oscilloscope is set to the ac mode when measuring ripple, since this blocks the dc output of the power supply. Normally, ripple voltage is small in relation to the power-supply voltage. Therefore, if the oscilloscope gain is set to display the ripple, the power-supply dc voltage would drive the display off the screen.

1. Connect the equipment as shown in Fig. 6-30.
2. Apply power. If the test is to be made under load conditions, close switch S_1. Open switch S_1 for a no-load test. The value of R_1 must be chosen to provide the power supply with the desired load.

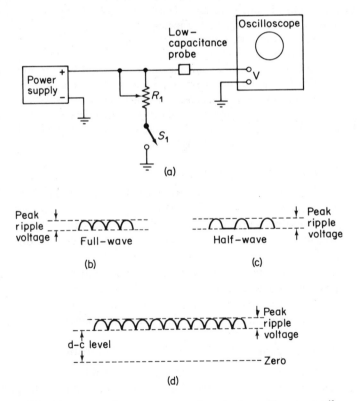

Fig. 6-30. Measuring power-supply ripple with an oscilloscope.

3. Adjust the oscilloscope sweep frequency and "sync" controls to produce two or three stationary cycles of each wave on the screen.

4. Measure the peak amplitude of the ripple on the voltage-calibrated vertical scale of the oscilloscope.

5. Measure the frequency of the ripple on the horizontal oscilloscope scale. (Refer to Sec. 5-9.1, steps 6 and 7.) When measuring ripple frequency, note that a full-wave power supply will produce two ripple "humps" per cycle (as shown in Fig. 6-30b), whereas a half-wave power supply will produce one "hump" per cycle (Fig. 6-30c).

6. If the power-supply dc output must be measured simultaneously with the ripple, set the oscilloscope for dc mode. The baseline should be deflected upward to the dc level, and the ripple should be displayed at that level (Fig. 6-30d). If this drives the display off the screen, reduce the vertical gain and measure the dc output. Then return the vertical gain to a level where the ripple can be measured, and use the vertical position control to bring the display back onto the screen for ripple measurement. This procedure will work with most laboratory oscilloscopes. On shop-type oscilloscopes where the vertical gain control must be set at a given "calibrate" position, the procedure will prove difficult, if not impossible. If this is so, it is easier to measure the dc voltage with a meter.

7. A study of the ripple waveform can reveal defects in the power supply (due to poor design or defective components.) For example, if the power supply is unbalanced (one rectifier passing more current than the others), the ripple "humps" will be unequal in amplitude. If there is noise or fluctuation in the power supply (particularly in Zener diodes, if used), the ripple "humps" will vary in amplitude or shape. If the ripple varies in frequency, the ac source is varying. If a full-wave power supply produces a half-wave output, one rectifier is not passing current.

6-10.5. Measuring Transformer Characteristics

If a regulator design does not prove satisfactory, the fault may be with the transformer. While it is usually simple to substitute other power-supply components (rectifiers, capacitors, etc.) during the breadboard test, it is more practical to test a transformer (unless another transformer is readily available for substitution).

The obvious test is to measure the transformer windings for opens, shorts, and the proper resistance value with an ohmmeter. In addition to basic resistance checks, it is possible to test a transformer's proper polarity markings, regulation, impedance ratio, and center-tap balance with a voltmeter.

6-10.5.1. Measuring Transformer Phase Relationships

When two supposedly identical transformers must be operated in parallel, and the transformers are not marked as to phase or polarity, the phase relationship of the transformers can be checked using a voltmeter and power source. The test circuit is shown in Fig. 6-31. For power transformers, the

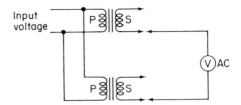

Input voltage

Fig. 6-31. Measuring transformer phase relationships.

source should be line voltage (115 V). Other transformers can be tested with a lower voltage dropped from a line source or (in extreme cases) from an audio generator.

The transformers are connected in proper phase relationship if the meter reading is zero or very low. The transformers are out of phase if the secondary output voltage is double that of the normal secondary output. This condition can be corrected by reversing either the primary or secondary leads (but not both) of one transformer (but not both transformers).

If the meter indicates some secondary voltage, it is possible that one transformer has a greater output than the other. This condition will result in considerable local current in the secondary winding and will produce a power loss (if not actual damage to the transformers).

6-10.5.2. Checking Transformer Polarity Markings

Many transformers are marked as to polarity or phase. These markings may consist of dots, color-coded wires, or a similar system. Unfortunately, transformer polarity markings are not always standard. This can prove very confusing during the breadboard stage of design.

Generally, transformer polarities are indicated on schematics as dots next to the terminals. When standard markings are used, the dots mean that if electrons are flowing *into* the terminal with the dot, the electrons will flow *out of* the secondary terminal with the dot. Therefore, the dots have the same polarity as far as the external circuits are concerned. No matter what system is used, the dots or other markings show *relative phase*, since instantaneous polarities are changing across the transformer windings.

From a practical design standpoint, there are only two problems of concern: the relationship of the primary to the secondary, and the relationship of markings on one transformer to those on another.

The phase relationship of primary to secondary can be found using the test circuit of Fig. 6-32. First check the voltage across terminals 1 and 3, then across 1 and 4 (or 1 and 2). Assume that there is 3 V across the primary, with 7 V across the secondary. If the windings are as shown in Fig. 6-32a, the 3 V will be added to the 7 V and will appear as 10 V across terminals 1 and 3. If the windings are as shown in Fig. 6-32b, the voltages will oppose each other, and will appear as 4 V (7 — 3) across terminals 1 and 3.

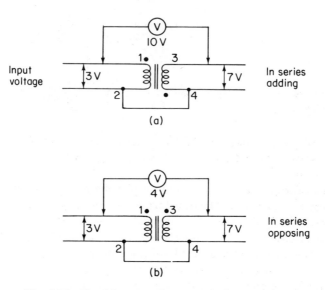

Fig. 6-32. Checking transformer polarity markings.

The phase relationship of one transformer marking to another can be found using the test circuit of Fig. 6-33. Assume that there is a 3-V output from the secondary of transformer A, and a 7-V output from transformer B. If the markings are consistent on both transformers, the two voltages will oppose, and 4 V will be indicated. If the markings are not consistent, the two voltages will add, resulting in a 10-V reading.

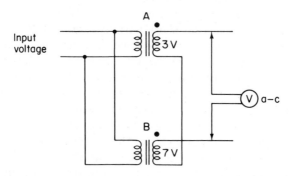

Fig. 6-33. Testing consistency of transformer polarities or phase markings.

6-10.5.3. Checking Transformer Regulation

All transformers have some regulating effect, even though not used in a power supply with a regulator circuit. That is, the output voltage of a transformer tends to remain constant with changes in load. Regulation is usually expressed as:

$$\text{percentage of regulation} = \frac{\text{no-load voltage} - \text{full-load voltage}}{\text{full-load voltage}} \times 100$$

However, some manufacturers rate their transformers by:

$$\text{percentage of regulation} = \frac{\text{no-load voltage} - \text{full-load voltage}}{\text{no-load voltage}} \times 100$$

Whichever factor is used, keep in mind that all transformers will not provide good regulation, even though designed for power-supply use. The regulation effect produced by a transformer is determined by core material, winding arrangement, wire size, etc. From a practical design standpoint a transformer with very poor regulation will require extra filtering and/or regulation. Often, an inexpensive transformer will show poor regulation. Thus, the savings on the transformer are offset by the increased cost of filter and regulator components.

Transformer regulation can be tested using the circuit of Fig. 6-34. The value of R_1 (load) should be selected to draw the maximum rated current from the secondary. For example, if the transformer is to be used with a power supply designed to deliver 100 W, and the secondary output voltage is 25 V, the value of R_1 should be such that 4 A will flow, or

$$R_1 = \frac{25}{4} = 6.25\ \Omega$$

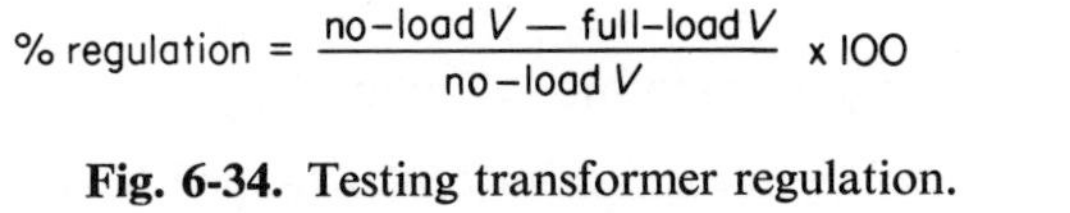

$$\% \text{ regulation} = \frac{\text{no-load } V - \text{full-load } V}{\text{full-load } V} \times 100$$

or

$$\% \text{ regulation} = \frac{\text{no-load } V - \text{full-load } V}{\text{no-load } V} \times 100$$

Fig. 6-34. Testing transformer regulation.

First, measure the secondary output voltage without a load, and then with a load. Use either equation to find the percentage of regulation.

6-10.5.4. Checking Transformer Impedance Ratio

The impedance ratio of a transformer is the square of the winding ratio. For example, if the winding ratio of a transformer is 15:1, the impedance ratio is $225:1$. Any impedance value placed across one winding will be reflected onto the other winding by a value equal to the impedance ratio. Assuming an impedance ratio of $225:1$ and an 1800-Ω impedance placed on the primary, the secondary would then have reflected impedance of $8\ \Omega$. Likewise, if a 10-Ω impedance were placed on the secondary, the primary would have a reflected impedance of $2250\ \Omega$.

Impedance ratio is related to turns ratio (primary to secondary). However, turns ratio information is not always available, so the ratio must be calculated using a test circuit as shown in Fig. 6-35.

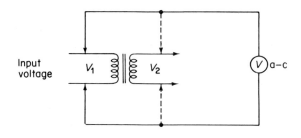

Turns ratio $= V_1/V_2$ or V_2/V_1

Impedance ratio $= (V_1/V_2)^2$ or $(V_2/V_1)^2$

Fig. 6-35. Measuring transformer turns and impedance ratios.

Measure both the primary and secondary voltage. Divide the larger voltage by the smaller, noting which is primary and which is secondary. For convenience, set either the primary or secondary to some exact voltage.

The *turns ratio* is equal to one voltage divided by the other.

The *impedance ratio* is the square of the turns ratio.

For example, assume that the primary shows 115 V, with 23 V at the secondary. This indicates a $5:1$ turns ratio, and a $25:1$ impedance ratio.

6-10.5.5. Checking Transformer-Winding Balance

There is always some imbalance in center-tapped transformers (such as used in dc–dc converters). That is, the turns ratio and impedance are not exactly the same on both sides of the center tap. A large imbalance can impair operation of any circuit, but especially a circuit such as used in dc–dc converters.

It is possible to find a large imbalance by measuring the dc resistance on either side of the center tap. However, a small imbalance might not show up.

It is usually more practical to measure the *voltage* on both sides of a center tap, as shown in Fig. 6-36. If the voltages are equal, the transformer winding is balanced. If a large imbalance is indicated by a large voltage difference, the winding should then be checked with an ohmmeter for shorted turns, poor design, etc.

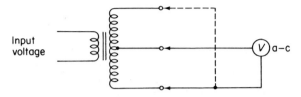

Fig. 6-36. Testing transformer-winding balance.

INDEX

INDEX